Torsion-Gravity and Spinors in Fundamental Theoretical Physics

Torsion-Gravity and Spinors in Fundamental Theoretical Physics

Editor

Luca Fabbri

MDPI • Basel • Beijing • Wuhan • Barcelona • Belgrade • Manchester • Tokyo • Cluj • Tianjin

Editor
Luca Fabbri
DIME, Sez. Metodi
Matematici
University of Genoa
Genoa
Italy

Editorial Office
MDPI
St. Alban-Anlage 66
4052 Basel, Switzerland

This is a reprint of articles from the Special Issue published online in the open access journal *Universe* (ISSN 2218-1997) (available at: www.mdpi.com/journal/universe/special_issues/Torsion_Gravity).

For citation purposes, cite each article independently as indicated on the article page online and as indicated below:

LastName, A.A.; LastName, B.B.; LastName, C.C. Article Title. *Journal Name* **Year**, *Volume Number*, Page Range.

ISBN 978-3-0365-7983-2 (Hbk)
ISBN 978-3-0365-7982-5 (PDF)

Contents

Editorial

Editorial for the Special Issue "Torsion-Gravity and Spinors in Fundamental Theoretical Physics"

Luca Fabbri

DIME, Sez. Metodi e Modelli Matematici, Università di Genova, Via all'Opera Pia 15, 16145 Genova, Italy; luca.fabbri@edu.unige.it

The Einsteinian theory of gravitation is among the best-established theories ever conceived in physics. Based on the geometrical idea that the energy density of matter could curve space-time, Einstein's field equations embrace conceptual simplicity, mathematical elegance and predictive power like few others.

One of the few cases of synthesis between simplicity, elegance and power is Maxwell's equations in electrodynamics, and they, too, are based on a similar geometrical idea, that the current density of matter curves the gauge potential. The electromagnetic tensor $F_{\mu\nu}$ in fact can be seen as the curvature of the gauge bundle in the same way in which the Riemann tensor $R_{\alpha\rho\mu\nu}$ is the curvature of the frame bundle.

However, now, the quick student or the attentive reader would raise the following question: given that the material Lagrangian generates energy, current and spin densities then, in the same spirit followed above, what does the spin density curve? The answer is that, hidden in the deep recesses of differential geometry, if care is taken to look attentively, there actually is an object whose curvature can be tied to the spin density: torsion.

Torsion, the antisymmetric part (in holonomic coordinates) of the most general connection, is a natural attribute of the geometric background in which Einsteinian gravity is built. With it, the torsional completion of Einsteinian gravity is the theory in which the space-time curvature couples to energy in the same manner in which the space-time torsion couples to spin, and in the same manner in which the gauge strength couples to the current.

Because the matter field that has both energy and spin, as well as current, is the spin fluid or the spinor field, use of the Weyssenhoff or Dirac theories becomes essential. As a consequence, the torsion field is expected to acquire a considerable prominence whenever spin effects might be important.

Of course this brings us, first of all, to quantum mechanics. But it may also bring us, perhaps not surprisingly, to the cosmology of black holes or of the early Universe, or in all those situations where the spin density is dominant.

The present collection brings together various experts in the field to make the present status of torsion in gravity and spin in quantum field theory a little clearer.

There are ten papers in this collection, split into two sections: The first is the review section and consists of three papers. The first, and largest of all, is Fundamental Theory of Torsion Gravity [1], written by myself with the aim of providing a basic reference for all following papers, as well as some historical introductions that readers might find interesting. The other two papers are Some Mathematical Aspects of $f(R)$-Gravity with Torsion: Cauchy Problem and Junction Conditions [2] by S. Vignolo and On the Mathematics of Coframe Formalism and Einstein-Cartan Theory–A Brief Review [3] by M. Tecchiolli, complementing the introduction with mathematical details on the structure of the differential equations of the theory and the presentation of the tetradic formulation.

The second is the research section and contains seven articles. Torsionally-Induced Stability in Spinors [4] deals with the problem of showing how torsion, creating a back-reaction between the spinor field and its spin, can quench the tendency to form singularities, whether they are gravitational, such as for the Hawking–Penrose theorem, or material, such

Citation: Fabbri, L. Editorial for the Special Issue "Torsion-Gravity and Spinors in Fundamental Theoretical Physics". *Universe* **2023**, *9*, 269. https://doi.org/10.3390/universe9060269

Received: 29 May 2023
Accepted: 31 May 2023
Published: 4 June 2023

as those encountered in quantum field theory. Detailed analyses are provided in Quantum Hydrodynamics of Spinning Particles in Electromagnetic and Torsion Fields [5] and Search for Manifestations of Spin-Torsion Coupling [6] by M. Trukhanova and co-workers, where axial vector spin–torsion coupling is investigated in general contexts, and additional effects of the torsion field are presented. A similarly general approach is also followed by C. Diether III and J. Christian in On the Role of Einstein-Cartan Gravity in Fundamental Particle Physics [7], where the authors examine the way in which torsion induces self-interactions for spinors that could balance the repulsive electrostatic forces of fermions. On the opposite side of the spectrum of applications, G. Milton in A Possible Explanation of Dark Matter and Dark Energy Involving a Vector Torsion Field [8] presents a model of torsion in gravity that could explain the dark sector of our Universe. Along the very same line, A. Ivanov and M. Wellenzohn tackle the problem of identifying the scalar field responsible for dark energy with the chameleon field in Can a Chameleon Field Be Identified with Quintessence? [9]. Finally, P. Asimakis and co-workers compare non-standard theories, such as types of teleparallel gravity against nucleosynthesis, in Big Bang Nucleosynthesis Constraints on $f(T, TG)$-Gravity [10].

With the last three papers on the dark sector and the previous three about the effects of torsion in particle physics, it is difficult to point to the direction in which torsion might make itself manifest. Every direction is open, and each is worthy of attention. For the purist, however, a spin density that is given by the Dirac field constitutes a very natural paradigm, and therefore I would be inclined to say that the phenomenologist might find the applications to quantum mechanics and particle physics to be those with higher discovery potential. However, as I said, every direction is open.

If, instead, you are a mathematical physicist, you may find Quantum Hydrodynamics of Spinning Particles in Electromagnetic and Torsion Fields [5] quite intriguing in its way of re-writing the full spinor theory in a form that is more suitable for formal manipulation, since in it the Dirac field is interpreted as a special type of spin fluid.

Those of you who are not new to torsion gravity may remember that a quarter of a century ago there was another Special Issue on torsion published by the Annales de la Fondation Louis de Broglie [11]. I authored a paper in that Special Issue and it pleases me immensely to be the Editor of the present Special Issue. However, for all the rest, the authorship has completely changed. Aside for me, not a single author from then has contributed now and this pleases me greatly too. It means that the torsion community is in constant renewal, with more ideas coming out afresh every year. It is my hope that younger generations will fuel the debate more than ever before.

But enough with the summary—you now know what you need to find the article that tickles your curiosity...

... Or just read them all to have a fair overview of the current state of torsion today.

Funding: This research received no external funding.

Conflicts of Interest: The authors declare no conflict of interest.

References

1. Fabbri, L. Fundamental Theory of Torsion Gravity. *Universe* **2021**, *7*, 305. [CrossRef]
2. Vignolo, S. Some Mathematical Aspects of $f(R)$-Gravity with Torsion: Cauchy Problem and Junction Conditions. *Universe* **2019**, *5*, 224. [CrossRef]
3. Tecchiolli, M. On the Mathematics of Coframe Formalism and Einstein-Cartan Theory—A Brief Review. *Universe* **2019**, *5*, 206. [CrossRef]
4. Fabbri, L. Torsionally-Induced Stability in Spinors. *Universe* **2023**, *9*, 73. [CrossRef]
5. Trukhanova, M.I.; Obukhov, Y.N. Quantum Hydrodynamics of Spinning Particles in Electromagnetic and Torsion Fields. *Universe* **2021**, *7*, 498. [CrossRef]
6. Trukhanova, M.I.; Andreev, P.; Obukhov, Y.N. Search for Manifestations of Spin-Torsion Coupling. *Universe* **2023**, *9*, 38. [CrossRef]
7. Diether, C.F., III; Christian, J. On the Role of Einstein-Cartan Gravity in Fundamental Particle Physics. *Universe* **2020**, *6*, 112. [CrossRef]

8. Milton, G.W. A Possible Explanation of Dark Matter and Dark Energy Involving a Vector Torsion Field. *Universe* **2022**, *8*, 298. [CrossRef]
9. Ivanov, A.N.; Wellenzohn, M. Can a Chameleon Field Be Identified with Quintessence? *Universe* **2020**, *6*, 221. [CrossRef]
10. Asimakis, P.; Saridakis, E.N.; Basilakos, S.; Yesmakhanova, K. Big Bang Nucleosynthesis Constraints on $f(T,TG)$-Gravity. *Universe* **2022**, *8*, 486. [CrossRef]
11. "Special Issue on Torsion" (Annales de la Fondation Louis de Broglie, Paris, 2007). Available online: https://fondationlouisdebroglie.org/AFLB-322/table322.htm (accessed on 28 May 2023)

 universe

MDPI

Review

Fundamental Theory of Torsion Gravity

Luca Fabbri

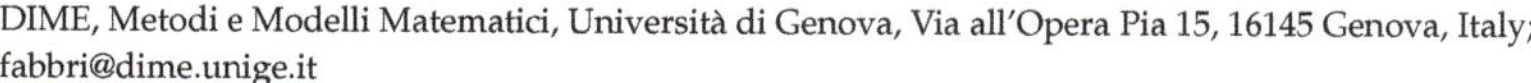
DIME, Metodi e Modelli Matematici, Università di Genova, Via all'Opera Pia 15, 16145 Genova, Italy; fabbri@dime.unige.it

Abstract: In this work, we present the general differential geometry of a background in which the space–time has both torsion and curvature with internal symmetries being described by gauge fields, and that is equipped to couple spinorial matter fields having spin and energy as well as gauge currents: torsion will turn out to be equivalent to an axial-vector massive Proca field and, because the spinor can be decomposed in its two chiral projections, torsion can be thought as the mediator that keeps spinors in stable configurations; we will justify this claim by studying some limiting situations. We will then proceed with a second chapter, where the material presented in the first chapter will be applied to specific systems in order to solve problems that seems to affect theories without torsion: hence the problem of gravitational singularity formation and positivity of the energy are the most important, and they will also lead the way for a discussion about the Pauli exclusion principle and the concept of macroscopic approximation. In a third and final chapter, we are going to investigate, in the light of torsion dynamics, some of the open problems in the standard models of particles and cosmology which would not be easily solvable otherwise.

Keywords: torsion-gravity; electrodynamics; spinors

Citation: Fabbri, L. Fundamental Theory of Torsion Gravity. *Universe* **2021**, 7, 305. https://doi.org/10.3390/universe7080305

Academic Editors: Antonino Del Popolo and Matteo Luca Ruggiero

Received: 8 July 2021
Accepted: 15 August 2021
Published: 18 August 2021

1. Introduction

In fundamental theoretical physics, there are a number of principles that are assumed, and, among them, one of the most important is the principle of covariance, stating that the form of physical laws must be independent from the coordinate system employed to write them. Covariance is mathematically translated into the instruction that such physical laws have to be written in tensorial forms.

On the other hand, because physical laws describe the shape and evolution of fields, differential operators must be used; because of covariance, all derivatives in the field equations have to be covariant: thus covariant derivatives must be defined. In its most general form, the covariant derivative of, say, a vector, is given by

$$D_{\alpha} V^{\nu} = \partial_{\alpha} V^{\nu} + V^{\sigma} \Gamma^{\nu}_{\sigma\alpha}$$

where the object $\Gamma^{\alpha}_{\nu\sigma}$ is called connection, it is defined in terms of the transformation law needed for the derivative to be fully covariant, and it has three indices: the upper index and the lower index on the left are the indices involved in the shuffling of the components of the vector, whereas the lower index on the right is the index related to the coordinate with respect to which the derivative is calculated eventually. Hence, there appears to be a clear distinction in the roles played by the left and the right of the lower indices, and therefore the connection cannot be taken to have any kind of symmetry property for indices transposition involving the two lower indices at all.

The fact that, in the most general case, the connection has no specific symmetry implies that the antisymmetric part of the connection is not zero, and it turns out to be a tensor: this is what is known as torsion tensor.

The circumstance for which the torsion tensor is not zero does not follow from arguments of generality alone, but also from explicit examples: for instance, torsion does describe some essential properties of Lie groups, as it was discussed by Cartan [1–4]. Cartan has been the first who pioneered into the study of torsion, and this is the reason why today torsion is also known as the Cartan tensor.

When back at the end of the 19th century, Ricci-Curbastro and Levi–Civita developed absolute differential calculus, or tensor calculus, they did it by assuming zero torsion to simplify computations, and the geometry they eventually obtained was entirely based on the existence of a Riemann metric: this is what we call Riemann geometry. Nothing in this geometry is spoiled by letting torsion take its place in it, the only difference being that now the metric would be accompanied by torsion as the fundamental objects of the geometry: the final setting is what is called Riemann–Cartan geometry.

Granted that, from a general mathematical perspective, torsion is present, one may wonder if there can be physical reasons for torsion to be zero. Physical arguments to prove that torsion must equal zero were indeed proposed in the past. However, none of them appeared to be free of fallacies or logical inconsistencies. A complete list with detailed reasons for their failure can be found in [5].

That torsion should not be equal to zero even in physical contexts is again quite general. In fact, by writing the RC geometry in anholonomic bases, the torsion can be seen as the strength of the potential arising from gauging the translation group, much in the same way in which the curvature is the strength of the potential arising from gauging the rotation group, as shown by Sciama and Kibble [6,7]. What Sciama and Kibble proved was that torsion is not just a tensor that could be added, but a tensor that must be added, besides curvature, in order to have the possibility to completely describe translations, besides rotations, in a full Poincaré gauge theory of physics [8].

At the beginning of the 20th century, when Einstein developed his theory of gravity, he did it by assuming zero torsion because, when torsion vanishes, the Ricci tensor is symmetric and therefore it can be consistently coupled to the symmetric energy tensor, realizing the identification between the space–time curvature, and its energy content expressed by Einstein field equations: this is the basic spirit of Einstein gravity. Today, we know that in physics there is also another quantity of interest called spin, and that, in its presence, the energy is no longer symmetric, so nowadays having a non-symmetric Ricci tensor besides a Cartan tensor would allow for a more exhaustive coupling in gravity, where the curvature would still be coupled to the energy but now torsion would be coupled to the spin: such a scheme would realize the identification between the space–time curvature and its energy content expressed by Einstein field equations and the identification between space–time torsion and its spin content expressed by the Sciama–Kibble field equations as the Einstein–Sciama–Kibble torsion gravity.

The ESK theory of gravity is thus the most complete theory describing the dynamics of the space–time, and, because torsion is coupled to the spin in the same spirit in which curvature is coupled to energy, then it is the theory of space–time in which the coupling to its matter content is achieved most exhaustively. The central point of the situation is therefore brought to the question asking whether there actually exists something possessing both spin and energy as a form of matter, which can profit from the setting that is provided by the ESK gravity.

As a matter of fact, such a theory not only exists, but it is also very well known, the Dirac spinorial field theory.

With so much insight, it is an odd circumstance that there be still such a controversy about the role of torsion besides that of curvature in gravity, and there may actually be several reasons for it. The single most important one may be that Einstein gravity was first published in the year 1916 when no spin was known and, despite being then insightful to set the torsion tensor to zero, when Dirac came with a theory of spinors comprising an intrinsic spin in 1928, the successes of Einstein theory of gravity were already too great to make anyone wonder about the possibility of modifying it.

Of course, this is no scientific reason to hinder research, but, sociologically, it can be easy to understand why one would not lightly go to look beyond something good, especially today that the successes of the Einstein theory of gravitation have become practically complete.

In the present report, we would like to change this tendency by considering torsion in gravity coupled to spinor fields and showing all advantages that we can get, from theoretical consistency, to phenomenological applications.

Thus, in a second chapter, we will investigate the theoretical advantages obtained from the torsion–spin interactions. These will span from the revision of the Hawking–Penrose theorem about the inevitability of gravitational singularity formation to some discussion about the positivity of the energy, passing through the Pauli exclusion principle and the concept of macroscopic approximation.

In a third and final chapter, we are going to employ the presented theory to assess some of the known open problems in the standard models of particles and cosmology.

One: Fundamental Theory

The first chapter will be about presenting the fundamental theory, and it will be divided into three sections: in the first section, we will define all the kinematic quantities and see how they can be dynamically coupled. It will be followed by a second section in which we will deepen the study about what is torsion and the spinor fields and the way they interact. A third section will be about studying limiting situations that can allow us to get even more information about the torsion–spin coupling.

2. Torsion Gravity for Spinor Fields

In this first section, we introduce the physical theory that will be our reference throughout the entire work: we start with the most general geometric introduction of the kinematic quantities. In addition, we will continue by establishing their link in terms of dynamical field equations.

2.1. Geometry and Its Matter Content

To build the geometric background on which to define kinematic fields, we start with the symmetry principle at the basis of any theory in physics: covariance under the most general transformation of coordinates. We will see in what way from such a general environment a natural definition of matter field will spontaneously emerge.

2.1.1. Tensor and Gauge Fields

The principle of covariance under the most general transformation of coordinates is possible one of the most self-evident principles in all of physics: it states that our way of writing equations might a priori be conditioned in principle by the coordinates we choose, but observable properties should not feel affected by coordinate artifacts brought by us. This means that of all possible manners we have to write physics, there must be one that is not influenced by any choice of coordinates, or in other words, this specific way of writing physics has to be invariant between different coordinate systems.

To see how this is possible, we start with the following definition. Suppose that a certain physical quantity can be described in terms of the object $T^{\alpha...\sigma}_{\rho...\zeta}$ such that it is written as $T^{\alpha...\sigma}_{\rho...\zeta}(x)$ with respect to coordinates x, and it is written as $T'^{\alpha...\sigma}_{\rho...\zeta}(x')$ with respect to coordinates x' in general, and suppose that

$$T'^{\alpha...\sigma}_{\rho...\zeta} = \frac{\partial x^{\beta}}{\partial x'^{\rho}} ... \frac{\partial x^{\theta}}{\partial x'^{\zeta}} \frac{\partial x'^{\alpha}}{\partial x^{\nu}} ... \frac{\partial x'^{\sigma}}{\partial x^{\tau}} T^{\nu...\tau}_{\beta...\theta}$$

where $x' = x'(x)$ determines the passage from the first to the second system of coordinates. If this happens, such a quantity is called a tensor. Now, suppose that one specific property of this quantity be described as

$$T^{\nu...\tau}_{\beta...\theta} = 0$$

in the first system of coordinates. According to the above definition, then, we have that

$$T'^{\alpha\ldots\sigma}_{\rho\ldots\zeta} = \frac{\partial x^{\beta}}{\partial x'^{\rho}} \cdots \frac{\partial x^{\theta}}{\partial x'^{\zeta}} \frac{\partial x'^{\alpha}}{\partial x^{\nu}} \cdots \frac{\partial x'^{\sigma}}{\partial x^{\tau}} T^{\nu\ldots\tau}_{\beta\ldots\theta} =$$
$$= \frac{\partial x^{\beta}}{\partial x'^{\rho}} \cdots \frac{\partial x^{\theta}}{\partial x'^{\zeta}} \frac{\partial x'^{\alpha}}{\partial x^{\nu}} \cdots \frac{\partial x'^{\sigma}}{\partial x^{\tau}} 0 \equiv 0$$

and thus

$$T'^{\alpha\ldots\sigma}_{\rho\ldots\zeta} = 0$$

showing that the same property pertains to that quantity also in the second system of coordinates as well. In this way, we have that, if the property of a quantity is encoded as the vanishing of a tensor, then we can be certain that such a property pertains to that quantity regardless of the system of coordinates. In addition, this is just covariance.

The principle of covariance is therefore implemented in the geometry by the straightforward requirement that this geometry be written in terms of tensors. Therefore, let be given two systems of coordinates as x and x' related by the most general coordinate transformation $x' = x'(x)$ and a set of functions of these coordinates written with respect to the first and the second system of coordinates as $T(x)$ and $T'(x')$ and such that, for a coordinate transformation, they are related by

$$T'^{\alpha\ldots\sigma}_{\rho\ldots\zeta} = \text{sign}\,\det\!\left(\frac{\partial x'}{\partial x}\right) \frac{\partial x^{\beta}}{\partial x'^{\rho}} \cdots \frac{\partial x^{\theta}}{\partial x'^{\zeta}} \frac{\partial x'^{\alpha}}{\partial x^{\nu}} \cdots \frac{\partial x'^{\sigma}}{\partial x^{\tau}} T^{\nu\ldots\tau}_{\beta\ldots\theta} \tag{1}$$

Then, this quantity is called *tensor* or *pseudo-tensor*, according to whether the sign of the determinant of such a transformation is positive or negative. For a tensor with at least two upper or two lower indices, we might switch the two indices obtaining a tensor called *transposition* of the original tensor in those two indices, and if it happens to be equal to the initial tensor up to the sign plus or minus, we say that the tensor is *symmetric* or *antisymmetric* in those two indices, respectively. Given a tensor with at least one upper and one lower index, we can consider one of the upper and one of the lower indices forcing them to have the same value and performing the sum over every possible value of those indices obtaining a tensor called *contraction* in those indices, and this process can be repeated until we reach a tensor whose contraction is zero, called *irreducible.* Particular cases are tensors having one index called *vectors*, while tensors without any index are called *scalars*. Tensors with the same index configuration can be *summed* and any two tensors can be *multiplied* in a component-by-component way, according to the usual rules of algebraic calculus as they are well known.

There is therefore no need to spend more time in the algebraic properties of tensors. However, differential properties of tensors need some deepening. The problem with differentiation applied to the case of tensors is that such an operation spoils the transformation law of a tensor in very general circumstances. Thus, if we want to construct an operation that is able to generalize the usual derivative up to a derivative that respects covariance, we must begin by noticing that, because a tensor is a set of fields, in general, it will have two types of variations: the first is due to the fact that tensors fields are fields, coordinates dependent, and so a local structure must be present as

$$_{\text{local}}\Delta T^{\alpha_1\ldots\alpha_j}_{\beta_1\ldots\beta_i} = T^{\alpha_1\ldots\alpha_j}_{\beta_1\ldots\beta_i}(x') - T^{\alpha_1\ldots\alpha_j}_{\beta_1\ldots\beta_i}(x) =$$
$$= \partial_{\mu} T^{\alpha_1\ldots\alpha_j}_{\beta_1\ldots\beta_i}(x)\delta x^{\mu}$$

at the first order infinitesimal; the second is due to the fact that tensors fields are tensors, so a system of components, and thus a re-shuffling of the different components must be allowed according to

$$\begin{aligned}
{}_{\text{structure}}\Delta T^{\alpha_1\ldots\alpha_j}_{\beta_1\ldots\beta_i} &= T'^{\alpha_1\ldots\alpha_j}_{\beta_1\ldots\beta_i} - T^{\alpha_1\ldots\alpha_j}_{\beta_1\ldots\beta_i} = \\
&= [(\delta\Gamma^{\alpha_1}_{\theta} T^{\theta\ldots\alpha_j}_{\beta_1\ldots\beta_i} + \ldots + \delta\Gamma^{\alpha_j}_{\theta} T^{\alpha_1\ldots\theta}_{\beta_1\ldots\beta_i}) - \\
&- (\delta\Gamma^{\theta}_{\beta_1} T^{\alpha_1\ldots\alpha_j}_{\theta\ldots\beta_i} + \ldots + \delta\Gamma^{\theta}_{\beta_i} T^{\alpha_1\ldots\alpha_j}_{\beta_1\ldots\theta})]
\end{aligned}$$

as the most general form in which this can be done while respecting the fact that the differential structure requires the linearity and the Leibniz rule, and again at the first order of infinitesimal. In full, we have

$$\begin{aligned}
\Delta T^{\alpha_1\ldots\alpha_j}_{\beta_1\ldots\beta_i} &= {}_{\text{local}}\Delta T^{\alpha_1\ldots\alpha_j}_{\beta_1\ldots\beta_i} + {}_{\text{structure}}\Delta T^{\alpha_1\ldots\alpha_j}_{\beta_1\ldots\beta_i} = \\
&= \partial_\mu T^{\alpha_1\ldots\alpha_j}_{\beta_1\ldots\beta_i}(x)\delta x^\mu + \\
&+ [(\delta\Gamma^{\alpha_1}_{\theta} T^{\theta\ldots\alpha_j}_{\beta_1\ldots\beta_i} + \ldots + \delta\Gamma^{\alpha_j}_{\theta} T^{\alpha_1\ldots\theta}_{\beta_1\ldots\beta_i}) - \\
&- (\delta\Gamma^{\theta}_{\beta_1} T^{\alpha_1\ldots\alpha_j}_{\theta\ldots\beta_i} + \ldots + \delta\Gamma^{\theta}_{\beta_i} T^{\alpha_1\ldots\alpha_j}_{\beta_1\ldots\theta})]
\end{aligned}$$

at the first order infinitesimal. Thus, defining $\delta\Gamma^{\alpha}_{\beta} = \Gamma^{\alpha}_{\beta\mu}\delta x^\mu$ and dividing by δx^μ, we obtain that

$$\begin{aligned}
D_\mu T^{\alpha_1\ldots\alpha_j}_{\beta_1\ldots\beta_i} &= \partial_\mu T^{\alpha_1\ldots\alpha_j}_{\beta_1\ldots\beta_i} + \\
&+ (\Gamma^{\alpha_1}_{\theta\mu} T^{\theta\ldots\alpha_j}_{\beta_1\ldots\beta_i} + \ldots + \Gamma^{\alpha_j}_{\theta\mu} T^{\alpha_1\ldots\theta}_{\beta_1\ldots\beta_i}) - \\
&- (\Gamma^{\theta}_{\beta_1\mu} T^{\alpha_1\ldots\alpha_j}_{\theta\ldots\beta_i} + \ldots + \Gamma^{\theta}_{\beta_i\mu} T^{\alpha_1\ldots\alpha_j}_{\beta_1\ldots\theta})
\end{aligned}$$

after taking the limit. This is the most general form of potential covariant derivative. To see that this derivative is indeed covariant, we have to require that $\Gamma^{\alpha}_{\beta\mu}$ transforms with a specific non-tensorial transformation law such as to compensate for the non-tensorial transformation law of the partial derivative. In the simplest case of one tensorial index, we have that the derivative is

$$D_\iota V^\alpha = \partial_\iota V^\alpha + V^\beta \Gamma^{\alpha}_{\beta\iota}$$

whose transformation law is given by

$$\begin{aligned}
\frac{\partial x^\beta}{\partial x'^{\beta'}}\frac{\partial x'^{\alpha'}}{\partial x^\alpha}(\partial_\beta V^\alpha + V^\rho \Gamma^{\alpha}_{\rho\beta}) &= \frac{\partial x^\beta}{\partial x'^{\beta'}}\frac{\partial x'^{\alpha'}}{\partial x^\alpha} D_\beta V^\alpha = \\
= (D_\beta V^\alpha)' = (\partial_\beta V^\alpha + V^\rho \Gamma^{\alpha}_{\rho\beta})' &= \partial_{\beta'} V^{\alpha'} + V^{\rho'} \Gamma'^{\alpha'}_{\rho'\beta'} = \\
&= \frac{\partial x^\theta}{\partial x'^{\beta'}}\frac{\partial}{\partial x^\theta}\left(\frac{\partial x'^{\alpha'}}{\partial x^\alpha} V^\alpha\right) + \frac{\partial x'^{\rho'}}{\partial x^\rho} V^\rho \Gamma'^{\alpha'}_{\rho'\beta'} = \\
&= \frac{\partial x^\theta}{\partial x'^{\beta'}}\frac{\partial x'^{\alpha'}}{\partial x^\alpha}\frac{\partial V^\alpha}{\partial x^\theta} + \frac{\partial x^\theta}{\partial x'^{\beta'}}\frac{\partial}{\partial x^\theta}\frac{\partial x'^{\alpha'}}{\partial x^\alpha} V^\alpha + \frac{\partial x'^{\rho'}}{\partial x^\rho} V^\rho \Gamma'^{\alpha'}_{\rho'\beta'}
\end{aligned}$$

in which terms with the derivatives disappear. Then,

$$\frac{\partial x^\beta}{\partial x'^{\beta'}}\frac{\partial x'^{\alpha'}}{\partial x^\alpha} V^\rho \Gamma^{\alpha}_{\rho\beta} = \frac{\partial x^\theta}{\partial x'^{\beta'}}\frac{\partial}{\partial x^\theta}\frac{\partial x'^{\alpha'}}{\partial x^\alpha} V^\alpha + \frac{\partial x'^{\rho'}}{\partial x^\rho} V^\rho \Gamma'^{\alpha'}_{\rho'\beta'}$$

and since this has to hold for any vector

$$\frac{\partial x^\beta}{\partial x'^{\beta'}}\frac{\partial x'^{\alpha'}}{\partial x^\alpha}\Gamma^{\alpha}_{\rho\beta} = \frac{\partial x^\theta}{\partial x'^{\beta'}}\frac{\partial}{\partial x^\theta}\frac{\partial x'^{\alpha'}}{\partial x^\rho} + \frac{\partial x'^{\rho'}}{\partial x^\rho}\Gamma'^{\alpha'}_{\rho'\beta'}$$

which is the non-tensorial transformation that the set of coefficients $\Gamma^{\alpha}_{\rho\beta}$ must have to ensure that the full derivative transforms as a tensor in this very specific case with a vector field. However, quite remarkably, the very same non-tensorial transformation of $\Gamma^{\alpha}_{\rho\beta}$ can be used for each term in the most general form of derivative for a generic tensor, and so the obtained result is completely general.

The set of coefficients $\Gamma^{\alpha}_{\rho\beta}$ have no specific symmetry properties in the lower indices, and consequently we have that we can write

$$\Gamma^{\alpha}_{\mu\nu} \equiv \frac{1}{2}(\Gamma^{\alpha}_{\mu\nu} + \Gamma^{\alpha}_{\nu\mu}) + \frac{1}{2}(\Gamma^{\alpha}_{\mu\nu} - \Gamma^{\alpha}_{\nu\mu})$$

where the transformation properties of the full object is inherited by the first part, which is symmetric in the two lower indices, and it can be indicated as

$$\Lambda^{\alpha}_{\mu\nu} = \frac{1}{2}(\Gamma^{\alpha}_{\mu\nu} + \Gamma^{\alpha}_{\nu\mu})$$

while the second part

$$Q^{\alpha}{}_{\mu\nu} = \Gamma^{\alpha}_{\mu\nu} - \Gamma^{\alpha}_{\nu\mu}$$

transforms as a tensor such that $Q^{\alpha}{}_{\mu\nu} = -Q^{\alpha}{}_{\nu\mu}$, meaning that it is antisymmetric in its second pair of indices. Thus,

$$\Gamma^{\alpha}_{\mu\nu} = \Lambda^{\alpha}_{\mu\nu} + \frac{1}{2}Q^{\alpha}{}_{\mu\nu}$$

in the most general case. As in the covariant derivatives, the connection enters linearly, and the splitting in symmetric and antisymmetric parts sums up to a linear combination of the tensor $Q^{\alpha}{}_{\mu\nu}$ plus the terms linear in the symmetric connection, which therefore forms yet another type of covariant derivative that is defined according to

$$\begin{aligned}\nabla_{\mu} T^{\alpha_1 \dots \alpha_j}_{\beta_1 \dots \beta_i} = \partial_{\mu} T^{\alpha_1 \dots \alpha_j}_{\beta_1 \dots \beta_i} + \\ +(\Lambda^{\alpha_1}_{\theta\mu} T^{\theta \dots \alpha_j}_{\beta_1 \dots \beta_i} + \dots + \Lambda^{\alpha_j}_{\theta\mu} T^{\alpha_1 \dots \theta}_{\beta_1 \dots \beta_i}) - \\ -(\Lambda^{\theta}_{\beta_1\mu} T^{\alpha_1 \dots \alpha_j}_{\theta \dots \beta_i} + \dots + \Lambda^{\theta}_{\beta_i\mu} T^{\alpha_1 \dots \alpha_j}_{\beta_1 \dots \theta})\end{aligned}$$

and in it the fact that the symmetric connection is indeed symmetric allows for particularly simplified expressions in some special cases. For instance, taking the symmetric covariant derivative of a tensor with all lower indices gives

$$\nabla_{\mu} T_{\beta_1 \dots \beta_i} = \partial_{\mu} T_{\beta_1 \dots \beta_i} - \Lambda^{\theta}_{\beta_1\mu} T_{\theta \dots \beta_i} - \dots - \Lambda^{\theta}_{\beta_i\mu} T_{\beta_1 \dots \theta}$$

which is particularly interesting because we see that the symmetric connection always saturates the same index in the upper position, so that, if we further specialize onto the case in which the tensor is completely antisymmetric, we obtain that

$$\begin{aligned}\nabla_{[\mu} T_{\beta \dots \rho]} = \nabla_{\mu} T_{\beta \dots \rho} - \nabla_{\beta} T_{\mu \dots \rho} + \dots - \nabla_{\rho} T_{\beta \dots \mu} = \\ = \partial_{\mu} T_{\beta \dots \rho} - \Lambda^{\sigma}_{\beta\mu} T_{\sigma \dots \rho} - \dots - \Lambda^{\sigma}_{\rho\mu} T_{\beta \dots \sigma} - \\ -\partial_{\beta} T_{\mu \dots \rho} + \Lambda^{\sigma}_{\mu\beta} T_{\sigma \dots \rho} + \dots + \Lambda^{\sigma}_{\rho\beta} T_{\mu \dots \sigma} + \dots \\ \dots - \partial_{\rho} T_{\beta \dots \mu} + \Lambda^{\sigma}_{\beta\rho} T_{\sigma \dots \mu} + \dots + \Lambda^{\sigma}_{\mu\rho} T_{\beta \dots \sigma} = \\ = \partial_{\mu} T_{\beta \dots \rho} - \partial_{\beta} T_{\mu \dots \rho} + \dots - \partial_{\rho} T_{\beta \dots \mu} = \partial_{[\mu} T_{\beta \dots \rho]}\end{aligned}$$

where all symmetric connections cancelled off leaving an expression written only in terms of partial derivatives but that is a completely antisymmetric covariant derivative in the most general case. This is a very peculiar property of tensors having all lower indices and being completely antisymmetric in all of these indices, and there is an entire domain related to this type of tensors and covariant derivatives, in which tensors are known as forms and the covariant derivatives are part of what is known as exterior calculus. Nevertheless, we will not discuss it here because we do not want to introduce even further mathematical concepts and after all forms and exterior derivatives are nothing but a specific type of tensors. We encourage the interested readers to study this domain on their own.

Thus, to summarize what we have done, we have that the set of functions $\Gamma^{\rho}_{\alpha\beta}$ transforming as

$$\Gamma'^{\rho}_{\sigma\tau} = \left(\Gamma^{\alpha}_{\mu\nu} - \frac{\partial x^{\alpha}}{\partial x'^{\kappa}} \frac{\partial^2 x'^{\kappa}}{\partial x^{\nu} \partial x^{\mu}}\right) \frac{\partial x'^{\rho}}{\partial x^{\alpha}} \frac{\partial x^{\mu}}{\partial x'^{\sigma}} \frac{\partial x^{\nu}}{\partial x'^{\tau}} \tag{2}$$

is called *connection*, and it can be decomposed as

$$\Gamma^{\rho}_{\alpha\beta} = \Lambda^{\rho}_{\alpha\beta} + \frac{1}{2} Q^{\rho}{}_{\alpha\beta} \tag{3}$$

where $\Lambda^{\rho}_{\alpha\beta}$ is a set of functions transforming according to the law of a connection but which are symmetric in the two lower indices called *symmetric connection* and

$$Q^{\rho}{}_{\alpha\beta} = \Gamma^{\rho}_{\alpha\beta} - \Gamma^{\rho}_{\beta\alpha} \tag{4}$$

which is a tensor antisymmetric in the two lower indices called *torsion tensor*. In terms of the connection, we may write the *covariant derivative* in the most general case as

$$\begin{aligned} D_{\mu} T^{\alpha_1 \dots \alpha_j}_{\beta_1 \dots \beta_i} = \partial_{\mu} T^{\alpha_1 \dots \alpha_j}_{\beta_1 \dots \beta_i} + \textstyle\sum_{k=1}^{k=j} \Gamma^{\alpha_k}_{\sigma\mu} T^{\alpha_1 \dots \sigma \dots \alpha_j}_{\beta_1 \dots \beta_i} - \\ - \textstyle\sum_{k=1}^{k=i} \Gamma^{\sigma}_{\beta_k \mu} T^{\alpha_1 \dots \alpha_j}_{\beta_1 \dots \sigma \dots \beta_i} \end{aligned} \tag{5}$$

decomposing as

$$\begin{aligned} D_{\mu} T^{\alpha_1 \dots \alpha_j}_{\beta_1 \dots \beta_i} = \nabla_{\mu} T^{\alpha_1 \dots \alpha_j}_{\beta_1 \dots \beta_i} + \tfrac{1}{2} \textstyle\sum_{k=1}^{k=j} Q^{\alpha_k}_{\sigma\mu} T^{\alpha_1 \dots \sigma \dots \alpha_j}_{\beta_1 \dots \beta_i} - \\ - \tfrac{1}{2} \textstyle\sum_{k=1}^{k=i} Q^{\sigma}_{\beta_k \mu} T^{\alpha_1 \dots \alpha_j}_{\beta_1 \dots \sigma \dots \beta_i} \end{aligned} \tag{6}$$

with spurious terms linear in the torsion tensor and

$$\begin{aligned} \nabla_{\mu} T^{\alpha_1 \dots \alpha_j}_{\beta_1 \dots \beta_i} = \partial_{\mu} T^{\alpha_1 \dots \alpha_j}_{\beta_1 \dots \beta_i} + \textstyle\sum_{k=1}^{k=j} \Lambda^{\alpha_k}_{\sigma\mu} T^{\alpha_1 \dots \sigma \dots \alpha_j}_{\beta_1 \dots \beta_i} - \\ - \textstyle\sum_{k=1}^{k=i} \Lambda^{\sigma}_{\beta_k \mu} T^{\alpha_1 \dots \alpha_j}_{\beta_1 \dots \sigma \dots \beta_i} \end{aligned} \tag{7}$$

which is the *covariant derivative calculated with respect to the symmetric connection*. If we apply the last definition to the particular case of tensors with all lower indices and being completely antisymmetric, we get

$$\nabla_{[\nu} T_{\alpha \dots \sigma]} = \partial_{[\nu} T_{\alpha \dots \sigma]} \equiv (\partial T)_{\nu\alpha \dots \sigma}, \tag{8}$$

which is still a tensor and such that it is completely antisymmetric called a *covariant curl* of the tensor field. When in the covariant derivative of a tensor with at least one upper index we contract the index of derivation with an upper index of the tensor field, we get what is known as *covariant divergence* in that index of the tensor field.

When we have introduced the concept of tensor, it naturally emerged that, in the definition, two types of indices were present, upper and lower, reflecting the fact that a tensor could transform according to two type of transformations, direct and inverse. However, these two types of transformation are two different forms of the same transformation, and so one should expect that the two types of indices be two different arrangements of the same system of components. Thus, there should be no difference in content if we move a given index up or down at will.

What this implies is that it should be possible to move indices up and down without losing or adding anything to the information content: this can be done by considering the

Kronecker tensor δ^{α}_{ν} and postulating the existence of two tensors $g_{\alpha\nu}$ and $g^{\alpha\nu}$ in general. Then, we can define the operation of raising and lowering of tensorial indices by considering that $A^{\pi}g_{\pi\nu}$ and $A_{\pi}g^{\pi\nu}$ are tensors that are related to the initial ones but with the index lowered and raised, respectively, and so we may define these two tensors as $A_{\pi}g^{\pi\nu} \equiv A^{\nu}$ and $A^{\pi}g_{\pi\nu} \equiv A_{\nu}$ as the same tensors but with the index moved in a different position with respect to the initial one. While it is certainly useful to have the possibility to perform such an operation, we also have to consider that such an operation has a two-fold ambiguity concerning the fact that, besides the contractions $A_{\pi}g^{\pi\nu} \equiv A^{\nu}$ and $A^{\pi}g_{\pi\nu} \equiv A_{\nu}$, we may have the contractions $A_{\pi}g^{\nu\pi} \equiv A^{\nu}$ and $A^{\pi}g_{\nu\pi} \equiv A_{\nu}$ too. In addition, we may decide to raise the previously lowered index to the initial position or lower the previously raise index to the initial position, so that the above ambiguity becomes four-fold with $A_{\pi}g^{\pi\nu}g_{\sigma\nu} \equiv A_{\sigma}$ and $A_{\pi}g^{\nu\pi}g_{\sigma\nu} \equiv A_{\sigma}$ as well as $A_{\pi}g^{\pi\nu}g_{\nu\sigma} \equiv A_{\sigma}$ and $A_{\pi}g^{\nu\pi}g_{\nu\sigma} \equiv A_{\sigma}$ as equally good possibilities that may be considered. On the other hand, requiring that raising one index up and then lowering that index down give back the initial tensor in all of the four possibilities leads to the following relationships

$$A_{\mu}(g^{\mu\sigma}g_{\sigma\kappa} - \delta^{\mu}_{\kappa}) = 0 \qquad A_{\mu}(g^{\sigma\mu}g_{\sigma\kappa} - \delta^{\mu}_{\kappa}) = 0$$
$$A_{\mu}(g^{\mu\sigma}g_{\kappa\sigma} - \delta^{\mu}_{\kappa}) = 0 \qquad A_{\mu}(g^{\sigma\mu}g_{\kappa\sigma} - \delta^{\mu}_{\kappa}) = 0$$

for any possible tensor A_{μ}, so that

$$(g^{\mu\sigma}g_{\sigma\kappa} - \delta^{\mu}_{\kappa}) = 0 \qquad (g^{\sigma\mu}g_{\sigma\kappa} - \delta^{\mu}_{\kappa}) = 0$$
$$(g^{\mu\sigma}g_{\kappa\sigma} - \delta^{\mu}_{\kappa}) = 0 \qquad (g^{\sigma\mu}g_{\kappa\sigma} - \delta^{\mu}_{\kappa}) = 0$$

identically. Taking the differences

$$g^{\mu\sigma}(g_{\sigma\kappa} - g_{\kappa\sigma}) = 0 \qquad (g^{\sigma\mu} - g^{\mu\sigma})g_{\sigma\kappa} = 0$$
$$g^{\sigma\mu}(g_{\sigma\kappa} - g_{\kappa\sigma}) = 0 \qquad (g^{\sigma\mu} - g^{\mu\sigma})g_{\kappa\sigma} = 0,$$

we may work out that

$$g_{\alpha\kappa} = g_{\kappa\alpha}$$
$$g^{\alpha\kappa} = g^{\kappa\alpha}$$

together with the condition

$$g^{\sigma\mu}g_{\kappa\sigma} = \delta^{\mu}_{\kappa}$$

meaning that, seen as matrices, they are symmetric and one the inverse of the other, and so, in particular, they are non-degenerate, as it has been demonstrated in [9]. This implies that what has been introduced to raise lower or lower upper indices has all the features of a metric and therefore these two tensors can also be identified with the metric of the space–time. We remark that this is exactly the opposite to the normal approach, where the metric is postulated, and then it is realized that it can be used to move up and down indices of tensors. The equivalence of these two a priori unrelated operations looks profound.

The metric determinant $\det(g_{\mu\nu}) = g$ can never be zero, but it follows the transformation law

$$g' = \det\left|\frac{\partial x}{\partial x'}\right|^{2} g$$

which is not the transformation law for a tensor. However, it can still be used to form a very important tensor as in the following. Consider in fact the non-tensorial quantity that is given by $\epsilon_{i_1 i_2 i_3 i_4}$ such that it is equal to the unity for an even permutation of (1234) and minus the unity for an odd permutation of (1234) and zero for a sequence that is not a permutation of (1234) at all. As this set of coefficients is completely antisymmetric with a number of indices that is equal to the dimension, we have that it has only one independent component, transforming as

$$\frac{\partial x^{i_1}}{\partial x'^{i'_1}}\frac{\partial x^{i_2}}{\partial x'^{i'_2}}\frac{\partial x^{i_3}}{\partial x'^{i'_3}}\frac{\partial x^{i_4}}{\partial x'^{i'_4}}\epsilon_{i_1 i_2 i_3 i_4} = \epsilon_{i'_1 i'_2 i'_3 i'_4}\alpha$$

for a given α function to be determined. In addition, because the determinant of any generic matrix can always be written in terms of these coefficients according to the expression given by $\det M = \Sigma_{i_j}\epsilon_{i_1 i_2 i_3 i_4}M^{1i_1}M^{2i_2}M^{3i_3}M^{4i_4}$, then

$$\det\frac{\partial x}{\partial x'} = \frac{\partial x^{i_1}}{\partial x'^1}\frac{\partial x^{i_2}}{\partial x'^2}\frac{\partial x^{i_3}}{\partial x'^3}\frac{\partial x^{i_4}}{\partial x'^4}\epsilon_{i_1 i_2 i_3 i_4} = \epsilon_{1234}\alpha = \alpha$$

furnishing the α function. Thus, we have

$$\epsilon_{i'_1 i'_2 i'_3 i'_4} = \det\frac{\partial x'}{\partial x}\frac{\partial x^{i_1}}{\partial x'^{i'_1}}\frac{\partial x^{i_2}}{\partial x'^{i'_2}}\frac{\partial x^{i_3}}{\partial x'^{i'_3}}\frac{\partial x^{i_4}}{\partial x'^{i'_4}}\epsilon_{i_1 i_2 i_3 i_4}$$

which is non-tensorial, but its non-tensoriality perfectly matches that of the determinant of the metric. Therefore, we have that they compensate in the combined form

$$(g^{\frac{1}{2}}\epsilon_{\alpha\nu\sigma\tau})' = \text{sign det}\left|\frac{\partial x'}{\partial x}\right|\frac{\partial x^{\beta}}{\partial x'^{\alpha}}\frac{\partial x^{\mu}}{\partial x'^{\nu}}\frac{\partial x^{\theta}}{\partial x'^{\sigma}}\frac{\partial x^{\rho}}{\partial x'^{\tau}}(g^{\frac{1}{2}}\epsilon_{\beta\mu\theta\rho}),$$

which is in fact the transformation that defines a pseudo-tensorial field. Notice, however, that, if we were to define the tensor with all lower indices as

$$\varepsilon_{\alpha\nu\sigma\tau} = \epsilon_{\alpha\nu\sigma\tau}|g|^{\frac{1}{2}}$$

the correspondent tensor with all upper indices would be given according to the following expression:

$$\varepsilon^{\alpha\nu\sigma\tau} = \epsilon^{\alpha\nu\sigma\tau}|g|^{-\frac{1}{2}}$$

in order for it to be consistently defined. This difference is necessary, as it can be seen from the fact that the quantity

$$\varepsilon^{i_1 i_2 i_3 i_4}\varepsilon_{j_1 j_2 j_3 j_4} = -\det\begin{vmatrix} \delta^{i_1}_{j_1} & \delta^{i_2}_{j_1} & \delta^{i_3}_{j_1} & \delta^{i_4}_{j_1} \\ \delta^{i_1}_{j_2} & \delta^{i_2}_{j_2} & \delta^{i_3}_{j_2} & \delta^{i_4}_{j_2} \\ \delta^{i_1}_{j_3} & \delta^{i_2}_{j_3} & \delta^{i_3}_{j_3} & \delta^{i_4}_{j_3} \\ \delta^{i_1}_{j_4} & \delta^{i_2}_{j_4} & \delta^{i_3}_{j_4} & \delta^{i_4}_{j_4} \end{vmatrix}$$

as it is very easy to check by performing a straightforward substitution and making all the direct calculations.

To summarize, the object δ^{β}_{α} that is unity or zero according to whether the value of its indices is equal or not is the *unity tensor* mentioned. We assume the existence of two tensors $g_{\alpha\kappa}$ and $g^{\alpha\kappa}$ symmetric and such that

$$g^{\sigma\mu}g_{\kappa\sigma} = \delta^{\mu}_{\kappa} \tag{9}$$

called *metric tensors*. In addition, we define

$$\delta^{i_0 i_1 i_2 i_3}_{j_0 j_1 j_2 j_3} = \det\begin{vmatrix} \delta^{i_0}_{j_0} & \delta^{i_1}_{j_0} & \delta^{i_2}_{j_0} & \delta^{i_3}_{j_0} \\ \delta^{i_0}_{j_1} & \delta^{i_1}_{j_1} & \delta^{i_2}_{j_1} & \delta^{i_3}_{j_1} \\ \delta^{i_0}_{j_2} & \delta^{i_1}_{j_2} & \delta^{i_2}_{j_2} & \delta^{i_3}_{j_2} \\ \delta^{i_0}_{j_3} & \delta^{i_1}_{j_3} & \delta^{i_2}_{j_3} & \delta^{i_3}_{j_3} \end{vmatrix} \tag{10}$$

as a *completely antisymmetric unity tensor*. The quantity given by $\epsilon_{i_0 i_1 i_2 i_3}$ equal to the unity, minus unity, or zero according to whether $(i_0 i_1 i_2 i_3)$ is an even, odd, or not a permutation of (0123) can be taken with the determinant of the metric $\det(g_{\mu\nu}) = g$ to define

$$\varepsilon^{\alpha\nu\sigma\tau} = \epsilon^{\alpha\nu\sigma\tau}|g|^{-\frac{1}{2}} \tag{11}$$

as well as

$$\varepsilon_{\alpha\nu\sigma\tau} = \epsilon_{\alpha\nu\sigma\tau}|g|^{\frac{1}{2}} \tag{12}$$

which are completely antisymmetric and such that

$$\varepsilon^{i_0 i_1 i_2 i_3}\varepsilon_{j_0 j_1 j_2 j_3} = -\delta^{i_0 i_1 i_2 i_3}_{j_0 j_1 j_2 j_3} \tag{13}$$

called *completely antisymmetric pseudo-tensors*. When a tensor with at least one index is multiplied by the metric tensor and the index is contracted with one index of the metric tensor, the result is a tensor in which the index has been *vertically moved*. In particular, if a tensor that is completely antisymmetric in k indices is multiplied by the completely antisymmetric pseudo-tensors and the k indices of the tensor are contracted with k indices of the completely antisymmetric pseudo-tensors, the result is a pseudo-tensor antisymmetric in $(4-k)$ indices called *dual*.

We are now at a point where we have defined for tensors a covariant operation that respects all rules of differentiation as well as the tensorial structure and an operation for the vertical re-configuration of tensorial indices, and we may wonder what happens when both operations are taken in parallel. More precisely, if the vertical index configuration cannot change the information content of a tensor, then this must be true for any tensor, and, in particular, if the tensor is the covariant derivative of some other tensor. Consequently, it must be possible to define

$$g^{\alpha\beta} D_\mu T^{\nu\ldots\zeta}_{\beta\rho\sigma\ldots\theta} = D_\mu T^{\alpha\nu\ldots\zeta}_{\rho\sigma\ldots\theta}$$

which therefore implies

$$D_\mu T^{\alpha\nu\ldots\zeta}_{\rho\sigma\ldots\theta} = D_\mu (g^{\alpha\beta} T^{\nu\ldots\zeta}_{\beta\rho\sigma\ldots\theta}) = D_\mu g^{\alpha\beta} T^{\nu\ldots\zeta}_{\beta\rho\sigma\ldots\theta} + \\ + g^{\alpha\beta} D_\mu T^{\nu\ldots\zeta}_{\beta\rho\sigma\ldots\theta}$$

so that we are left with the equation

$$D_\mu g^{\alpha\beta} T^{\nu\ldots\zeta}_{\beta\rho\sigma\ldots\theta} = 0$$

for any tensor, implying $D_\mu g^{\alpha\beta} = 0$ as well. This means that the metric tensor is covariantly constant. Conditions of vanishing of the covariant derivative of the metric tensor mean that the irrelevance of the indices disposition must be valid regardless of the differential order of the tensor. If we were to follow the common approach defining the metric first, these conditions would mean that the metric structure and the local structure will have to be independent. This is reasonable since, if a vector is constant, its norm should be constant too. It is interesting to notice that, since we have two types of covariant derivatives and because the present arguments hold, regardless of the specific covariant derivative, then we have to assume that both covariant derivatives of the metric tensor vanish as $D_\mu g_{\alpha\beta} = \nabla_\mu g_{\alpha\beta} = 0$ in general. In particular, we have that $D_\theta \varepsilon_{\alpha\beta\mu\nu} = \nabla_\theta \varepsilon_{\alpha\beta\mu\nu} = 0$ hold as well. If we are insisting that this happen, then there are very remarkable consequences that follow. To see this, expand

$$0 = D_\rho g_{\alpha\beta} = \partial_\rho g_{\alpha\beta} - g_{\alpha\mu}\Gamma^\mu_{\beta\rho} - g_{\mu\beta}\Gamma^\mu_{\alpha\rho}$$

and take the three different indices permutations combined together with the definition of torsion to get

$$\Gamma^{\rho}_{\alpha\beta}=\tfrac{1}{2}Q^{\rho}{}_{\alpha\beta}+\tfrac{1}{2}(Q_{\alpha\beta}{}^{\rho}+Q_{\beta\alpha}{}^{\rho})+ \\ +\tfrac{1}{2}g^{\rho\mu}(\partial_{\beta}g_{\alpha\mu}+\partial_{\alpha}g_{\mu\beta}-\partial_{\mu}g_{\alpha\beta})$$

in which $Q_{\rho\alpha\sigma}$ is the torsion tensor antisymmetric in the two lower indices, while $(Q_{\alpha\beta\rho}+Q_{\beta\alpha\rho})$ is a tensor symmetric in those indices, whereas the remaining coefficients written in terms of the partial derivatives of the metric tensor transform as a connection and they are symmetric in those very indices. This expression shows that the most general connection can be decomposed in terms of the torsion plus a symmetric connection, as we already knew from expression (3), but, in addition, it tells us the explicit form of $\Lambda^{\rho}_{\alpha\beta}$ as given by a symmetric combination of two torsions plus a symmetric connection entirely written in terms of the metric. It is essential to note that, if we want all possible connections to give rise to covariant derivatives, which, once applied onto the metric, give zero, then we have to restrict the torsion to verify

$$Q_{\alpha\beta\rho}=-Q_{\beta\alpha\rho}$$

spelling its complete antisymmetry [10]. The condition of metric-compatible connection extended to all connections implies the torsion to be completely antisymmetric, once again establishing a link between two structures that are a priori unrelated. The complete antisymmetry of torsion is equivalent to the existence of a single symmetric part of the connection, and therefore to the existence of a unique connection writable in terms of the metric alone. It is a remarkable fact that the torsion tensor could be reduced to be completely antisymmetric by employing a number of unrelated arguments as those presented in [11–14] and, although torsion might well not display such a symmetry, it is certainly intriguing to argue what the consequences are of this condition. We will see that some of these consequence are of paramount importance next.

Thus, we summarize by saying that the torsion tensor with all lower indices is taken to be completely antisymmetric and therefore it is possible to write it according to

$$Q_{\alpha\sigma\nu}=\tfrac{1}{6}W^{\mu}\varepsilon_{\mu\alpha\sigma\nu} \tag{14}$$

in terms of the W^{μ} pseudo-vector, therefore called the torsion pseudo-vector, while the connection

$$\Lambda^{\rho}_{\alpha\beta}=\tfrac{1}{2}g^{\rho\mu}(\partial_{\beta}g_{\alpha\mu}+\partial_{\alpha}g_{\mu\beta}-\partial_{\mu}g_{\alpha\beta}) \tag{15}$$

is symmetric and written entirely in terms of the partial derivatives of the metric tensor and, for this reason called the *metric connection*, so that

$$\Gamma^{\rho}_{\alpha\beta}=\tfrac{1}{2}g^{\rho\mu}\Big[(\partial_{\beta}g_{\alpha\mu}+\partial_{\alpha}g_{\mu\beta}-\partial_{\mu}g_{\alpha\beta})+\tfrac{1}{6}W^{\nu}\varepsilon_{\nu\mu\alpha\beta}\Big] \tag{16}$$

is the most general connection and such a decomposition is equivalent to the validity of the following conditions:

$$\nabla_{\theta}\varepsilon_{\alpha\beta\mu\nu}\equiv D_{\theta}\varepsilon_{\alpha\beta\mu\nu}=0 \tag{17}$$

$$\nabla_{\mu}g_{\alpha\beta}\equiv D_{\mu}g_{\alpha\beta}=0 \tag{18}$$

called *metric-compatibility conditions for the connection.*

Thus far, we have defined tensors and the properties compatible with the derivation. It is now the time to see what happens when we go to a following order derivative.

We may proceed to calculate the commutator of two derivatives, which in the particular case of vectors is

$$[D_\alpha, D_\beta]T^\sigma = (\Gamma^\rho_{\alpha\beta} - \Gamma^\rho_{\beta\alpha})D_\rho T^\sigma + \\ +(\partial_\alpha \Gamma^\sigma_{\kappa\beta} - \partial_\beta \Gamma^\sigma_{\kappa\alpha} + \Gamma^\rho_{\kappa\beta}\Gamma^\sigma_{\rho\alpha} - \Gamma^\rho_{\kappa\alpha}\Gamma^\sigma_{\rho\beta})T^\kappa$$

with no second derivatives. The only derivative term left is proportional to the torsion tensor $Q^\rho_{\mu\alpha}$ plus another

$$G^\sigma_{\kappa\alpha\beta} = \partial_\alpha \Gamma^\sigma_{\kappa\beta} - \partial_\beta \Gamma^\sigma_{\kappa\alpha} + \Gamma^\rho_{\kappa\beta}\Gamma^\sigma_{\rho\alpha} - \Gamma^\rho_{\kappa\alpha}\Gamma^\sigma_{\rho\beta}$$

which, although written in terms of the connection alone, is a tensor. With these expressions, we have

$$[D_\alpha, D_\beta]T^\sigma = Q^\rho_{\alpha\beta}D_\rho T^\sigma + G^\sigma_{\kappa\alpha\beta}T^\kappa$$

giving the commutator of vectors in particular. As it has been done for the connection and the most general covariant derivative, the interesting thing is that the definition of tensor $G^\sigma_{\kappa\alpha\beta}$ can be used in the most general case of the commutator of covariant derivatives. We also have

$$(\partial\partial T)_{\alpha\beta\rho\ldots\mu} = \partial_{[\alpha}(\partial T)_{\beta\rho\ldots\mu]} = \partial_{[\alpha}\partial_{[\beta}T_{\rho\ldots\mu]]} = \\ = \partial_{[\alpha}\partial_\beta T_{\rho\ldots\mu]} = 0$$

because partial derivatives always commute and therefore their commutator is always zero. Before we have had the opportunity to briefly talk about external calculus, where the external derivatives are used to calculate the border of a manifold, and the above expression refers to the fact that the border has a border that vanishes, or that there is no border of a border. Once again, apart from curiosity, there is no need to deepen these concepts in the following.

To summarize, from the connection, we may calculate

$$G^\sigma{}_{\kappa\alpha\beta} = \partial_\alpha \Gamma^\sigma_{\kappa\beta} - \partial_\beta \Gamma^\sigma_{\kappa\alpha} + \Gamma^\sigma_{\rho\alpha}\Gamma^\rho_{\kappa\beta} - \Gamma^\sigma_{\rho\beta}\Gamma^\rho_{\kappa\alpha} \tag{19}$$

which is a tensor antisymmetric in the last two indices and verifying the following cyclic permutation condition

$$D_\kappa Q^\rho{}_{\mu\nu} + D_\nu Q^\rho{}_{\kappa\mu} + D_\mu Q^\rho{}_{\nu\kappa} + \\ + Q^\pi{}_{\nu\kappa}Q^\rho{}_{\mu\pi} + Q^\pi{}_{\mu\nu}Q^\rho{}_{\kappa\pi} + Q^\pi{}_{\kappa\mu}Q^\rho{}_{\nu\pi} - \\ - G^\rho{}_{\kappa\nu\mu} - G^\rho{}_{\mu\kappa\nu} - G^\rho{}_{\nu\mu\kappa} \equiv 0 \tag{20}$$

called the *curvature tensor* and decomposable as

$$G^\sigma{}_{\kappa\alpha\beta} = R^\sigma{}_{\kappa\alpha\beta} + \tfrac{1}{2}(\nabla_\alpha Q^\sigma{}_{\kappa\beta} - \nabla_\beta Q^\sigma{}_{\kappa\alpha}) + \\ + \tfrac{1}{4}(Q^\sigma{}_{\rho\alpha}Q^\rho{}_{\kappa\beta} - Q^\sigma{}_{\rho\beta}Q^\rho{}_{\kappa\alpha}) \tag{21}$$

in terms of torsion and

$$R^\sigma{}_{\kappa\alpha\beta} = \partial_\alpha \Lambda^\sigma_{\kappa\beta} - \partial_\beta \Lambda^\sigma_{\kappa\alpha} + \Lambda^\sigma_{\rho\alpha}\Lambda^\rho_{\kappa\beta} - \Lambda^\sigma_{\rho\beta}\Lambda^\rho_{\kappa\alpha} \tag{22}$$

as a tensor antisymmetric in the last two indices and such that it verifies the cyclic permutation condition

$$R^\rho{}_{\kappa\nu\mu} + R^\rho{}_{\mu\kappa\nu} + R^\rho{}_{\nu\mu\kappa} \equiv 0 \tag{23}$$

called *metric curvature tensor*. By employing torsion and curvature, it is possible to demonstrate that we have

$$\begin{aligned}[D_{\mu},D_{\nu}]T^{\alpha_1\dots\alpha_j}_{\beta_1\dots\beta_i}=Q^{\eta}{}_{\mu\nu}D_{\eta}T^{\alpha_1\dots\alpha_j}_{\beta_1\dots\beta_i}+\\+\sum_{k=1}^{k=j}G^{\alpha_k}{}_{\sigma\mu\nu}T^{\alpha_1\dots\sigma\dots\alpha_j}_{\beta_1\dots\beta_i}-\\-\sum_{k=1}^{k=i}G^{\sigma}{}_{\beta_k\mu\nu}T^{\alpha_1\dots\alpha_j}_{\beta_1\dots\sigma\dots\beta_i}\end{aligned} \tag{24}$$

as the expression for *commutator of covariant derivatives* of the tensor field. In particular, we have that

$$\partial\partial T=0 \tag{25}$$

which is valid in the most general circumstance.

We have the validity of the following decomposition

$$\begin{aligned}R_{\kappa\rho\alpha\mu}=\tfrac{1}{2}(\partial_{\alpha}\partial_{\rho}g_{\mu\kappa}-\partial_{\mu}\partial_{\rho}g_{\kappa\alpha}+\partial_{\mu}\partial_{\kappa}g_{\alpha\rho}-\partial_{\kappa}\partial_{\alpha}g_{\mu\rho})+\\+\tfrac{1}{4}g^{\sigma\nu}[(\partial_{\rho}g_{\alpha\nu}+\partial_{\alpha}g_{\rho\nu}-\partial_{\nu}g_{\rho\alpha})(\partial_{\kappa}g_{\mu\sigma}+\partial_{\mu}g_{\kappa\sigma}-\partial_{\sigma}g_{\kappa\mu})-\\-(\partial_{\rho}g_{\mu\nu}+\partial_{\mu}g_{\rho\nu}-\partial_{\nu}g_{\rho\mu})(\partial_{\kappa}g_{\alpha\sigma}+\partial_{\alpha}g_{\kappa\sigma}-\partial_{\sigma}g_{\kappa\alpha})]\end{aligned} \tag{26}$$

showing the antisymmetry also in the first two indices as well as the symmetry involving all four indices

$$R_{\rho\kappa\mu\nu}=R_{\mu\nu\rho\kappa} \tag{27}$$

and, as a consequence, the metric curvature tensor has one independent contraction $R_{\mu\sigma}=R^{\rho}{}_{\mu\rho\sigma}$, which is symmetric and called *Ricci metric curvature tensor* with contraction $R=R_{\mu\sigma}g^{\mu\sigma}$ called *Ricci metric curvature scalar*, so that, with torsion, we can write

$$\begin{aligned}G_{\kappa\rho\alpha\mu}=\tfrac{1}{2}(\partial_{\alpha}\partial_{\rho}g_{\mu\kappa}-\partial_{\mu}\partial_{\rho}g_{\kappa\alpha}+\partial_{\mu}\partial_{\kappa}g_{\alpha\rho}-\partial_{\kappa}\partial_{\alpha}g_{\mu\rho})+\\+\tfrac{1}{4}g^{\sigma\nu}[(\partial_{\rho}g_{\alpha\nu}+\partial_{\alpha}g_{\rho\nu}-\partial_{\nu}g_{\rho\alpha})(\partial_{\kappa}g_{\mu\sigma}+\partial_{\mu}g_{\kappa\sigma}-\partial_{\sigma}g_{\kappa\mu})-\\-(\partial_{\rho}g_{\mu\nu}+\partial_{\mu}g_{\rho\nu}-\partial_{\nu}g_{\rho\mu})(\partial_{\kappa}g_{\alpha\sigma}+\partial_{\alpha}g_{\kappa\sigma}-\partial_{\sigma}g_{\kappa\alpha})]+\\+\tfrac{1}{12}\nabla^{\eta}W^{\sigma}(g_{\alpha\eta}\varepsilon_{\sigma\kappa\rho\mu}-g_{\mu\eta}\varepsilon_{\sigma\kappa\rho\alpha})+\tfrac{1}{144}[W_{\sigma}W^{\sigma}(g_{\mu\rho}g_{\alpha\kappa}-g_{\mu\kappa}g_{\alpha\rho})+\\+(W_{\alpha}W_{\rho}g_{\mu\kappa}-W_{\mu}W_{\rho}g_{\alpha\kappa}+W_{\mu}W_{\kappa}g_{\alpha\rho}-W_{\alpha}W_{\kappa}g_{\mu\rho})]\end{aligned} \tag{28}$$

showing the antisymmetry in the first two indices, and, as a consequence, it has one independent contraction chosen as $G_{\mu\sigma}=G^{\rho}{}_{\mu\rho\sigma}$ called *Ricci curvature tensor* whose contraction $G=G_{\mu\sigma}g^{\mu\sigma}$ is called *Ricci curvature scalar*.

In addition, finally, we may consider the cyclic permutation of commutator of commutators of covariant derivatives and see that the results are geometric identities.

In general, we have that we can write

$$\begin{aligned}D_{\mu}G^{\nu}{}_{\iota\kappa\rho}+D_{\kappa}G^{\nu}{}_{\iota\rho\mu}+D_{\rho}G^{\nu}{}_{\iota\mu\kappa}+\\+G^{\nu}{}_{\iota\beta\mu}Q^{\beta}{}_{\rho\kappa}+G^{\nu}{}_{\iota\beta\kappa}Q^{\beta}{}_{\mu\rho}+G^{\nu}{}_{\iota\beta\rho}Q^{\beta}{}_{\kappa\mu}\equiv 0\end{aligned} \tag{29}$$

for torsion and curvature valid as a geometric identity.

Thus far, we have introduced the concept of tensor and the way to move its indices, which we recall were coordinate indices. Coordinate indices are important since they are the type of indices involved in differentiation. However, on the other hand, tensors in coordinate indices always feel the specificity of the coordinate system. Tensorial equations do remain formally the same in all coordinate system, but the tensors themselves change in content while changing the coordinate system. The only types of tensors which, also in content, remain the same in all of the coordinate systems are the tensors that are identically equal to zero and the scalars. Zero tensors offer little information, but scalars can be used to build a formalism in which tensors can be rendered, both in form and in content, completely invariant. This formalism is known as Lorentz formalism.

In Lorentz formalism, the idea is that of introducing a basis of vectors ξ_a^α having two types of indices: one type of indices (Greek) is the usual coordinate index referring to the component of the vector, whereas the other type of indices (Latin) is a new Lorentz index referring to which vector of the basis we are considering. Under the point of view of coordinate transformations, the coordinate index ensures the transformation law of a vector, but clearly the other index ensures some different type of transformation that we will next find to be a Lorentz transformation.

Consider, for example, the tensor given by $T_{\alpha\sigma}$ and multiply it by two of the vectors ξ_a^α of the basis contracting the coordinate indices together: so $T_{\alpha\sigma}\xi_a^\alpha\xi_s^\sigma = T_{as}$ is an object that according to a coordinate transformation law does not transform at all, thus it is completely invariant, and this is exactly what we wanted. For one tensor with upper indices, the procedure would be the same but just made in terms of the covectors ξ_α^a as clear. Converting a coordinate index to a Lorentz index and then back to the coordinate index requires that $\xi_b^\alpha\xi_\alpha^c = \delta_b^c$ and $\xi_k^\alpha\xi_\sigma^k = \delta_\sigma^\alpha$ as a simple consistency condition. Finally, the operation for moving Lorentz indices is performed in terms of the metric tensor in Lorentz form $g_{\alpha\sigma}\xi_a^\alpha\xi_s^\sigma = g_{as}$, but, because we can always ortho-normalize the basis, the metric tensor in Lorentz form is just the Minkowskian matrix $g_{as} = \eta_{as}$ as it is well known indeed. Once the basis ξ_a^σ is assigned, we may pass to another basis $\xi_a^{\prime\sigma}$ linked to the initial according to the transformation $\xi_a^{\prime\sigma} = \Lambda_a^b\xi_b^\sigma$ with Λ_a^b chosen as to preserve the structure of the Minkowskian matrix and so such that it has to yield $\eta = \Lambda\eta\Lambda^T$ known as Lorentz transformation and justifying the name of the formalism.

In conclusion, after that, the coordinate tensors are converted into the Lorentz tensors, they are scalars under a general coordinate transformation but tensors under the Lorentz transformations. In doing so, we have converted the most general formalism into an equivalent formalism in which, however, the structure of the transformation now is very specific, and it can be made explicit. It is, in fact, known from the theory of Lie groups that any continuous transformation is writable according to

$$\Lambda = e^{\frac{1}{2}\sigma^{ab}\theta_{ab}}$$

in which $\theta_{ab} = -\theta_{ba}$ are the parameters while $\sigma_{ab} = -\sigma_{ba}$ are the generators and which verify specific commutation relationships that depend on the specific transformation alone. In the case of Lorentz transformation, it is known that we have six parameters and six generators given by

$$(\sigma_{ab})^i_j = \delta^i_a\eta_{jb} - \delta^i_b\eta_{ja}$$

and verifying

$$[\sigma_{ab}, \sigma_{cd}] = \eta_{ad}\sigma_{bc} - \eta_{ac}\sigma_{bd} + \eta_{bc}\sigma_{ad} - \eta_{bd}\sigma_{ac}$$

in general. While the generators are peculiar of this so-called real representation, the commutations relationship are meant to be a general character of the Lorentz transformation. As such, they will always be the same for any representation of Lorentz transformations. This shall be the Lorentz transformation that we will employ next.

We may condense everything into the following statements, starting from the fact that given a Lorentz transformation Λ the set of functions $T^{a_1 \dots a_i}_{r_1 \dots r_j}$ transforming as

$$T'^{a'_1 \dots a'_m}_{r'_1 \dots r'_n} = (\Lambda^{-1})^{r_1}_{r'_1} \dots (\Lambda^{-1})^{r_n}_{r'_n} (\Lambda)^{a'_1}_{a_1} \dots (\Lambda)^{a'_m}_{a_m} T^{a_1 \dots a_m}_{r_1 \dots r_n} \tag{30}$$

is a *tensor in Lorentz formalism*. Compared to the coordinate formalism, symmetry properties and contractions, as well as all algebraic operations, are given analogously.

However, again, Lorentz transformations can be local and so differential operations must be defined by introducing a connection. As we have done before, the connection must be introduced in general in terms of its transformation.

Therefore, once again, we summarize by saying that the set of functions $\Omega^{a}_{b\mu}$ such that, under a general coordinate transformation, transforming as a lower Greek index vector and under a Lorentz transformation transforming as

$$\Omega'^{a'}_{b'\nu} = \Lambda^{a'}_{a}\left[\Omega^{a}_{b\nu} - (\Lambda^{-1})^{a}_{k}(\partial_{\nu}\Lambda)^{k}_{b}\right](\Lambda^{-1})^{b}_{b'} \tag{31}$$

is called *spin connection*, and no decomposition nor in particular any torsion can be defined as no transposition of indices of different types is defined. With it, we have

$$\begin{aligned} D_{\mu}T^{a_1\ldots a_i}_{r_1\ldots r_j} = \partial_{\mu}T^{a_1\ldots a_i}_{r_1\ldots r_j} + \textstyle\sum_{k=1}^{k=i}\Omega^{a_k}_{p\mu}T^{a_1\ldots p\ldots a_i}_{r_1\ldots r_j} - \\ - \textstyle\sum_{k=1}^{k=j}\Omega^{p}_{r_k\mu}T^{a_1\ldots a_i}_{r_1\ldots p\ldots r_j} \end{aligned} \tag{32}$$

as *covariant derivative* of tensors in Lorentz formalism.

As we have anticipated, the passage to this formalism is done with the ξ^{a}_{σ} and ξ^{σ}_{a} vectors while the vertical movement of Latin indices is done with the η^{ab} matrix.

Thus, the passage from general coordinate formalism to the Lorentz formalism is made via the introduction of the bases of vectors ξ^{a}_{σ} and ξ^{σ}_{a} dual of one another

$$\xi^{a}_{\mu}\xi^{\mu}_{r} = \delta^{a}_{r} \tag{33}$$

$$\xi^{a}_{\mu}\xi^{\rho}_{a} = \delta^{\rho}_{\mu} \tag{34}$$

called *tetrad fields* and such that they verify the pair of ortho-normality conditions given by

$$g^{\alpha\sigma}\xi^{a}_{\alpha}\xi^{b}_{\sigma} = \eta^{ab} \tag{35}$$

$$g_{\alpha\sigma}\xi^{\alpha}_{a}\xi^{\sigma}_{b} = \eta_{ab} \tag{36}$$

as η are the *Minkowskian matrices*, preserved by Lorentz transformations. With the dual bases, ortho-normal with respect to the Minkowskian matrices, we can take a tensor in coordinate formalism with at least one Greek index and multiply it by the basis contracting one Greek index with the Greek index of the bases therefore obtaining the tensor in Lorentz formalism with a Latin index, and with a vertical movement of Latin indices which is performed in terms of the Minkowskian matrix as it is expected.

Notice that, if these two formalisms are perfectly equivalent, then their covariant derivatives should be equivalent and in particular we should be able from the most general connection to derive the spin connection. Upon requiring that $D_{\mu}\xi^{\alpha}_{a}=0$ as well as $D_{\mu}\eta_{ab}=0$, we have exactly this.

In fact, in terms of the most general coordinate connection and tetrad fields, we can always write

$$\Omega^{a}_{b\mu} = \xi^{\nu}_{b}\xi^{a}_{\rho}\left(\Gamma^{\rho}_{\nu\mu} - \xi^{\rho}_{k}\partial_{\mu}\xi^{k}_{\nu}\right) \tag{37}$$

antisymmetric in the Lorentz indices and such that, from it, we can derive the torsion tensor according to

$$Q^{a}{}_{\mu\nu} = -(\partial_{\mu}\xi^{a}_{\nu} - \partial_{\nu}\xi^{a}_{\mu} + \xi^{b}_{\nu}\Omega^{a}_{b\mu} - \xi^{b}_{\mu}\Omega^{a}_{b\nu}) \tag{38}$$

as it is easy to see, and we have that conditions (37) and $\Omega_{ab\mu} = -\Omega_{ba\mu}$ are respectively equivalent to

$$D_{\mu}\xi^{r}_{\alpha} = 0 \tag{39}$$

$$D_{\mu}\eta_{ab} = 0 \tag{40}$$

as general *coordinate-Lorentz compatibility conditions.*

In Lorentz formalism, from the spin connection, we get

$$G^{a}{}_{b\alpha\beta} = \partial_{\alpha}\Omega^{a}_{b\beta} - \partial_{\beta}\Omega^{a}_{b\alpha} + \Omega^{a}_{k\alpha}\Omega^{k}_{b\beta} - \Omega^{a}_{k\beta}\Omega^{k}_{b\alpha} \tag{41}$$

as the *curvature tensor*. Then, we have that

$$[D_{\mu}, D_{\nu}]T^{r_1...r_j} = Q^{\eta}{}_{\mu\nu}D_{\eta}T^{r_1...r_j} + \\ + \sum_{k=1}^{k=j} G^{r_k}{}_{p\mu\nu}T^{r_1...p...r_j} \tag{42}$$

is the general coordinate covariant *commutator of covariant derivatives* of the tensor field in Lorentz formalism.

As it should be expected by now, we have that

$$G^{a}{}_{b\mu\nu} = \xi^{a}_{\alpha}\xi^{\beta}_{b}G^{\alpha}{}_{\beta\mu\nu} \tag{43}$$

showing that the curvature tensor in Lorentz formalism is antisymmetric both in coordinate indices and in Lorentz indices, and so as a consequence the curvature also in this formalism has the same independent contractions which are therefore $G_{b\sigma} = G^{a}{}_{b\rho\sigma}\xi^{\rho}_{a}$ for the *Ricci curvature tensor* and $G = G_{a\sigma}\xi^{\sigma}_{p}\eta^{ap}$ for the *Ricci curvature scalar.*

After index renaming, we get

$$D_{\mu}G^{a}{}_{j\kappa\rho} + D_{\kappa}G^{a}{}_{j\rho\mu} + D_{\rho}G^{a}{}_{j\mu\kappa} + \\ + G^{a}{}_{j\beta\mu}Q^{\beta}{}_{\rho\kappa} + G^{a}{}_{j\beta\kappa}Q^{\beta}{}_{\mu\rho} + G^{a}{}_{j\beta\rho}Q^{\beta}{}_{\kappa\mu} \equiv 0 \tag{44}$$

with curvature in Lorentz form as a geometric identity.

In this way, we conclude the introduction of the most general covariant formalism with the further conversion into the specific Lorentz formalism, in which the Lorentz transformation has been made explicit in terms of its real representation. We will soon see that another representation is possible. However, before this, we introduce gauge fields.

Our main goal is going to be focusing on the fact that fields may be complex, and so it makes sense to ask what symmetries can be established for these fields: if a field is complex, there arises the issue of phase transformations and, correspondingly, it is possible to construct a calculus that is in all aspects analogous to the one we just built.

Thus, given a real function α, we have that a complex field that transforms according to the transformation

$$\phi' = e^{iq\alpha}\phi \tag{45}$$

is called *gauge field* of q *charge*, with algebraic operations defined as for geometric tensors.

Let it be given a covector field A_{ν} such that, for a phase transformation, it transforms according to the law

$$A'_{\nu} = A_{\nu} - \partial_{\nu}\alpha \tag{46}$$

then this vector is called *gauge potential*. With it,

$$D_{\mu}\phi = \partial_{\mu}\phi + iqA_{\mu}\phi \tag{47}$$

is said to be the *gauge derivative* of the gauge field.

For the gauge fields, we may introduce the operation of complex conjugation without the necessity of introducing any additional structure, and hence, for a gauge field of q charge, the complex conjugate gauge field has $-q$ charge.

There is no decomposition of the gauge potential into more fundamental elements. In fact, complex conjugation is compatible with gauge derivatives automatically.

From the gauge connection, we define

$$F_{\alpha\beta} = \partial_\alpha A_\beta - \partial_\beta A_\alpha \tag{48}$$

that is such that $F = \partial A$ and so it is a tensor which is antisymmetric and invariant by a gauge transformation called *gauge strength*. With it, we have that

$$[D_\mu, D_\nu]\phi = iqF_{\mu\nu}\phi \tag{49}$$

is the *commutator of gauge derivatives* of gauge fields.

Clearly, the gauge strength cannot be decomposed in terms of more fundamental underlying structures.

Furthermore, we have that

$$\partial_\nu F_{\alpha\sigma} + \partial_\sigma F_{\nu\alpha} + \partial_\alpha F_{\sigma\nu} = 0 \tag{50}$$

or equivalently $\partial F = 0$ as a gauge geometric identity.

There is a point that needs to be elucidated regarding the definition of the Maxwell strength. As this expression can be generalized up to

$$F_{\alpha\beta} = \nabla_\alpha A_\beta - \nabla_\beta A_\alpha,$$

then one may wonder if some non-minimal coupling could be invoked to write it as

$$F'_{\alpha\beta} = D_\alpha A_\beta - D_\beta A_\alpha = \nabla_\alpha A_\beta - \nabla_\beta A_\alpha + Q_{\alpha\beta\rho}A^\rho$$

which would violate gauge invariance. Therefore, which one between $F_{\alpha\beta}$ and $F'_{\alpha\beta}$ should be considered? The answer is actually quite simple conceptually, and it is that, in a theory of electrodynamics established within a purely geometric context, the Maxwell strength is not just a curl of a vector but the specific curl of a vector that comes as the formal expression of the curvature of two covariant derivatives. In this sense, it is clear that $F'_{\alpha\beta}$ as compared to $F_{\alpha\beta}$ has a lesser geometric meaning. Moreover, the form $F_{\alpha\beta}$ is also the one for which the geometric identities (50) called Cauchy identities are valid. In addition, so this is the only form that will interest us in the following.

This concludes the introduction of gauge fields, based on a parallel with geometric tensor. We shall now move to a following part in which these two formalisms will be merged into a single one known as spinorial formalism.

2.1.2. Spinorial Fields

In the previous parts, we have introduced tensor fields and the way to pass from coordinate into Lorentz indices, specifying that, with such a conversion, we also had the conversion of the most general coordinate transformation into the specific Lorentz transformation: the advantage of this specific Lorentz transformation is that, although it had been introduced in real representation, nevertheless, it can also be written in other representations like most notably the complex representation. In such representation, we will see that gauge fields find place naturally.

In order to find a Lorentz transformation in complex representation, we specify that these transformations are classified by semi-integer labels known as spin, and here we consider the simplest $\frac{1}{2}$-spin case: so, for the complex generators, we select those whose irreducible form is given in terms of two-dimensional matrices. General results from the theory of Lie groups tell us that the Lorentz transformation can be written according to the following form:

$$\boldsymbol{\Lambda} = e^{\frac{1}{2}\sigma^{ab}\theta_{ab}}$$

where $\theta^{ab} = -\theta^{ba}$ are the parameters as given above and $\sigma_{ab} = -\sigma_{ba}$ are the generators verifying

$$[\sigma_{ab}, \sigma_{cd}] = \eta_{ad}\sigma_{bc} - \eta_{ac}\sigma_{bd} + \eta_{bc}\sigma_{ad} - \eta_{bd}\sigma_{ac}$$

as commutation relationships. The actual form for these Lorentz generators in the case of the complex irreducible two-dimensional matrices is known to be given in terms of the Pauli matrices

$$\sigma^1 = \begin{pmatrix} 0 & 1 \\ 1 & 0 \end{pmatrix} \quad \sigma^2 = \begin{pmatrix} 0 & -i \\ i & 0 \end{pmatrix} \quad \sigma^3 = \begin{pmatrix} 1 & 0 \\ 0 & -1 \end{pmatrix}$$

according to

$$\sigma^{0A}_{\pm} = \pm\tfrac{1}{2}\sigma^A$$
$$\sigma_{AB} = -\tfrac{i}{2}\varepsilon_{ABC}\sigma^C$$

as a straightforward check would demonstrate. We notice that, in the passage from real to complex representation, a two-fold multiplicity has arisen since two opposite expressions are possible for the boosts and thus for the Lorentz transformation in full. This ambiguity can be overcome by having these two irreducible two-dimensional generators merged into a single reducible four-dimensional generators

$$\sigma^{0A} = \tfrac{1}{2}\begin{pmatrix} -\sigma^A & 0 \\ 0 & \sigma^A \end{pmatrix}$$
$$\sigma_{AB} = -\tfrac{i}{2}\varepsilon_{ABC}\begin{pmatrix} \sigma^C & 0 \\ 0 & \sigma^C \end{pmatrix}$$

which still verify the Lorentz commutation algebra. Such a merging also has the advantage that, with four-dimensional matrices, it is possible to introduce

$$\begin{pmatrix} 0 & \mathbb{I} \\ \mathbb{I} & 0 \end{pmatrix} = \gamma^0$$
$$\begin{pmatrix} 0 & \sigma^K \\ -\sigma^K & 0 \end{pmatrix} = \gamma^K$$

in terms of which the four-dimensional generators are

$$\sigma^{ab} = \tfrac{1}{4}\left[\gamma^a, \gamma^b\right]$$

and where

$$\{\gamma^a, \gamma^b\} = 2\eta^{ab}\mathbb{I}$$

in terms of the Minkowskian matrix. This way of writing four-dimensional matrices constitutes an advantage because we can see a manifest (1 + 3)-dimensional space–time form in the last two expressions. Then, it is possible to employ these last two expressions to derive a whole list of useful identities involving these matrices. To begin, we have

$$\sigma_{ab} = -\tfrac{i}{2}\varepsilon_{abcd}\pi\sigma^{cd}$$

which implicitly defines the π matrix. This matrix is the one usually indicated like a gamma with an index five as originally it was used to study five-dimensional theories, but, because we will always be in the space–time, the index five for us has no meaning and so we will use a notation with no index at all. Notice that, with this definition, we have extinguished all possible matrices since the matrices

$$\mathbb{I} \qquad \gamma^a \qquad \sigma^{ab} \qquad \gamma^a\pi \qquad \pi$$

are 16 linearly independent matrices spanning the space of four-dimensional matrices, and so they form a basis for such a space. These matrices are called Clifford matrices and they will have great importance. We have that

$$\gamma_0\gamma_a^\dagger\gamma_0 = \gamma_a$$
$$\gamma_0\sigma_{ab}^\dagger\gamma_0 = -\sigma_{ab}$$
$$\pi^\dagger = \pi$$

specifying the behavior of the Clifford matrices under conjugation. By direct inspection, one can easily see that

$$\gamma_a\gamma_b = \eta_{ab}\mathbb{I} + 2\sigma_{ab}$$

as well as

$$\gamma_i\gamma_j\gamma_k = \gamma_i\eta_{jk} - \gamma_j\eta_{ik} + \gamma_k\eta_{ij} + i\varepsilon_{ijkq}\pi\gamma^q$$

showing that products of, however, many gamma matrices can always be reduced to the product of at most two of them. Therefore, there is no need to compute the product of three or more gamma matrices. Because $\varepsilon_{0123} = 1$, we have $\pi = i\gamma^0\gamma^1\gamma^2\gamma^3$ and so

$$\{\pi, \gamma_a\} = 0$$
$$[\pi, \sigma_{ab}] = 0$$

as expected. In fact, this representation is reducible, and then Schur's lemma ensures us that there must exist one matrix different from the identity commuting with all the generators of the group. By working with all the previous identities, one can find

$$[\gamma_i, \sigma_{jk}] = \gamma_k\eta_{ij} - \gamma_j\eta_{ik}$$
$$\{\gamma_i, \sigma_{jk}\} = i\varepsilon_{ijkq}\pi\gamma^q$$

and similarly

$$\{\sigma_{ab}, \sigma_{cd}\} = \tfrac{1}{2}(\eta_{ad}\eta_{bc}\mathbb{I} - \eta_{ac}\eta_{bd}\mathbb{I} + i\varepsilon_{abcd}\pi)$$
$$[\sigma_{ab}, \sigma_{cd}] = \eta_{ad}\sigma_{bc} - \eta_{ac}\sigma_{bd} + \eta_{bc}\sigma_{ad} - \eta_{bd}\sigma_{ac}$$

as other fundamental identities. The list may go on, but, for our purposes, there is no need to reach products with more gamma matrices. The last identity tells us that the σ_{ab} matrices are the generators of the Lorentz algebra as expected. As already said, the parameters are the same we had in the Lorentz formalism since real and complex representations are merely two different forms of the same transformation. This transformation is thus given by

$$\Lambda = e^{\frac{1}{2}\sigma^{ab}\theta_{ab}}$$

in its most general form. However, in view of studying complex fields, we know that the complex phase transformation $e^{iq\alpha}$ must also be introduced. Therefore, we have

$$\Lambda e^{iq\alpha} = e^{(\frac{1}{2}\sigma^{ab}\theta_{ab} + iq\alpha\mathbb{I})} = S$$

as the Lorentz-phase transformation in its most complete form possible. This form is also called spinorial transformation. It is what we will employ to define the spinorial fields ψ as the column of four complex functions that are scalars for coordinate transformations while transforming according to $\psi' = S\psi$ under the spinorial transformations.

We may now summarize by saying that, given the most general spinorial transformation S, the column and row of complex scalars ψ and $\overline{\psi}$ transforming as

$$\psi' = S\psi \qquad \overline{\psi}' = \overline{\psi}S^{-1} \tag{51}$$

are called *spinorial fields*. Operations of sum and product respect spinor transformation.

As above, the transformation S is local and so we have to introduce the spinorial connection defined in terms of the transformation law that guarantees the derivative to be covariant for general spinorial transformations.

Therefore, we have that the coefficients $\boldsymbol{\Omega}_\nu$ transforming according to

$$\boldsymbol{\Omega}'_\nu = S(\boldsymbol{\Omega}_\nu - S^{-1}\partial_\nu S)S^{-1} \tag{52}$$

are called *spinorial connection*. Once the spinorial connection is assigned, we have that

$$\boldsymbol{D}_\mu\psi = \partial_\mu\psi + \boldsymbol{\Omega}_\mu\psi \qquad \boldsymbol{D}_\mu\overline{\psi} = \partial_\mu\overline{\psi} - \overline{\psi}\boldsymbol{\Omega}_\mu \tag{53}$$

are the *covariant derivatives* of the spinorial fields.

We now give a list of properties of the Clifford matrices.

We have that the *Clifford matrices* $\boldsymbol{\gamma}^a$ such that

$$\boldsymbol{\Lambda}\boldsymbol{\gamma}^b\boldsymbol{\Lambda}^{-1} \equiv (\Lambda^{-1})^b_a\boldsymbol{\gamma}^a \tag{54}$$

verify the anticommutation relationships

$$\{\boldsymbol{\gamma}_a, \boldsymbol{\gamma}_b\} = 2\eta_{ab}\mathbb{I} \tag{55}$$

so that we can define the matrices $\boldsymbol{\sigma}_{ab}$ as

$$\boldsymbol{\sigma}_{ab} = \tfrac{1}{4}[\boldsymbol{\gamma}_a, \boldsymbol{\gamma}_b] \tag{56}$$

and

$$\boldsymbol{\sigma}_{ab} = -\tfrac{i}{2}\varepsilon_{abcd}\boldsymbol{\pi}\boldsymbol{\sigma}^{cd} \tag{57}$$

for the $\boldsymbol{\pi}$ matrix to be implicitly defined. Then,

$$\boldsymbol{\gamma}_0\boldsymbol{\gamma}_a^\dagger\boldsymbol{\gamma}_0 = \boldsymbol{\gamma}_a \tag{58}$$

$$\boldsymbol{\gamma}_0\boldsymbol{\sigma}_{ab}^\dagger\boldsymbol{\gamma}_0 = -\boldsymbol{\sigma}_{ab} \tag{59}$$

$$\boldsymbol{\pi}^\dagger = \boldsymbol{\pi} \tag{60}$$

alongside the square properties

$$\boldsymbol{\gamma}_a\boldsymbol{\gamma}^a = 4\mathbb{I} \tag{61}$$

$$\boldsymbol{\sigma}_{ab}\boldsymbol{\sigma}^{ab} = -3\mathbb{I} \tag{62}$$

$$\boldsymbol{\pi}^2 = \mathbb{I} \tag{63}$$

together with the anticommutation properties

$$\{\boldsymbol{\pi}, \boldsymbol{\gamma}_a\} = 0 \tag{64}$$

$$\{\boldsymbol{\gamma}_i, \boldsymbol{\sigma}_{jk}\} = i\varepsilon_{ijkq}\boldsymbol{\pi}\boldsymbol{\gamma}^q \tag{65}$$

and the commutation properties

$$[\pi, \sigma_{ab}] = 0 \tag{66}$$
$$[\gamma_a, \sigma_{bc}] = \eta_{ab}\gamma_c - \eta_{ac}\gamma_b \tag{67}$$
$$[\sigma_{ab}, \sigma_{cd}] = \eta_{ad}\sigma_{bc} - \eta_{ac}\sigma_{bd} + \eta_{bc}\sigma_{ad} - \eta_{bd}\sigma_{ac} \tag{68}$$

as well as

$$\gamma_a\gamma_b = \eta_{ab}\mathbb{I} + 2\sigma_{ab} \tag{69}$$
$$\gamma_i\gamma_j\gamma_k = \gamma_i\eta_{jk} - \gamma_j\eta_{ik} + \gamma_k\eta_{ij} + i\varepsilon_{ijkq}\pi\gamma^q \tag{70}$$

all being spinorial identities. Employing γ_0, we can define

$$\overline{\psi} = \psi^\dagger\gamma_0 \qquad \gamma_0\overline{\psi}^\dagger = \psi \tag{71}$$

as the *spinor conjugation*. In particular, we have

$$\pi_L = \tfrac{1}{2}(\mathbb{I} - \pi) \tag{72}$$
$$\pi_R = \tfrac{1}{2}(\mathbb{I} + \pi) \tag{73}$$

as *left-handed/right-handed chiral projectors*. They verify

$$\pi_L^\dagger = \pi_L \tag{74}$$
$$\pi_R^\dagger = \pi_R \tag{75}$$

alongside

$$\pi_L^2 = \pi_L \tag{76}$$
$$\pi_R^2 = \pi_R \tag{77}$$

together with

$$\pi_L\pi_R = \pi_R\pi_L = 0 \tag{78}$$

and such that

$$\pi_L + \pi_R = \mathbb{I} \tag{79}$$

in general. We can also define

$$\pi_L\psi = \psi_L \qquad \overline{\psi}\pi_R = \overline{\psi}_L \tag{80}$$
$$\pi_R\psi = \psi_R \qquad \overline{\psi}\pi_L = \overline{\psi}_R \tag{81}$$

and

$$\overline{\psi}_L + \overline{\psi}_R = \overline{\psi} \qquad \psi_L + \psi_R = \psi \tag{82}$$

as *left-handed/right-handed chiral parts*. With the pair of conjugate spinors, we define the *bi-linear spinorial quantities* according to

$$2\overline{\psi}\sigma^{ab}\pi\psi = \Sigma^{ab} \tag{83}$$
$$2i\overline{\psi}\sigma^{ab}\psi = M^{ab} \tag{84}$$
$$\overline{\psi}\gamma^{a}\pi\psi = S^{a} \tag{85}$$
$$\overline{\psi}\gamma^{a}\psi = U^{a} \tag{86}$$
$$i\overline{\psi}\pi\psi = \Theta \tag{87}$$
$$\overline{\psi}\psi = \Phi \tag{88}$$

such that they are all real tensor quantities. From them,

$$\psi\overline{\psi} \equiv \tfrac{1}{4}\Phi\mathbb{I} + \tfrac{1}{4}U_{a}\gamma^{a} + \tfrac{i}{8}M_{ab}\sigma^{ab} - -\tfrac{1}{8}\Sigma_{ab}\sigma^{ab}\pi - \tfrac{1}{4}S_{a}\gamma^{a}\pi - \tfrac{i}{4}\Theta\pi \tag{89}$$

from which we get the relationships

$$2U_{\mu}S_{\nu}\sigma^{\mu\nu}\pi\psi + U^{2}\psi = 0 \tag{90}$$
$$i\Theta S_{\mu}\gamma^{\mu}\psi + \Phi S_{\mu}\gamma^{\mu}\pi\psi + U^{2}\psi = 0 \tag{91}$$

and

$$U_{a}\gamma^{a}\psi = -S_{a}\gamma^{a}\pi\psi = (\Phi\mathbb{I} + i\Theta\pi)\psi \tag{92}$$

as well as the relationships

$$\Sigma^{ab} = -\tfrac{1}{2}\varepsilon^{abij}M_{ij} \tag{93}$$
$$M^{ab} = \tfrac{1}{2}\varepsilon^{abij}\Sigma_{ij} \tag{94}$$

and

$$M_{ab}\Phi - \Sigma_{ab}\Theta = U^{j}S^{k}\varepsilon_{jkab} \tag{95}$$
$$M_{ab}\Theta + \Sigma_{ab}\Phi = U_{[a}S_{b]} \tag{96}$$

with

$$M_{ik}U^{i} = \Theta S_{k} \tag{97}$$
$$\Sigma_{ik}U^{i} = \Phi S_{k} \tag{98}$$
$$M_{ik}S^{i} = \Theta U_{k} \tag{99}$$
$$\Sigma_{ik}S^{i} = \Phi U_{k} \tag{100}$$

and also

$$\tfrac{1}{2}M_{ab}M^{ab} = -\tfrac{1}{2}\Sigma_{ab}\Sigma^{ab} = \Phi^{2} - \Theta^{2} \tag{101}$$
$$U_{a}U^{a} = -S_{a}S^{a} = \Theta^{2} + \Phi^{2} \tag{102}$$
$$\tfrac{1}{2}M_{ab}\Sigma^{ab} = -2\Theta\Phi \tag{103}$$
$$U_{a}S^{a} = 0 \tag{104}$$

called *Fierz re-arrangements* of spinor fields. If both scalars Θ and Φ do not vanish identically, we can always find a special frame where the most general spinor is written as

$$\psi=\phi e^{-\frac{i}{2}\beta\pi}e^{-i\alpha}\begin{pmatrix}1\\0\\1\\0\end{pmatrix} \tag{105}$$

up to the reversal of the third axis and up to the discrete transformation $\psi\rightarrow\pi\psi$ and called *polar form*. From this, we can write

$$\Sigma^{ab}=2\phi^2(\cos\beta u^{[a}s^{b]}-\sin\beta u_j s_k\varepsilon^{jkab}) \tag{106}$$

$$M^{ab}=2\phi^2(\cos\beta u_j s_k\varepsilon^{jkab}+\sin\beta u^{[a}s^{b]}) \tag{107}$$

in terms of

$$S^a=2\phi^2 s^a \tag{108}$$

$$U^a=2\phi^2 u^a \tag{109}$$

and

$$\Theta=2\phi^2\sin\beta \tag{110}$$

$$\Phi=2\phi^2\cos\beta \tag{111}$$

showing that the fields ϕ and β are a scalar and a pseudo-scalar, respectively. Then,

$$u_a u^a=-s_a s^a=1 \tag{112}$$

$$u_a s^a=0 \tag{113}$$

showing that the normalized velocity vector u^a and the normalized spin axial-vector s^a possess three independent components each. This means that ϕ and β are the only true real scalar degrees of freedom and called *module* and *Yvon–Takabayashi angle*. The reader interested in details for all these statements can have a look at [15].

The conditions of compatibility now read $\boldsymbol{D}_\mu\boldsymbol{\gamma}_a=0$ in general: if the spinorial matrix also has a tensorial index, the covariant derivative is to be completed to the form

$$\boldsymbol{D}_\mu\boldsymbol{B}_a=\partial_\mu\boldsymbol{B}_a-\boldsymbol{B}_b\Omega^b{}_{a\mu}+[\boldsymbol{\Omega}_\mu,\boldsymbol{B}_a]$$

which can be taken for the gamma matrix and hence implementing the above condition, and recalling that these matrices in Lorentz indices are constants, yields

$$-\boldsymbol{\gamma}_b\Omega^b{}_{a\mu}+[\boldsymbol{\Omega}_\mu,\boldsymbol{\gamma}_a]=0$$

as a relation among connections. By writing a general

$$\boldsymbol{\Omega}_\mu=a\Omega^{ij}{}_\mu\boldsymbol{\sigma}_{ij}+\boldsymbol{A}_\mu$$

and plugging it into the above relation, we obtain that

$$-\boldsymbol{\gamma}_b\Omega^b{}_{k\mu}+a\Omega^{ij}{}_\mu[\boldsymbol{\sigma}_{ij},\boldsymbol{\gamma}_k]+[\boldsymbol{A}_\mu,\boldsymbol{\gamma}_k]=0$$

and with $[\boldsymbol{\sigma}_{ij},\boldsymbol{\gamma}_k]=\eta_{kj}\boldsymbol{\gamma}_i-\eta_{ki}\boldsymbol{\gamma}_j$ we get $a=1/2$ and

$$[\boldsymbol{A}_\mu,\boldsymbol{\gamma}_s]=0$$

telling that $\boldsymbol{A}_\mu$ must commute with all gamma matrices, and thus, with all possible matrices, implying that it must be proportional to the identity matrix. Writing it as

$$\boldsymbol{A}_\mu = (pC_\mu + ibA_\mu)\mathbb{I},$$

it is possible to see that, for $b=q$, it is possible to interpret the vector A_μ as the gauge potential. Because the other term is related to conformal transformations, which are not symmetries in our case, we set $p=0$ in general. Then, we have that, all considered, we may write the expression

$$\boldsymbol{\Omega}_\mu = \tfrac{1}{2}\Omega^{ij}{}_\mu \boldsymbol{\sigma}_{ij} + iqA_\mu\mathbb{I}$$

as the most general form of spinorial connection.

To summarize, we have that the most general spinorial connection is given by

$$\boldsymbol{\Omega}_\mu = \tfrac{1}{2}\Omega_{ab\mu}\boldsymbol{\sigma}^{ab} + iqA_\mu\mathbb{I} \tag{114}$$

in terms of the generator-valued spin connection and the gauge potential, and this is equivalent to the fact that the spinorial covariant derivatives of the gamma matrices are

$$\boldsymbol{D}_\mu\boldsymbol{\gamma}_a = 0 \tag{115}$$

vanishing identically, as it is quite straightforward to see.

We have that, from the spinorial connection, we define

$$\boldsymbol{F}_{\alpha\beta} = \partial_\alpha\boldsymbol{\Omega}_\beta - \partial_\beta\boldsymbol{\Omega}_\alpha + [\boldsymbol{\Omega}_\alpha, \boldsymbol{\Omega}_\beta] \tag{116}$$

as the *spinorial curvature*. With it,

$$[\boldsymbol{D}_\mu, \boldsymbol{D}_\nu]\psi = Q^\alpha{}_{\mu\nu}\boldsymbol{D}_\alpha\psi + \boldsymbol{F}_{\mu\nu}\psi \tag{117}$$

as *commutator of covariant derivatives* of spinor fields.

Correspondingly, the curvature is decomposable as

$$\boldsymbol{F}_{\mu\nu} = \tfrac{1}{2}G_{ab\mu\nu}\boldsymbol{\sigma}^{ab} + iqF_{\mu\nu}\mathbb{I} \tag{118}$$

with the curvature tensor and gauge strength.

For a final step, we have

$$\begin{aligned} &D_\mu\boldsymbol{F}_{\kappa\rho} + D_\kappa\boldsymbol{F}_{\rho\mu} + D_\rho\boldsymbol{F}_{\mu\kappa} + \\ &+\boldsymbol{F}_{\beta\mu}Q^\beta{}_{\rho\kappa} + \boldsymbol{F}_{\beta\kappa}Q^\beta{}_{\mu\rho} + \boldsymbol{F}_{\beta\rho}Q^\beta{}_{\kappa\mu} \equiv 0 \end{aligned} \tag{119}$$

as spinorial geometrical identities holding in general.

We conclude with some fundamental comments: the first and most important one is about the fact that so far we have encountered three types of transformation laws: the first type was the most general coordinate transformation; the second type was the gauge transformation; the third type was the specific Lorentz transformation, which was given in real representation for tensors and complex representation for spinors. The coordinate transformation is known as *passive transformation*; the Lorentz transformation in real representation as well as the Lorentz transformation in complex representation merged with the gauge transformation that is the spinor transformation, are known altogether as *active transformations*. Because they have the very same parameters, we then have that both active transformations have to be performed simultaneously.

Another interesting comment is on the connections and how they are built: the torsion tensor, when the metric tensor is used, gives the connection (16); this connection, when the dual bases of tetrad fields are employed, gives the spin connection (37); this spin connection, when the gamma matrices and their commutators are considered, with the gauge potential, when multiplied by the identity matrix, give the spinorial connection (114). Remarkably, all fields fit within the most general spinorial connection, with no room for anything else: this circumstance can be seen as a sort of geometric unification of all the physical fields that are involved. On the other hand, however, in order to see it that way, we have to wait until we interpret these geometric quantities.

A final comment regards the structure of the covariant commutator (117), in which, by interpreting the covariant derivative as the covariant generators of translations, one sees that the completely antisymmetric torsion plays the role that in Lie group theory is played by the completely antisymmetric structure coefficients; we also recall to the reader that, in the curvature, there appear sigma matrices which are the generators of the Lorentz transformations and therefore of the space–time rotations. An additional interpretation that can be assigned to the covariant commutator is that, when a field is moved around, it would fail to go back to the starting point and have the initial orientation. A position mismatch is measured by torsion and a directional mismatch is measured by curvature, and this is why torsion is also said to describe the dislocations while curvature is also said to describe the disclinations of a round trip. This shows intuitively that both torsion and curvature have to be accounted for the most general description of space–time.

For some introduction to the general theory of spinors and their classifications, we refer the readers to [16,17].

Geometry and Matter in Interaction

Now that, in terms of general symmetry arguments, we have completed the definitions of all geometric quantities for the kinematic background, the next step is to have them coupled to one another in order to assign their dynamics.

2.1.3. Covariant Field Equations

When in 1916 Einstein wanted to construct the theory of gravitation, the idea he wished to follow was inspired geometrically, based on the principle of equivalence.

The principle of equivalence states the equivalence at a local level between inertia and gravitation, in the sense that locally inertial and gravitational forces can simulate one another so well that, when both present, their effects can be made to cancel: it can be stated by saying that one can always find a system of coordinates in which locally the accelerations due to gravitation are negligible.

On the other hand, one can demonstrate a theorem originally due to Weyl whose statement sounds analogous: it states that one can always find a system of coordinates in which in a point the symmetric part of the connection vanishes.

In the previous sections, we have discussed in what way the condition of complete antisymmetry of torsion gives rise to a unique symmetric part of the connection, thus removing any possible ambiguity in the implementation of Weyl theorem: hence, for a completely antisymmetric torsion, the Weyl theorem is the mathematical implementation of the principle of equivalence insofar as the acceleration due to gravitation is encoded within the symmetric connection. A unique symmetric connection corresponds to a uniquely defined gravitational field as our physical intuition would suggest. Furthermore, the single symmetric connection is entirely written in terms of the derivatives of the metric, and therefore, if the gravitational field is encoded within the symmetric connection, then the gravitational potential is encoded within the metric tensor.

The metric tensor is a tensor, but it cannot vanish and none of its derived scalar is non-trivial, and the connection is not a tensor, so they will always depend on the choice of coordinates: hence, the information about gravity will always be intertwined with inertial information, which is not a surprise, since after all we know, they are locally indistinguishable. On the other hand, we wish to have a way to tell gravity apart from inertial information, and, to do that, it is necessary to take a less local level, then considering the Riemann curvature tensor: if gravity is contained in the metric tensor as well as in the connection, then it is contained in the Riemann curvature tensor too, but the Riemann curvature tensor is a tensor from which non-trivial scalars can be derived or which can be vanished, and this is what makes it able to discriminate gravity from inertial forces. If the metric is Minkowskian and the connection is zero, we cannot know whether this is because gravity is absent or compensated by inertial forces, and, similarly, if the metric is not Minkowskian and the connection is not zero, we cannot know whether this is because gravity is present or simulated by inertial forces as above. However, if the Riemann curvature tensor is zero, we know it is because gravity is absent, and, if the Riemann curvature tensor is not zero, we know gravity is present in general terms. This has to be so, as there can not be any compensation due to inertial forces since there can be no inertial forces, within the Riemann curvature tensor.

Therefore, the principle of equivalence is the manifestation of the interpretative principle telling that gravitation is geometrized, and this is so as a consequence of the fact that gravity alone is contained in the Riemann curvature.

This statement has to be taken into account together with the parallel fact that, in Einstein relativity, the mass is a form of energy, as it is very well known indeed.

Putting the two things together, it becomes clear that the gravitational field equations that were given in terms of a second-order differential operator of the gravitational potential proportional to the mass density have to be considered as an approximated form of a more general set of gravitational field equations given by a certain linear combination of the curvature proportional to the energy.

The energy density is a tensor having two indices and therefore the curvature we are looking for must have two indices as well, which tells that we need the contraction of the Riemann curvature given by the Ricci curvature.

In 1916, all matter forms that were known consisted of macroscopic fluids, scalars, and electro-dynamic fields, all of which have an energy density symmetric in the two indices. This may be a problem as the Ricci curvature is not symmetric.

In addition, this is where Einstein assumption of the vanishing torsion came about: assuming torsion to be equal to zero meant that a specific linear combination of the Ricci curvatures were symmetric, and thus proportional to the energy.

To see this, consider identity (29) in the case in which torsion vanishes. Its full contraction gives, in the most general case, the following identity:

$$\nabla_{\mu}(R^{\mu\nu}-\tfrac{1}{2}g^{\mu\nu}R-g^{\mu\nu}\Lambda)=0$$

where the object in parenthesis is symmetric indeed, and so it can be taken to be proportional to the energy density.

Now, Einstein geometrical insight is expressed by the gravitational field equations

$$R^{\mu\nu}-\tfrac{1}{2}g^{\mu\nu}R-g^{\mu\nu}\Lambda=\tfrac{1}{2}kE^{\mu\nu} \tag{120}$$

called Einstein field equations: from them, it follows that the energy density verifies $E^{\mu\nu}=E^{\nu\mu}$ and $\nabla_{\mu}E^{\mu\nu}=0$ as is well known.

Therefore, Einstein field equations are the most general linear combination of curvatures for which geometric identities imply the validity of the symmetry and conservation law for the energy of matter. In this sense, the field equations are established on the bases of their conservation laws, themselves obtained from geometric identities, and this is what represents the Einstein spirit of geometrization—at least in the most general case without torsion.

Then, one might wonder what happens if torsion were not neglected.

The first thing we would have to keep in mind is that, in this case, geometry would provide both a curvature and a torsion tensor. The second point to be retained is that the Einsteinian gravitational theory is based on the fact that the curvature tensor is sourced by the energy. Putting things together, we should expect in the presence of torsion that there be another conserved quantity in parallel to the energy and another field equation coupling such a conserved quantity to the torsion tensor itself.

Such a quantity, however, is already at hand.

In 1928, Dirac was the first to describe a system of matter fields, named spinors, which possessed an energy together with a spin, and this is the quantity we are seeking.

In a torsional completion of the theory of gravitation, matter fields described by both an energy and a spin can naturally find a place when the spin is coupled to torsion much in the same way in which the energy is coupled to curvature. For such a theory, the full system of field equations is given by the spin–torsion field equations, which simply spell the proportionality between torsion and spin, called Sciama–Kibble field equations, alongside the curvature–energy field equations, which are formally the same as in Einstein gravity, and therefore still called Einstein field equations. Altogether, they are known under the name of Einstein–Sciama–Kibble ESK field equations [18–20].

However, contrary to what is believed, the ESK field equations are actually not the most general either because, while torsion and gravitation are independent, their field equations have the same coupling constant, and this accounts for an arbitrary restriction.

If we want independent fields to have independent coupling to their independent sources, we must find a way to obtain the ESK field equations generalized so that the two coupling constants are different.

We will not spend time on the mathematical details of this generalization, but the interested reader can find such generalized system of field equations in the case of two different coupling constants in [21].

However, then again, this is not still the most general system of field equations because the torsion tensor enters algebraically in its coupling to the spin density tensor.

As mentioned, the above system of field equations has the feature that torsion and spin are algebraically related and this constitutes a conceptual problem because in the case in which the spin density were to vanish, then torsion would vanish too, with no possibility to propagate, and hence the torsion tensor would be unphysical.

That the torsion–spin coupling is algebraic might not be seen as a problem because also the curvature–energy coupling is algebraic, but there are reasons for this situation not be to entirely analogous: the most important is that the torsion that enters in the field equations is the general Cartan torsion, with the consequence that, if the spin density were to be vanishing everywhere the Cartan torsion would be vanishing as well, but the curvature that enters the field equations is the Ricci curvature and not the Riemann curvature, with the consequence that, even if the energy density were to be vanishing everywhere, the Ricci curvature would also be vanishing, but this would not imply that the Riemann curvature would be equal to zero, and gravity may still be present.

In addition, the curvature has an internal structure given in terms of first-order derivatives of the connection and thus in terms of second-order derivatives of the metric tensor, so that there exists a dynamics for the gravitational field, unlike for torsion.

If we desire that the torsion dynamics be implemented in the theory, then we have to look for dynamical terms in the torsion–spin field equations, and also for torsional contribution in all of the other field equations as well.

We specify that our main goal is following the Einstein spirit of geometrization, and, in order to do so, we are going to obtain the field equations for the theory in a genuinely geometric way by finding the most general form of the field equations that is compatible with the constraints given by underlying geometric identities.

In order to construct the most general system of field equations, we are going to start by distinguishing them into two different types: the field equations for the geometry–matter coupling, which shall be written in the form of second-order derivatives of the metric and torsion and also gauge potentials equal to sources given by the energy and spin and also the current of fields; and the matter field equations, which will be written in the form of a first-order differential operator containing metric and torsion and gauge potentials acting on the spinor field and equalling the spinor field itself. This discrimination comes from the fact that, on the one hand, it is possible to employ spinors to construct sources for the tensor and gauge field equations, but, on the other hand, it is not possible to use tensor and gauge fields to build sources of the spinorial field equations. In the spinorial field equations, the derivatives of the spinor field must be proportional to the spinor field itself. This discrimination between the form of geometric and matter field equations is therefore intrinsic to the structure of the fields we use.

We start by considering the fact that field equations for the metric have to be in the form of some derivative of the metric equal to some source: because the covariant derivative of the metric tensor vanishes identically, then any dynamics of the metric can only be described in terms of the partial derivatives of the metric, or, equivalently, by the metric connection (15). Again, the metric connection is not a tensor, and the only way we have from the symmetric connection to form a tensor is to take another partial derivative, therefore forming the metric curvature tensor as given by (22). As Equation (27) shows, the metric curvature tensor is one peculiar combination of second-order partial derivatives of the metric, that is, arguments of symmetry under the most general coordinate transformations force at least second-order derivatives of the metric in the differential field equations. Then, arguments of simplicity would require that we do not take any further differential structure. In the following, we will see that second-order derivatives in the metric field equations endow them with a character that no other field equation will have, rendering them somewhat peculiar indeed.

For the moment, what we have established is that the metric field equations will have to be given in the form of some combination of the metric curvature tensor, and to see what combination, we start from considering that, if the leading term were to be given by the Riemann metric curvature tensor $R^{\alpha\tau\sigma\nu}$, then the vacuum equations would reduce to the condition of vanishing of Riemann metric curvature tensor, so that they would imply that there only be the trivial metric. Hence, if we want non-trivial metrics to be possible in vacuum, then the Riemann metric curvature tensor must appear contracted as the Ricci metric curvature tensor $R^{\alpha\mu}$ for leading term, and of course we may have contractions such as the Ricci metric curvature scalar $Rg_{\alpha\mu}$ or even $\Lambda g_{\alpha\mu}$ as sub-leading terms in general: as we have already seen above, the most general form of linear combination of curvatures in the field equations is given by (120), in which the only constant Λ is still undetermined, and it will remain undetermined since there is no way to fix it on geometrical grounds. Thus, we might well think of it as a generic integration constant, which can always be added and whose value cannot be fixed.

We now turn our attention to the other field equations, for which the covariant derivatives of the fields will not be identically zero.

The field equations for the torsion have to be in the form of covariant derivatives of the torsion axial-vector equal to some source: taking covariant derivatives of the torsion axial-vector implies that we will have to write the field equation in the form of the covariant divergence of the torsion axial-vector equal to a source constituted by a pseudo-scalar field, but the temporal derivative will be specified for the temporal component of the torsion axial-vector solely. In addition, thus, we must take two covariant derivatives of the torsion axial-vector as a leading term.

To assess what are the most general field equations for the torsion axial-vector, we consider that the leading term given in the form of two covariant derivatives of the torsion axial-vector $\nabla_\sigma\nabla_\alpha W_\rho$ is to be such that one of the indices of the derivatives has to be contracted yielding the two forms $\nabla_\sigma\nabla^\sigma W_\rho$ and $\nabla_\rho\nabla_\sigma W^\sigma$ as leading terms: sub-leading terms may be added eventually and so we may establish the most general form of field equations as

$$\begin{aligned}&2\Pi\nabla_\sigma\nabla^\sigma W^\eta - 2H\nabla^\eta\nabla_\rho W^\rho -\\&-V\nabla_\alpha W_\nu W_\rho \varepsilon^{\alpha\nu\rho\eta} - UW^\alpha W_\alpha W^\eta -\\&-2LR^{\eta\rho}W_\rho + 2NRW^\eta + PW^\eta = \kappa S^\eta\end{aligned}$$

where S^{α} will have to be fixed on general grounds.

This general field equation can be restricted with the Velo–Zwanziger method [22,23]. Thus, taking its divergence

$$\begin{gathered}
2(\Pi-H)\nabla_{\eta}\nabla^{\eta}\nabla_{\rho}W^{\rho}+\\
+V\nabla_{\eta}\nabla_{\alpha}W_{\nu}W_{\rho}\varepsilon^{\eta\alpha\nu\rho}+\\
+V\nabla_{\alpha}W_{\nu}\nabla_{\eta}W_{\rho}\varepsilon^{\eta\alpha\nu\rho}-\\
-2[UW^{\rho}W^{\eta}+(L-\Pi)R^{\eta\rho}]\nabla_{\eta}W_{\rho}+\\
+(2N-L+\Pi)\nabla_{\eta}RW^{\eta}-\\
-(UW^{\alpha}W_{\alpha}-2NR-P)\nabla_{\eta}W^{\eta}=\kappa\nabla_{\eta}S^{\eta},
\end{gathered}$$

it becomes possible to see that there appears a third-order time derivative for the temporal component of the torsion axial-vector implying that the constraint obtained from the field equations would actually determine the time evolution of some components of the torsion axial-vector field. Since this would spoil a balance between the number of independent field equations and the amount of degrees of freedom of a given field, then no higher-order derivative terms must be produced in the constraints and thus we set $\Pi=H$ identically. Once this is done, there is no second-order derivative in time for any components of the field in the constraint, which is thus a true constraint, which substituted back into the field equations gives

$$\begin{gathered}
2H\nabla_{\sigma}\nabla^{\sigma}W^{\eta}-2H(UW^{\alpha}W_{\alpha}-2NR-P)^{-1}\cdot\\
\cdot\nabla^{\eta}[V\nabla_{\tau}\nabla_{\alpha}W_{\nu}W_{\rho}\varepsilon^{\tau\alpha\nu\rho}+V\nabla_{\alpha}W_{\nu}\nabla_{\tau}W_{\rho}\varepsilon^{\tau\alpha\nu\rho}-\\
-2[UW^{\rho}W^{\tau}+(L-H)R^{\tau\rho}]\nabla_{\tau}W_{\rho}+\\
+(2N-L+H)\nabla_{\tau}RW^{\tau}-\kappa\nabla_{\tau}S^{\tau}]+\\
+2H\nabla^{\eta}(UW^{\alpha}W_{\alpha}-2NR)(UW^{\alpha}W_{\alpha}-2NR-P)^{-2}\cdot\\
\cdot[V\nabla_{\tau}\nabla_{\alpha}W_{\nu}W_{\rho}\varepsilon^{\tau\alpha\nu\rho}+V\nabla_{\alpha}W_{\nu}\nabla_{\tau}W_{\rho}\varepsilon^{\tau\alpha\nu\rho}-\\
-2[UW^{\rho}W^{\tau}+(L-H)R^{\tau\rho}]\nabla_{\tau}W_{\rho}+\\
+(2N-L+H)\nabla_{\tau}RW^{\tau}-\kappa\nabla_{\tau}S^{\tau}]-\\
-V\nabla_{\alpha}W_{\nu}W_{\rho}\varepsilon^{\alpha\nu\rho\eta}-UW^{\alpha}W_{\alpha}W^{\eta}-\\
-2LR^{\eta\rho}W_{\rho}+2NRW^{\eta}+PW^{\eta}=\kappa S^{\eta}
\end{gathered}$$

which contains second-order time derivatives of all components of the torsion axial-vector, and therefore this is a true field equation. To check the propagation properties of the field, we consider its characteristic determinant

$$(UW^{\alpha}W_{\alpha}-2NR-P)n^{2}+2[UW^{\tau}W^{\nu}+(L-H)R^{\tau\nu}]n_{\tau}n_{\nu}=0$$

and, by following the general discussion of Velo and Zwanziger, one can see that, in general, acausality may be possible unless we have $L=H$ and $N=U=0$ identically, in which case $n^{2}=0$ and thus causality is ensured. Notice that there are no constraints on V, which remains a free parameter.

Placing all constraints together gives field equations

$$4\nabla_{\rho}(\partial W)^{\rho\eta}-VW_{\rho}(\partial W)_{\alpha\nu}\varepsilon^{\rho\alpha\nu\eta}+2PW^{\eta}=2\kappa S^{\eta}$$

because H can be reabsorbed within a redefinition of all the other constants.

To proceed, we notice that, for the metric field equations, the source contribution from the torsion axial-vector field has to be built with no quartic torsion term because they would correspond to what in the torsion field equations are cubic torsion terms, which are absent, and no second derivatives of torsion because they would give rise to curvatures, which cannot be present since they are already addressed. Thus, it is possible to come to the most general form of this contribution as the one given by

$$E^{\mu\nu} = aW^{\mu}W^{\nu} + bW^{2}g^{\mu\nu} + z(W^{\nu}W_{\rho}(\partial W)_{\alpha\sigma}\varepsilon^{\rho\alpha\sigma\mu} + $$
$$+ W^{\mu}W_{\rho}(\partial W)_{\alpha\sigma}\varepsilon^{\rho\alpha\sigma\nu}) + y(\nabla_{\sigma}W^{\mu}(\partial W)^{\sigma\nu} + \nabla_{\sigma}W^{\nu}(\partial W)^{\sigma\mu}) + x\nabla^{\mu}W_{\sigma}\nabla^{\nu}W^{\sigma} + $$
$$+ w\nabla_{\sigma}W^{\mu}\nabla^{\sigma}W^{\nu} + v\nabla_{\alpha}W_{\sigma}\nabla^{\alpha}W^{\sigma}g^{\mu\nu} + u(\partial W)^{\nu\sigma}(\partial W)^{\mu}{}_{\sigma} + t(\partial W)^{2}g^{\mu\nu}$$

in terms of ten constants: because we know that $\nabla_{\nu}E^{\nu\mu} = 0$ and because in vacuum the divergence of the torsion field equations gives

$$4P\nabla\cdot W + V(\partial W)_{\eta\rho}(\partial W)_{\alpha\nu}\varepsilon^{\eta\rho\alpha\nu} = 0,$$

then one can easily see that it must be $V = z = v = 0$ with $x = y = -w$ and $x + u = -4t$ and together with $a = -2b = 2tP$ which must hold identically.

We also notice that we must have $P = 2M^{2}$ because this is just the mass term of the torsion axial-vector field as it is well known.

The field equations for the gauge field are also in the form of covariant derivatives of the gauge potential equal to some source: nevertheless, taking derivatives of the gauge potential means that that we have to consider the gauge strength because this is the only term that is differential in the potential and which is still gauge invariant, but, since this is irreducible, any contraction of the gauge strength vanishes and therefore these terms alone cannot be not enough. Hence, we have to take one more covariant derivative of the gauge strength as a leading term.

The most general field equations for the gauge fields have a leading term in the form $\nabla_{\sigma}F_{\alpha\rho}$ and, after contraction, we get $\nabla_{\sigma}F^{\sigma\rho}$ as the leading term: then, we get

$$\nabla_{\sigma}F^{\sigma\eta} - \tfrac{1}{12}BF_{\alpha\nu}W_{\rho}\varepsilon^{\alpha\nu\rho\eta} = qJ^{\eta}$$

in which the source J^{α} will have to be fixed as well.

The contribution from the gauge field is similarly built in terms of squares of the gauge strength strength, since any other term would violate gauge symmetry. Thus,

$$E^{\mu\nu} = \alpha F^{\mu\rho}F^{\nu}{}_{\rho} + \beta F^{\alpha\pi}F_{\alpha\pi}g^{\mu\nu}$$

in terms of two constants: again, because $\nabla_{\nu}E^{\nu\mu} = 0$ and using the form of the electrodynamic field equations, we can see that $B = 0$ and $\alpha = -4\beta$ identically.

In the metric field equation, the contributions due to torsion and gauge fields are analogous, and torsion and gauge fields are independent, so we may normalize torsion and gauge fields with no loss of generality in order to have the two constants t and β with the same value, and it is still without losing generality that they can be reabsorbed in the k constant. We notice that, in reabsorbing within a renaming of the constant k the values of the constants t and β, we did not lose any generality in their absolute value, but, in order not to lose any generality also for the sign, all constants would have to be positive, and this in general may not be the case: the reason why we did it anyway is that those constants are in front of torsion and gauge fields' energy contributions, which are positive defined. Of course, we might have assumed those constants to possess a generic sign, but, in the final form of the field equations, we would have discovered that those signs were positive, and thus we can assume this immediately with no loss of generality.

To proceed with the inclusion of matter fields, it is fundamental to notice that spinor fields are defined in terms of gamma matrices that can also be used in building fundamental quantities, whose employment allows for lowering the order of derivatives in all such quantities because every time covariance demands for a single covariant index to be present, and one gamma matrix can be used instead of one spinorial derivative.

The most general field equations for the spinor field have a leading term containing $\nabla_{\mu}\psi$ so that, after multiplying by the matrix γ^{ν}, it is possible to contract the indices getting $\gamma^{\mu}\nabla_{\mu}\psi$ as a leading term: therefore, we may establish the most general form of field equations as

$$i\gamma^{\mu}\nabla_{\mu}\psi - XW_{\sigma}\gamma^{\sigma}\pi\psi - m\psi = 0$$

where the imaginary unit has been placed because, in free cases, $i\gamma^{\mu}\nabla_{\mu}\psi - m\psi = 0$ so that taking the square of the derivative gives $\nabla^{2}\psi + m^{2}\psi = 0$, and m can be interpreted as the mass term, which is what is expected. That is, the imaginary unit has to be interpreted as what ensures the mass of the field will behave as to provide non-imaginary contribution to the dynamics of the free field equations.

Then, we have to write the general form of their contribution in the metric field equations, and this can be constructed by employing no more than one spinorial derivative of the spinor field, since gamma matrices can be used to saturate indices: eventually,

$$\begin{aligned} E^{\rho\sigma} = &\zeta[\nabla^{\rho}(\overline{\psi}\gamma^{\sigma}\psi) + \nabla^{\sigma}(\overline{\psi}\gamma^{\rho}\psi)] + \\ &+ i\xi(\overline{\psi}\gamma^{\rho}\nabla^{\sigma}\psi - \nabla^{\sigma}\overline{\psi}\gamma^{\rho}\psi + \overline{\psi}\gamma^{\sigma}\nabla^{\rho}\psi - \nabla^{\rho}\overline{\psi}\gamma^{\sigma}\psi) + \\ &+ \chi\nabla_{\alpha}(\overline{\psi}\gamma^{\alpha}\psi)g^{\rho\sigma} + \lambda i(\overline{\psi}\gamma^{\alpha}\nabla_{\alpha}\psi - \nabla_{\alpha}\overline{\psi}\gamma^{\alpha}\psi)g^{\rho\sigma} + \\ &+ \tau(W^{\sigma}\overline{\psi}\gamma^{\rho}\pi\psi + W^{\rho}\overline{\psi}\gamma^{\sigma}\pi\psi) + \upsilon W_{\alpha}\overline{\psi}\gamma^{\alpha}\pi\psi g^{\sigma\rho} + \mu\overline{\psi}\psi g^{\rho\sigma} \end{aligned}$$

in general; the contribution as a source of the torsion field equations is the spin density of the material field, and it can be taken without any spinorial derivative at all when gamma matrices are considered, therefore obtaining that

$$S^{\mu} = \omega\overline{\psi}\gamma^{\mu}\pi\psi$$

also in general; the contribution as source of the gauge field equations is the current density of the material field, and similarly it is given according to

$$J^{\rho} = p\overline{\psi}\gamma^{\rho}\psi$$

again in the most general case: by considering again the divergences of all field equations and with the same reasoning as before, one can eventually see that $\zeta = 0$ as well as $\mu = -2\lambda m$ and $p = 4\xi$ with $\tau = -2\xi X$ and $\upsilon = -2\lambda X$ and also $\kappa\omega = 2\xi X$ identically.

Finally, we notice that, without affecting the metric or the torsion or the gauge fields, the spinor field may be renormalized in such a way that, without losing generality, we can always set $4\xi = 1$ and, as a consequence, it is possible to see that the full system of field equations has been completely determined.

It is constituted by the metric field equations given according to the expression

$$\begin{aligned} R^{\rho\sigma} - \tfrac{1}{2}Rg^{\rho\sigma} - \tfrac{k}{2}[\tfrac{1}{4}(\partial W)^{2}g^{\rho\sigma} - (\partial W)^{\sigma\alpha}(\partial W)^{\rho}{}_{\alpha}] - \\ -\tfrac{k}{2}(\tfrac{1}{4}F^{2}g^{\rho\sigma} - F^{\rho\alpha}F^{\sigma}{}_{\alpha}) - \tfrac{k}{2}M^{2}(W^{\rho}W^{\sigma} - \tfrac{1}{2}W^{2}g^{\rho\sigma}) - \\ -\Lambda g^{\rho\sigma} = \tfrac{1}{2}k[\tfrac{i}{4}(\overline{\psi}\gamma^{\rho}\nabla^{\sigma}\psi - \nabla^{\sigma}\overline{\psi}\gamma^{\rho}\psi + \overline{\psi}\gamma^{\sigma}\nabla^{\rho}\psi - \nabla^{\rho}\overline{\psi}\gamma^{\sigma}\psi) - \\ -\tfrac{1}{2}X(W^{\sigma}\overline{\psi}\gamma^{\rho}\pi\psi + W^{\rho}\overline{\psi}\gamma^{\sigma}\pi\psi)] \end{aligned}$$

and the torsion field equations given according to

$$\nabla_{\rho}(\partial W)^{\rho\mu} + M^{2}W^{\mu} = X\overline{\psi}\gamma^{\mu}\pi\psi$$

with gauge field equations given by

$$\nabla_{\sigma}F^{\sigma\mu} = q\overline{\psi}\gamma^{\mu}\psi$$

as the form that is usually known, while the matter field equations are

$$i\gamma^{\mu}\nabla_{\mu}\psi - XW_{\sigma}\gamma^{\sigma}\boldsymbol{\pi}\psi - m\psi = 0$$

with parameters Λ and M and also m describing intrinsic properties of metric and torsion and also spinor fields, while parameters k, X, and q are the constants that measure the strength with which metric, torsion, and gauge fields couple to energy, spin, and current.

It is possible to write the above system of coupled field equations into the system of coupled field equations with respect to which all the torsionless derivatives and curvatures are the torsionful derivatives and curvatures.

Thus, we can give the full system of field equations as the torsion–spin and the curvature–energy field equations as

$$D_{[\rho}D^{\sigma}Q_{\mu\nu]\sigma} + Q_{\eta[\mu\nu}G_{\rho]\sigma}g^{\sigma\eta} - G_{\sigma[\rho}Q_{\mu\nu]\eta}g^{\sigma\eta} + M^2 Q_{\rho\mu\nu} = \tfrac{1}{12}S_{\rho\mu\nu} \tag{121}$$

and

$$\begin{aligned}
&G^{\rho\sigma} - \tfrac{1}{2}Gg^{\rho\sigma} - 18k[\tfrac{1}{3}D_{\alpha}D^{[\alpha}D_{\pi}Q^{\rho\sigma]\pi} - \tfrac{1}{3}D_{\alpha}D_{\eta}Q^{\eta\pi[\alpha}Q^{\rho\sigma]\nu}g_{\nu\pi} - \\
&-\tfrac{1}{3}Q^{\rho\eta\varphi}D^{[\sigma}D_{\pi}Q^{\eta\varphi]\pi} - \tfrac{1}{3}Q^{\sigma\eta\varphi}D^{[\rho}D_{\pi}Q^{\eta\varphi]\pi} + \tfrac{1}{2}D^{\pi}D_{\tau}Q^{\tau\mu\nu}Q_{\pi\mu\nu}g^{\rho\sigma} + \\
&+\tfrac{1}{4}D_{\pi}Q^{\pi\mu\nu}D^{\tau}Q_{\tau\mu\nu}g^{\rho\sigma} - D_{\pi}Q^{\pi\mu\rho}D^{\tau}Q_{\tau\mu}{}^{\sigma} - \tfrac{1}{3}D_{\eta}Q^{\eta\pi\alpha}D_{\alpha}Q^{\rho\sigma}{}_{\pi} + \\
&+\tfrac{1}{3}(Q^{\rho\eta\varphi}D_{\tau}Q^{\tau\pi\sigma} + Q^{\sigma\eta\varphi}D_{\tau}Q^{\tau\pi\rho})Q^{\eta\varphi}{}_{\pi}] - \tfrac{1}{2}k(\tfrac{1}{4}F^2 g^{\rho\sigma} - F^{\rho\alpha}F^{\sigma}{}_{\alpha}) - \\
&-(12kM^2+1)(\tfrac{1}{2}D_{\alpha}Q^{\alpha\rho\sigma} - \tfrac{1}{4}Q^{\rho\alpha\pi}Q^{\sigma}{}_{\alpha\pi} + \tfrac{1}{8}Q^2 g^{\rho\sigma}) - \Lambda g^{\rho\sigma} = \tfrac{1}{2}kT^{\rho\sigma}
\end{aligned} \tag{122}$$

called *Sciama–Kibble field equations* and *Einstein field equations* and they come alongside the gauge-current field equations

$$D_{\sigma}F^{\sigma\mu} + \tfrac{1}{2}F_{\alpha\nu}Q^{\alpha\nu\mu} = J^{\mu} \tag{123}$$

called *Maxwell field equations*, where the sources are given by the spin and the energy

$$S^{\rho\mu\nu} = -8X\tfrac{i}{4}\overline{\psi}\{\gamma^{\rho}, \sigma^{\mu\nu}\}\psi \tag{124}$$

and

$$\begin{aligned}
T^{\rho\sigma} = &\tfrac{i}{2}(\overline{\psi}\gamma^{\rho}\boldsymbol{D}^{\sigma}\psi - \boldsymbol{D}^{\sigma}\overline{\psi}\gamma^{\rho}\psi) + (8X+1)D_{\alpha}(\tfrac{i}{4}\overline{\psi}\{\gamma^{\alpha}, \sigma^{\rho\sigma}\}\psi) + \\
&+\tfrac{1}{2}(8X+1)Q^{\rho\mu\nu}\tfrac{i}{4}\overline{\psi}\{\gamma^{\sigma}, \sigma_{\mu\nu}\}\psi - (8X+1)Q^{\sigma\mu\nu}\tfrac{i}{4}\overline{\psi}\{\gamma^{\rho}, \sigma_{\mu\nu}\}\psi
\end{aligned} \tag{125}$$

alongside the current

$$J^{\mu} = q\overline{\psi}\gamma^{\mu}\psi \tag{126}$$

given in terms of the matter field. They come alongside the spinorial field equation

$$i\gamma^{\mu}\boldsymbol{D}_{\mu}\psi - i(X+\tfrac{1}{8})Q_{\nu\tau\alpha}\gamma^{\nu}\gamma^{\tau}\gamma^{\alpha}\psi - m\psi = 0 \tag{127}$$

called *Dirac spinorial field equations*, which decompose according to

$$\tfrac{i}{2}(\overline{\psi}\gamma^{\mu}\boldsymbol{D}_{\mu}\psi - \boldsymbol{D}_{\mu}\overline{\psi}\gamma^{\mu}\psi) - (X+\tfrac{1}{8})Q^{\pi\tau\eta}S^{\sigma}\varepsilon_{\pi\tau\eta\sigma} - m\Phi = 0 \tag{128}$$

$$D_{\mu}U^{\mu} = 0 \tag{129}$$

$$\tfrac{i}{2}(\overline{\psi}\gamma^{\mu}\boldsymbol{\pi}\boldsymbol{D}_{\mu}\psi - \boldsymbol{D}_{\mu}\overline{\psi}\gamma^{\mu}\boldsymbol{\pi}\psi) - (X+\tfrac{1}{8})Q^{\pi\tau\eta}U^{\sigma}\varepsilon_{\pi\tau\eta\sigma} = 0 \tag{130}$$

$$D_\mu S^\mu - 2m\Theta = 0 \quad (131)$$

$$i(\overline{\psi}\boldsymbol{D}^\alpha\psi - \boldsymbol{D}^\alpha\overline{\psi}\psi) - D_\mu M^{\mu\alpha} + (2X + \tfrac{1}{4})\varepsilon_{\pi\tau\eta\sigma}Q^{\pi\tau\eta}\Sigma^{\sigma\alpha} - 2mU^\alpha = 0 \quad (132)$$

$$D_\alpha\Phi - 2(\overline{\psi}\sigma_{\mu\alpha}\boldsymbol{D}^\mu\psi - \boldsymbol{D}^\mu\overline{\psi}\sigma_{\mu\alpha}\psi) + (2X + \tfrac{1}{4})\Theta Q^{\pi\tau\eta}\varepsilon_{\pi\tau\eta\alpha} = 0 \quad (133)$$

$$D_\nu\Theta - 2i(\overline{\psi}\sigma_{\mu\nu}\boldsymbol{\pi}\boldsymbol{D}^\mu\psi - \boldsymbol{D}^\mu\overline{\psi}\sigma_{\mu\nu}\boldsymbol{\pi}\psi) - (2X + \tfrac{1}{4})\Phi Q^{\pi\tau\eta}\varepsilon_{\pi\tau\eta\nu} + 2mS_\nu = 0 \quad (134)$$

$$(\boldsymbol{D}_\alpha\overline{\psi}\boldsymbol{\pi}\psi - \overline{\psi}\boldsymbol{\pi}\boldsymbol{D}_\alpha\psi) + D^\mu\Sigma_{\mu\alpha} + (2X + \tfrac{1}{4})\varepsilon^{\pi\tau\eta\mu}Q_{\pi\tau\eta}M_{\mu\alpha} = 0 \quad (135)$$

$$D^\mu S^\rho\varepsilon_{\mu\rho\alpha\nu} + i(\overline{\psi}\gamma_{[\alpha}\boldsymbol{D}_{\nu]}\psi - \boldsymbol{D}_{[\nu}\overline{\psi}\gamma_{\alpha]}\psi) + (2X + \tfrac{1}{4})Q^{\pi\tau\eta}\varepsilon_{\pi\tau\eta[\alpha}S_{\nu]} = 0 \quad (136)$$

$$D^{[\alpha}U^{\nu]} - i\varepsilon^{\alpha\nu\mu\rho}(\boldsymbol{D}_\mu\overline{\psi}\gamma_\rho\boldsymbol{\pi}\psi - \overline{\psi}\gamma_\rho\boldsymbol{\pi}\boldsymbol{D}_\mu\psi) - (12X + \tfrac{3}{2})Q^{\alpha\nu\rho}U_\rho - 2mM^{\alpha\nu} = 0 \quad (137)$$

which are altogether equivalent to the Dirac spinor field equations and called *Gordon decompositions*. Spin, energy, and current verify

$$D_\rho S^{\rho\mu\nu} + \tfrac{1}{2}T^{[\mu\nu]} \equiv 0 \quad (138)$$

and

$$D_\mu T^{\mu\nu} + T_{\rho\beta}Q^{\rho\beta\nu} - S_{\mu\rho\beta}G^{\mu\rho\beta\nu} + J_\rho F^{\rho\nu} \equiv 0 \quad (139)$$

alongside

$$D_\rho J^\rho = 0 \quad (140)$$

satisfied in the most general case.

Intriguingly, we notice that, in the spinor field equations, the mass appears linearly, and thus it may be positive as well as negative, and therefore it is possible to have two different types of spinor field equations. Such possibility is clear because, if $m \to -m$ is accompanied by the discrete transformation $\psi \to \boldsymbol{\pi}\psi$, then the system of field equations is invariant, and, consequently, any solutions of the first are also a solution of the second. The fact that we may have two different spinor field equations is translated into the fact that we may have two different solutions linked by $\psi \to \boldsymbol{\pi}\psi$ in general.

The full system of field equations is invariant under the transformation of parity reflection [24], and it is the most general under the restriction of being at the least-order differential form [25]. What this means is that arguments of compatibility with covariance, generality, and having field equations at their least-order derivative are enough to lead to the above physical field equations. In addition, this is true regardless of the principle of equivalence. The principle of equivalence might have been a guide for Einstein from a historical perspective, but mathematically there is no need for it. Its role is reduced to that of an interpretative principle telling us that the metric is what encodes the information about gravitation. Moreover, it is common knowledge of Einsteinian gravity that, when Einstein field equations are linearized and taken in the static case and for small velocities, they reduce to Newton equations, in which the time–time component of the metric is witnessed to be the Newtonian gravitational potential. Henceforth, the principle of equivalence can be abandoned. Certainly, this principle may give important insights, but it can be equally well disregarded, as the interpretation of gravity within the metric tensor naturally emerges from specific limits of the Einstein field equations and these come from arguments of simplicity, generality, and compatibility with identities proper to the underlying geometric structure.

3. Torsion-Spin Interactions

In this second section, we will consider the above physical field equations in order to investigate their properties: the idea will be to write them in an equivalent but somewhat clearer manner. We will end with general remarks about the interaction of geometry with its material content.

3.1. Torsion and Spinor Decomposition

To have the physical field equations converted in more manageable forms, we will decompose all quantities that can be decomposed into more fundamental ones: we will separate torsion from all torsionless quantities in all the covariant derivatives and curvature tensors. Finally, the spinor field will also be decomposed into its two irreducible chiral projections and elementary degrees of freedom.

3.1.1. Torsion as Axial-Vector Massive Field

Among all geometric fields, torsion has a special property indeed. The gauge potential is a gauge field for phase transformations, and the metric tensor can be considered a gauge field for coordinate transformations, so both are always depending on the phase or the coordinate system, while torsion is a tensor that does not have any relation with such properties. Thus, torsion can be split from gauge and metric connections, with all the covariant derivatives and curvatures being written as covariant derivatives and curvatures with no torsion but with all the torsion terms appearing as independent.

To have the most general connection decomposed into the simplest symmetric connection plus torsion terms, we only need to substitute (16) in (37), and this in (114).

Thus, the system of field equations reduces to

$$\nabla_{\rho}(\partial W)^{\rho\mu}+M^{2}W^{\mu}=X\overline{\psi}\gamma^{\mu}\pi\psi \tag{141}$$

and

$$\begin{aligned}R^{\rho\sigma}-\tfrac{1}{2}Rg^{\rho\sigma}-\Lambda g^{\rho\sigma}=\tfrac{k}{2}[\tfrac{1}{4}F^{2}g^{\rho\sigma}-F^{\rho\alpha}F^{\sigma}{}_{\alpha}+\\+\tfrac{1}{4}(\partial W)^{2}g^{\rho\sigma}-(\partial W)^{\sigma\alpha}(\partial W)^{\rho}{}_{\alpha}+M^{2}(W^{\rho}W^{\sigma}-\tfrac{1}{2}W^{2}g^{\rho\sigma})+\\+\tfrac{i}{4}(\overline{\psi}\gamma^{\rho}\nabla^{\sigma}\psi-\nabla^{\sigma}\overline{\psi}\gamma^{\rho}\psi+\overline{\psi}\gamma^{\sigma}\nabla^{\rho}\psi-\nabla^{\rho}\overline{\psi}\gamma^{\sigma}\psi)-\\-\tfrac{1}{2}X(W^{\sigma}\overline{\psi}\gamma^{\rho}\pi\psi+W^{\rho}\overline{\psi}\gamma^{\sigma}\pi\psi)]\end{aligned} \tag{142}$$

for the torsion–spin and curvature–energy coupling, and

$$\nabla_{\sigma}F^{\sigma\mu}=q\overline{\psi}\gamma^{\mu}\psi \tag{143}$$

for the gauge–current coupling. Then, we also have that

$$i\gamma^{\mu}\nabla_{\mu}\psi-XW_{\sigma}\gamma^{\sigma}\pi\psi-m\psi=0 \tag{144}$$

for the spinor field equations.

If we take the divergence of (141) and contract (142), we obtain the constraints

$$M^{2}\nabla_{\mu}W^{\mu}=2Xmi\overline{\psi}\pi\psi \tag{145}$$

and

$$\tfrac{2}{k}R+\tfrac{8}{k}\Lambda-M^{2}W^{2}=-m\overline{\psi}\psi \tag{146}$$

where (144) has been used.

It is now possible to interpret torsion: just a quick look at the torsion–spin and curvature–energy field equations simply reveals that *torsion is an axial-vector massive field* verifying Proca field equations with corresponding energy and torsional-spin coupling within the gravitational field equations [26]. With this insight, one might now wonder if there really was the necessity to go through the trouble of insisting on the presence of torsion if all comes to the presence of an axial-vector massive field, asking why we could not simply impose torsion equal to zero and then allowing an axial-vector massive field to be included into the theory. The answer is that, although mathematically it is equivalent to follow both approaches, conceptually the former approach is the most straightforward construction in which all quantities are defined and all relationships are built in the most general manner, while, on the other hand, the latter approach would be afflicted by a number of arbitrary assumptions. If this latter approach were the one to be followed, we would have to justify why torsion albeit in general present should be removed, why, among all fields that could be included, we pick precisely a vector field with pseudo-tensorial properties, and why it would have to be massive, and hence resulting into an approach having three unjustified assumptions in alternative to the other approach in which assumptions are either justified or not assumed at all. In order to avoid this high degree of arbitrariness, we prefer to follow the approach that we actually followed here. This leads after all to the presence of an axial-vector massive field. Then, if in some part of the theory, there were to appear new physics that could somehow be reconducted to an axial-vector massive field, we would know that these effects would emerge from the existence of torsion. In fact, such effects might be something that we have already observed, even if we ignored that they could come from the torsion tensor.

Spinors as Sum of Chiral Parts

Analogously to the covariant decomposition of torsion, there is also a perfectly covariant split of the spinor field into its two chiral parts according to (80) and (81) and therefore in its degrees of freedom as expressed by (105).

When (105) is plugged into the Gordon decompositions, we obtain the polar forms of the Gordon decompositions, among which we find the following two equations:

$$-XW_{\mu}-\tfrac{1}{4}g_{\mu\nu}\varepsilon^{\nu\rho\sigma\alpha}\partial_{\rho}\xi^{k}_{\sigma}\xi^{j}_{\alpha}\eta_{jk}-(\nabla\alpha-qA)^{\iota}u_{[\iota}s_{\mu]}+s_{\mu}m\cos\beta+\tfrac{1}{2}\nabla_{\mu}\beta=0$$

and

$$s_{\mu}m\sin\beta-(\nabla\alpha-qA)^{\rho}u^{\nu}s^{\alpha}\varepsilon_{\mu\rho\nu\alpha}+\tfrac{1}{2}|\xi|^{-1}\xi^{k}_{\mu}\partial_{\alpha}(|\xi|\xi^{\alpha}_{k})+\nabla_{\mu}\ln\phi=0$$

which are very special since we can show that these two expressions imply the spinor field Equations (144): in fact, by employing the above pair of equations, we have that

$$\begin{aligned}i\gamma^{\mu}\nabla_{\mu}\psi-XW_{\sigma}\gamma^{\sigma}\boldsymbol{\pi}\psi-m\psi=\\=(\nabla\alpha-qA)^{\iota}(i\gamma^{\mu}u^{\nu}s^{\alpha}\varepsilon_{\mu\iota\nu\alpha}+u_{[\iota}s_{\mu]}\gamma^{\mu}\boldsymbol{\pi}+\gamma_{\iota})\psi-\\-m(is_{\mu}\gamma^{\mu}\sin\beta+s_{\mu}\gamma^{\mu}\boldsymbol{\pi}\cos\beta+\mathbb{I})\psi\end{aligned}$$

and then, using (90, 91), we get

$$i\gamma^{\mu}\nabla_{\mu}\psi-XW_{\sigma}\gamma^{\sigma}\boldsymbol{\pi}\psi-m\psi=0$$

which is the Dirac spinor field Equation (144) as expected.

Therefore, we may summarize by saying that the Dirac spinorial field equation are equivalent to the equations

$$\nabla_{\mu}\beta-2XW_{\mu}-\tfrac{1}{2}g_{\mu\nu}\varepsilon^{\nu\rho\sigma\alpha}\partial_{\rho}\xi^{k}_{\sigma}\xi^{j}_{\alpha}\eta_{jk}-2(\nabla\alpha-qA)^{\iota}u_{[\iota}s_{\mu]}+2ms_{\mu}\cos\beta=0 \tag{147}$$

and

$$\nabla_{\mu}\ln\phi^{2}+|\xi|^{-1}\xi^{k}_{\mu}\partial_{\alpha}(|\xi|\xi^{\alpha}_{k})-2(\nabla\alpha-qA)^{\rho}u^{\nu}s^{\alpha}\varepsilon_{\mu\rho\nu\alpha}+2ms_{\mu}\sin\beta=0 \tag{148}$$

in the most general case that is possible.

Thus, we interpret spinor fields in this way: despite being fundamental, *spinor fields are reducible, constituted from two chiral parts. Their independence is measured by the Yvon–Takabayashi angle describing internal dynamics of the spinor field. The module describes the overall matter distribution*. These are the two degrees of freedom of the spinor field, with all space–time derivatives of these two degrees of freedom specified by (147) and (148) [27].

3.2. Torsion–Spinor Interactions

We now have all elements to deepen the investigation about the interaction between geometry and matter.

3.2.1. Torsion–Spinor Binding

In the recent parts, we have seen that (145) and (146) provide very simply links between geometrical structures and the bi-linear spinorial scalars. In addition, (147) and (148) constitute some form of dynamical conditions upon such spinorial scalars.

We have in fact that (145) and (146) can be written as

$$M^2\nabla_\mu W^\mu = 4Xm\phi^2 \sin\beta \tag{149}$$

and

$$\tfrac{2}{k}R + \tfrac{8}{k}\Lambda - M^2W^2 = -2m\phi^2 \cos\beta \tag{150}$$

linking torsion and curvature to Yvon–Takabayashi angle and module, these last being subject to

$$\nabla_\mu\beta - 2XW_\mu - \tfrac{1}{2}g_{\mu\nu}\varepsilon^{\nu\rho\sigma\alpha}\partial_\rho\xi^k_\sigma\xi^j_\alpha\eta_{jk} - 2(\nabla\alpha - qA)^\iota u_{[\iota}s_{\mu]} + 2ms_\mu\cos\beta = 0 \tag{151}$$

and

$$\nabla_\mu \ln\phi^2 + |\xi|^{-1}\xi^k_\mu\partial_\alpha(|\xi|\xi^\alpha_k) - 2(\nabla\alpha - qA)^\rho u^\nu s^\alpha\varepsilon_{\mu\rho\nu\alpha} + 2ms_\mu\sin\beta = 0 \tag{152}$$

as dynamical conditions. Therefore, by solving these last equations, *we can always integrate spinor fields as all their degrees of freedom can be tied to torsion and curvature*.

This is remarkable because it shows that formally the spinorial degrees of freedom can always be replaced by quantities related to the underlying geometric structure.

To conclude this part, we will give a few more results starting from the introduction of the potentials

$$K_\mu = 2XW_\mu + \tfrac{1}{2}g_{\mu\nu}\varepsilon^{\nu\rho\sigma\alpha}\partial_\rho\xi^k_\sigma\xi^j_\alpha\eta_{jk} + 2(\nabla\alpha - qA)^\iota u_{[\iota}s_{\mu]} \tag{153}$$

and

$$G_\mu = -|\xi|^{-1}\xi^k_\mu\partial_\alpha(|\xi|\xi^\alpha_k) + 2(\nabla\alpha - qA)^\rho u^\nu s^\alpha\varepsilon_{\mu\rho\nu\alpha}, \tag{154}$$

in terms of which we have

$$\nabla_\mu\beta - K_\mu + s_\mu 2m\cos\beta = 0 \tag{155}$$

and

$$\nabla_\mu \ln\phi^2 - G_\mu + s_\mu 2m\sin\beta = 0 \tag{156}$$

as the Dirac equations in polar form. From these, we get

$$\left|\nabla\tfrac{\beta}{2}\right|^2 - m^2 - \phi^{-1}\nabla^2\phi + \tfrac{1}{2}(\nabla G + \tfrac{1}{2}G^2 - \tfrac{1}{2}K^2) = 0 \tag{157}$$

and

$$\nabla_\mu(\phi^2\nabla^\mu\tfrac{\beta}{2})-\tfrac{1}{2}(\nabla K+KG)\phi^2=0 \tag{158}$$

as a Hamilton–Jacobi equation and a continuity equation.

Alternatively, we may define

$$\tfrac{1}{2}(\nabla_\mu\beta-2XW_\mu-\tfrac{1}{2}g_{\mu\nu}\varepsilon^{\nu\rho\sigma\alpha}\partial_\rho\xi^k_\sigma\xi^j_\alpha\eta_{jk})=Y_\mu \tag{159}$$

and

$$-\tfrac{1}{2}[\nabla_\mu\ln\phi^2+|\xi|^{-1}\xi^k_\mu\partial_\alpha(|\xi|\xi^\alpha_k)]=Z_\mu \tag{160}$$

in terms of which

$$Y_\mu-(\nabla\alpha-qA)^\iota u_{[\iota}s_{\mu]}+ms_\mu\cos\beta=0 \tag{161}$$

and

$$Z_\mu-(\nabla\alpha-qA)^\rho u^\nu s^\alpha\varepsilon_{\mu\rho\nu\alpha}+ms_\mu\sin\beta=0 \tag{162}$$

as Dirac equations in polar form. Then, defining

$$P_\nu=\nabla_\nu\alpha-qA_\nu \tag{163}$$

as the momentum, we have that

$$P^\nu=m\cos\beta u^\nu+Y_\mu u^{[\mu}s^{\nu]}+Z_\mu s_\rho u_\sigma\varepsilon^{\mu\rho\sigma\nu} \tag{164}$$

giving its explicit form in terms of mass and velocity but also in terms of the Yvon–Takabayashi angle and spin as well as the potentials given in the (159) and (160) above.

There is a very important point to be clarified regarding the spinorial active transformations acting on spinorial fields. Consider the rotations around the third axis and the spinors in polar form (105): despite the fact that these spinors are aligned along the axis around which we perform the above rotation, that rotation does not leave them unchanged (as we have for vectors). This might already sound problematic, but, in addition, we also have that, when such a rotation is given for an angle $\theta=2\pi$, it is $\boldsymbol{\Lambda}=-\mathbb{I}$ implying that the spinor would not go back to the initial configuration (as we have when we perform a passive rotation). This too sounds peculiar. Thus, we might ask, is there any intuitive way to see things under which these odd behaviors would look natural? First of all, we have to take into account the fact that the rotation is an active rotation, and therefore an operation that, keeping fixed the space–time, moves the spinor. Then, we have to keep in mind that spinors are more sensitive than vectors to the structure of the space–time, as if anchored instead of being free to slide in it. Thus, for a given active rotation around a certain axis, a vector behaves like a pole, and, if aligned to the axis of rotation, it would be left unchanged as a whole. For the same active rotation, however, spinors would behave as a pole with a flag, so, even if aligned to the axis of rotation, they would be left unchanged almost fully but not quite entirely. They would indeed behave as if the rotation was taking place on a Möbius band. One way to picture them would be that of the belt trick, or the spinning plates, as described in [28].

We have shown that there is a duplicity in the spinorial structure made clear from the fact that spinors were defined up to the $\psi\rightarrow\pi\psi$ discrete transformation, or from the fact that the combined $\psi\rightarrow\pi\psi$ and $m\rightarrow-m$ is one symmetry of the physical field equations. Such duplicity may suggest a form of matter/antimatter duality [29,30].

Physical effects and phenomenological implications provided by a torsion tensor with a dynamical axial-vector field have also been recently presented in [31].

4. Limiting Situations

In this third section, we will consider the above theory in some specific cases so to deepen their examination: we will first consider what happens as a consequence of the fact that torsion is an axial-vector field with mass and in addition we will discuss what happens as a consequence of the fact that also the spinor field is massive. Eventually, we will see what happens in the complementary situation in which masslessness will allow another symmetry.

4.1. Massive Cases

We start from the analysis of the consequences of massive torsion: assuming also that the torsion mass is quite large, we will study the effective approximation. Finally, some comments about low-speed conditions will be given from the perspective of the non-relativistic limit.

4.1.1. Effective Approximation

To begin our investigation, we remark that torsion had a first property that was unlike what any other space–time or gauge fields had, and that it comes as a general feature of the geometry and not from a symmetry principle, with the consequence that there is no symmetry protecting it from being massive. Thus, the torsional field Equations (141) are such that, in the presence of a massive field, they can be taken in the approximation in which the dynamical term is negligible compared to the mass term. Thus, we may write

$$M^2W^\mu \approx X\overline{\psi}\gamma^\mu\pi\psi \tag{165}$$

yielding an algebraic equation that can be used to have torsion substituted in all other field equations in terms of the spin of the spinor, so that all torsional contributions can effectively be converted into spin–spin interactions.

This is the so-called effective approximation.

Let us now move back to the physical field equations, which consist of expressions (141)–(144). These equations, by employing the variational formalism, can be derived from a dynamical action whose Lagrangian is

$$\begin{aligned}\mathscr{L} = & -\tfrac{1}{4}(\partial W)^2 + \tfrac{1}{2}M^2W^2 - \tfrac{1}{k}R - \tfrac{2}{k}\Lambda - \tfrac{1}{4}F^2 + \\ & + i\overline{\psi}\gamma^\mu\nabla_\mu\psi - X\overline{\psi}\gamma^\mu\pi\psi W_\mu - m\overline{\psi}\psi\end{aligned} \tag{166}$$

where torsion is already decomposed. Equivalently,

$$\begin{aligned}\mathscr{L} = & -\tfrac{1}{4}(\partial W)^2 + \tfrac{1}{2}M^2W^2 - \tfrac{1}{k}R - \tfrac{2}{k}\Lambda - \tfrac{1}{4}F^2 + \\ & + i\overline{\psi}_L\gamma^\mu\nabla_\mu\psi_L + i\overline{\psi}_R\gamma^\mu\nabla_\mu\psi_R + \\ & + X\overline{\psi}_L\gamma^\mu\psi_L W_\mu - X\overline{\psi}_R\gamma^\mu\psi_R W_\mu - \\ & - m\overline{\psi}_R\psi_L - m\overline{\psi}_L\psi_R\end{aligned} \tag{167}$$

in which the chiral split is already done.

In effective approximation, the Lagrangian becomes

$$\begin{aligned}\mathscr{L}_{\text{effective}} = & -\tfrac{1}{k}R - \tfrac{2}{k}\Lambda - \tfrac{1}{4}F^2 + \\ & + i\overline{\psi}\gamma^\mu\nabla_\mu\psi + \tfrac{1}{2}\tfrac{X^2}{M^2}\overline{\psi}\gamma^\mu\psi\overline{\psi}\gamma_\mu\psi - m\overline{\psi}\psi\end{aligned} \tag{168}$$

where (102) was used. Equivalently,

$$\begin{aligned}\mathscr{L}_{\text{effective}} = -\tfrac{1}{k}R - \tfrac{2}{k}\Lambda - \tfrac{1}{4}F^2 + \\ + i\overline{\psi}_L\gamma^\mu\nabla_\mu\psi_L + i\overline{\psi}_R\gamma^\mu\nabla_\mu\psi_R + \\ + \tfrac{X^2}{M^2}\overline{\psi}_L\gamma^\mu\psi_L\overline{\psi}_R\gamma_\mu\psi_R - \\ - m\overline{\psi}_R\psi_L - m\overline{\psi}_L\psi_R\end{aligned} \tag{169}$$

which is exactly the Lagrangian of the Nambu–Jona–Lasinio model [32,33].

As (102) shows, it is precisely the axial-vector nature of the field that produces the inversion of the sign of the potential, making the contact interaction attractive.

In addition, as it is clear, such an interaction takes place between two chiral projections.

In fact, general knowledge of the NJL model shows that the torsionally-induced spin–spin contact interaction is an attraction between the two chiral parts of the spinor.

We recall that the role of the Higgs boson is analogous.

This is not surprising since the torsion–spin coupling is the axial-vector analog of the scalar Yukawa coupling. In fact, if the effective Lagrangian (168) is further re-arranged in terms of (102), it can be put in the form

$$\begin{aligned}\mathscr{L}^{\text{spinor}}_{\text{effective}} = i\overline{\psi}\gamma^\mu\nabla_\mu\psi + \\ + \tfrac{1}{2}\tfrac{X^2}{M^2}(|\overline{\psi}\psi|^2 + |i\overline{\psi}\boldsymbol{\pi}\psi|^2) - m\overline{\psi}\psi\end{aligned} \tag{170}$$

as the Lagrangian of the spinor field complemented with the torsionally-induced spin-contact interactions. On the other hand, in the standard model of particle physics [34], we might take into account the Lagrangian for the electron in the presence of the Higgs interaction alone. If the Higgs is taken in effective approximation, we have

$$M^2 H \approx -\tfrac{Y}{2}\bar{e}e \tag{171}$$

which is analogous to (165) in scalar form. Plugging it into the standard model Lagrangian gives

$$\mathscr{L}^{\text{electron}}_{\text{effective}} = i\bar{e}\gamma^\mu\nabla_\mu e + \tfrac{Y^2}{4M^2}|\bar{e}e|^2 - m\bar{e}e \tag{172}$$

for the electronic field with the Higgs-induced interaction. The comparison between (170) and (172) shows that

$$\mathscr{V}^{\text{spinor}}_{\text{effective}} = -\tfrac{1}{2}\tfrac{X^2}{M^2}(|\overline{\psi}\psi|^2 + |i\overline{\psi}\boldsymbol{\pi}\psi|^2) \tag{173}$$

$$\mathscr{V}^{\text{electron}}_{\text{effective}} = -\tfrac{Y^2}{4M^2}|\bar{e}e|^2 \tag{174}$$

meaning that torsion gives a self-interaction with a scalar part and a pseudo-scalar part, so spin dependent, while the Higgs gives rise to a scalar self-interaction only. Apart from this, they are both attractive and occur between the chiral parts.

From the Lagrangian (170), we extract the potential

$$\mathscr{V} = -\tfrac{X^2}{2M^2}(|\overline{\psi}\psi|^2 + |i\overline{\psi}\boldsymbol{\pi}\psi|^2) \tag{175}$$

which is negative, as expected for attractive interactions, and so the energy is the kinetic energy plus the potential energy, given by the general expression according to

$$\mathscr{E} = \mathscr{K} - \tfrac{X^2}{2M^2}(|\overline{\psi}\psi|^2 + |i\overline{\psi}\boldsymbol{\pi}\psi|^2) \tag{176}$$

and we recall that all quantities are densities. In fact, a straightforward dimensional analysis shows that we have

$$E = K - \tfrac{X^2}{2M^2}\tfrac{1}{V} \tag{177}$$

having interpreted $|\overline{\psi}\psi|^2+|i\overline{\psi}\pi\psi|^2=V^{-2}$ as inverse volume, which is reasonable at least on dimensional grounds. On the other hand, it is possible to compute what turns out to be the expression for the internal energy of a van der Waals gas with negative pressure, given by

$$U=T-C^2\tfrac{1}{V} \tag{178}$$

in terms of a generic constant C, as it is known from general thermodynamic arguments.

Because thermodynamically the kinetic energy can be interpreted as the temperature, and of course the energy is the internal energy, then the formal similarities of these two apparently unrelated expressions are striking.

In this thermodynamic analogy, we have that the single spinor field can be seen as a matter distribution behaving in the same way in which a van der Waals attractive gas with attractive intermolecular forces would [35].

Consider now the pair of second-order derivative Equations (157) and (158) with $K_\mu \approx 2XW_\mu$ and $G_\mu \approx 0$ and implement the torsion effective approximation: (157) becomes

$$\begin{aligned}\nabla^2\phi-4X^4M^{-4}\phi^5+2X^2M^{-2}K\cdot s\phi^3 -\\ -|\nabla\beta/2|^2\phi+m^2\phi =0\end{aligned} \tag{179}$$

with a quintic potential. We see that such a nonlinear potential is attractive.

Summarizing, in the effective approximation, torsional interactions give rise to a contact force much in the same way in which the Higgs field would, with these two forces being similarly attractive and chiral. In addition, we have seen that the torsional potential would also be analogous to the internal energy of an attractive van der Waals gas.

Consequently, insofar as this effective approximation holds, there is a clear indication that torsion is a sort of internal binding force, a tension, localizing the spinor.

4.1.2. Non-Relativistic Limit

In the initial section in which we introduced kinematic quantities, it was clear that tensors and gauge fields were characterized by general definition while spinors were defined in a way that was strongly dependent on the background being a (1 + 3)-dimensional space–time. Therefore, in such a space–time, the spinorial transformation law has a total of six parameters while spinor fields defined in terms of this transformation have a total of eight real components, and we have seen how to remove six components from the spinor field leaving it with two physical degrees of freedom.

However, now one might wonder what would happen when we consider the non-relativistic limit. In such a limit of small velocities, boosts can no longer be viable transformation laws and so time gets frozen, reducing the background to effectively be a three-dimensional space. In this case, spinorial transformation laws would possess a total of three parameters while spinor fields defined by this transformation would have four real components, so that we could remove three of the components from the spinor, hence leaving it with only one physical degree of freedom and nothing more.

To be mathematically precise, in the (1 + 3)-dimensional space–time, the spinor can always be written as (105) like

$$\psi=\phi e^{-i\alpha}\begin{pmatrix} e^{\frac{i}{2}\beta}\\ 0\\ e^{-\frac{i}{2}\beta}\\ 0\end{pmatrix} \tag{180}$$

in the representation we used throughout this presentation, called chiral representation, with Yvon–Takabayashi angle expressed in terms of imaginary exponentials. It is, however, possible to introduce another representation in which the Yvon–Takabayashi is expressed in terms of real circular functions, called standard representation, obtained via the unitary transformation

$$U = \frac{1}{\sqrt{2}}\begin{pmatrix} \mathbb{I} & \mathbb{I} \\ -\mathbb{I} & \mathbb{I} \end{pmatrix} \tag{181}$$

which operates on gamma matrices to give

$$\gamma^0 = \begin{pmatrix} \mathbb{I} & 0 \\ 0 & -\mathbb{I} \end{pmatrix} \tag{182}$$

$$\gamma^K = \begin{pmatrix} 0 & \sigma^K \\ -\sigma^K & 0 \end{pmatrix} \tag{183}$$

so that

$$\sigma^{0A} = \frac{1}{2}\begin{pmatrix} 0 & \sigma^A \\ \sigma^A & 0 \end{pmatrix} \tag{184}$$

$$\sigma_{AB} = -\frac{i}{2}\varepsilon_{ABC}\begin{pmatrix} \sigma^C & 0 \\ 0 & \sigma^C \end{pmatrix} \tag{185}$$

and on spinors to give

$$\psi = \sqrt{2}\phi e^{-i\alpha}\begin{pmatrix} \cos\frac{\beta}{2} \\ 0 \\ -i\sin\frac{\beta}{2} \\ 0 \end{pmatrix} \tag{186}$$

in general. In (1 + 3)-dimensional space–times, spinors can always be written as above.

In three-dimensional space, the spinor can always be written according to

$$\psi = \phi e^{-i\alpha}\begin{pmatrix} 1 \\ 0 \end{pmatrix} \tag{187}$$

and the representation is unique.

Upon comparison, it becomes easy to see that the non-relativistic limit requires a small spatial part of the velocity u^a but also a small Yvon–Takabayashi angle β and, when this is accomplished, we have that, in standard representation, the spinor reduces to the form

$$\psi = \sqrt{2}\phi e^{-i\alpha}\begin{pmatrix} 1 \\ 0 \\ 0 \\ 0 \end{pmatrix} \tag{188}$$

where the lower component has vanished, and the upper component has reduced to

$$\psi = \phi e^{-i\alpha}\begin{pmatrix} 1 \\ 0 \end{pmatrix} \tag{189}$$

up to an overall constant, which is irrelevant.

It is also worth noticing that, so far, we have been able to obtain a procedure of non-relativistic limit that involves no definition of momentum. However, if the momentum in (163) is considered, we would see that only in the case in which all spin contributions are negligible can the explicit form of the momentum (164) reduce to

$$P^{\nu} \approx m\cos\beta u^{\nu} \tag{190}$$

so that the non-relativistic limit is given as a small spatial part of P^a as commonly used.

Therefore, we have that the non-relativistic limit is implemented by the requirement that, when written in standard representation, the spinor loses its lower component

$$\psi \rightarrow \sqrt{2}\phi e^{-i\alpha}\begin{pmatrix}1\\0\\0\\0\end{pmatrix} \tag{191}$$

and this is why this component is called small component.

Equivalently, we have that the conditions

$$u^a \rightarrow \begin{pmatrix}1\\\hline 0\\0\\0\end{pmatrix} \tag{192}$$

$$\beta \rightarrow 0 \tag{193}$$

are what implements the non-relativistic limit.

In addition, additionally, if the spin is negligible, then

$$P^a \rightarrow \begin{pmatrix}m\\\hline 0\\0\\0\end{pmatrix} \tag{194}$$

is the final form of non-relativistic limit, and the one that is normally employed.

We notice that, because u^a is the velocity and, as we said, β is already linked to the internal dynamics, then, in a non-relativistic limit, the spinor loses both the overall and the internal motions, which is intuitive. In addition, it is remarkable that the spinorial lower component is connected to *Zitterbewegung* effects which are yet another signature of internal dynamics [36]. There seems to be a very tight relation linking the Yvon–Takabayashi angle with effects of *Zitterbewegung* as manifestations of internal dynamics for the Dirac spinorial matter fields in general [37].

4.2. Massless Case

In this part, we will study the complementary situation given when both torsion and spinors are massless.

Ultra-Relativistic Limit

Let us now consider what happens when torsion as well as the Dirac spinor are both massless. The torsional field Equations (141) become

$$\nabla_\rho(\partial W)^{\rho\mu} = X\overline{\psi}\gamma^\mu\boldsymbol{\pi}\psi \tag{195}$$

which are analogous to the electro-dynamic field equations apart from the fact that these above are parity-odd. This aside, both are vector field equations in a massless case, and, as such, we should expect some symmetry to be present. The full Lagrangian in the case of masslessness also for the spinor field is given by the following:

$$\begin{aligned}\mathscr{L}_{\text{massless}} = &-\tfrac{1}{4}(\partial W)^2 - \tfrac{1}{k}R - \tfrac{2}{k}\Lambda - \tfrac{1}{4}F^2 + \\ &+ i\overline{\psi}_L\gamma^\mu\nabla_\mu\psi_L + i\overline{\psi}_R\gamma^\mu\nabla_\mu\psi_R + \\ &+ X\overline{\psi}_L\gamma^\mu\psi_L W_\mu - X\overline{\psi}_R\gamma^\mu\psi_R W_\mu\end{aligned} \tag{196}$$

as it is straightforward to see.

This is invariant for the transformation

$$W'_\nu = W_\nu - \partial_\nu \omega \tag{197}$$

with

$$\psi'_L = e^{-iX\omega}\psi_L \qquad \psi'_R = e^{iX\omega}\psi_R \tag{198}$$

or in compact form

$$\psi' = e^{iX\omega\pi}\psi \tag{199}$$

known as chiral gauge transformation.

Additionally, expression (105) can also be written as

$$\psi = \phi e^{-i\alpha} e^{-\frac{i}{2}\beta\pi} \begin{pmatrix} 1 \\ 0 \\ 1 \\ 0 \end{pmatrix} \tag{200}$$

and, from this expression, it is clear that it is always possible to perform a chiral gauge transformation taking the local parameter to be $\beta = 2X\omega$ and leaving

$$\psi' = \phi e^{-i\alpha} \begin{pmatrix} 1 \\ 0 \\ 1 \\ 0 \end{pmatrix} \tag{201}$$

in terms of the module alone: this has to be expected, as symmetries come with redundant information that can be removed by reducing the fields, and because in this case the chiral symmetry is an additional symmetry with one parameter, we have to expect that one degree of freedom be removed. It is clear that the only degree of freedom to remain is the one that cannot be removed in any way whatsoever that is the module.

Because in the massless approximation the two chiral Lagrangians become separable, the two chiral projections are independent, and therefore the Yvon–Takabayashi angle can be vanished, since it carries no information.

This is yet another fact that supports the evidence for which the Yvon–Takabayashi angle can be related to internal dynamics and *Zitterbewegung* for spinor fields.

In addition, this is possible because of the attractiveness that characterizes the axial-vector massive torsion mediation of the chiral mutual interaction within the spinor field.

If torsion were not an axial-vector, the chiral interaction would not be attractive, and, if such an attraction were not massive enough, it would not be sufficiently strong to grant stability for the bound-state spinorial field itself.

4.3. Two: Basic Applications

This second chapter will be about applying the above theory to solve or discuss fundamental problems in modern physics: in the first section, we will tackle the problem of gravitational singularity formation. In the second section, we will discuss the problem of positivity of energy.

5. Consequences of Spin

In this first section, our main goal is to take into account the problem of the formation of gravitational singularities and face it in terms of the modifications brought by the presence of torsion interacting with spinors. Some comments on the Pauli exclusion principle will be made.

5.1. Singularity Avoidance

If we consider Einstein gravity on its own, it is remarkably difficult to overestimate its success. From planetary precession, through gravitational waves, to black holes, there is not a single prediction that has not been corroborated yet. In fact, if Dark Matter is just another form of matter, there is not a single effect, whether predicted or not, that has never been confirmed so far. Nevertheless, there is a black spot, theoretically.

The Hawking–Penrose theorem is a very general result showing how, under very general conditions on energy, gravitationally-induced singularities form. If true, such a result would constitute an indication that Einstein gravitation has to be generalized, or at least included in an extended framework. There are, in fact, several attempts at extended models, whether they are simple extensions of Einstein gravity, or major revisions of all Einstein concepts of a geometric theory in itself. All these models and theories are certainly worth our attention. However, at times, the solution to a given puzzle might well be much closer than expected. If we wish to try a solution that is based on the physics we already have, the most straightforward possibility is to use the torsion tensor.

Employing torsion to solve this problem has already been done [38]. However, contrary to the expectation that torsion could solve or at least alleviate this issue, Kerlick found that the issue was actually worsened. This way was then abandoned.

Nevertheless, to a more attentive examination, we may find a possible way out. A closer look at the reasons why torsion would enhance the formation of singularities will reveal that the gravitational field is increased because, in the energy density, there are positive contributions coming from the fact that torsional effects for the spin contact interaction of spinors are taken to be repulsive.

This happens to be the case because Kerlick considers the simplest generalization of Einstein gravity, the original Einstein–Sciama–Kibble theory, where torsion is tied to the spin in terms of the Newton gravitational constant.

However, as discussed above, a more general theory of torsion would, first of all, involve a torsion–spin coupling that is not the Newton gravitational constant, but which can be any possible constant and in particular a constant with the opposite sign. In addition, secondly, in the most general case in which torsion propagates, in the effective approximation, the torsion–spin coupling constant has an opposite sign necessarily.

In fact, in this case, in effective approximation, we found that we do have an attractive torsion effect, resulting in a negative potential in the energy density, decreasing gravitation and making the singularity formation avoidable.

Indeed, the torsional contribution could provide such a negative potential that the whole energy may turn negative, the gravitational field may turn repulsive, and singularity formation would be avoided necessarily.

To put words into expressions, take (142) contracted as

$$-R-4\Lambda=\frac{k}{2}(-M^2W^2+m\Phi) \tag{202}$$

and plug this back into the original equations to get

$$\begin{aligned}R^{\rho\sigma}+\Lambda g^{\rho\sigma}=\frac{k}{2}[&\tfrac{1}{4}F^2g^{\rho\sigma}-F^{\rho\alpha}F^{\sigma}{}_{\alpha}+\tfrac{1}{4}(\partial W)^2g^{\rho\sigma}-(\partial W)^{\sigma\alpha}(\partial W)^{\rho}{}_{\alpha}+M^2W^\rho W^\sigma+\\&+\tfrac{i}{4}(\overline{\psi}\gamma^\rho\nabla^\sigma\psi-\nabla^\sigma\overline{\psi}\gamma^\rho\psi+\overline{\psi}\gamma^\sigma\nabla^\rho\psi-\nabla^\rho\overline{\psi}\gamma^\sigma\psi)-\\&-\tfrac{1}{2}X(W^\sigma S^\rho+W^\rho S^\sigma)-\tfrac{1}{2}m\Phi g^{\rho\sigma}]\end{aligned} \tag{203}$$

equivalent to those in the original form. For the singularity theorem in Einstein gravity, we have that the condition

$$R^{\rho\sigma}u_\rho u_\sigma\geqslant 0 \tag{204}$$

must be verified, and, when this is the case, then singularity formation will become inevitable. With no cosmological constant and neglecting electro-dynamics, we obtain that the condition to have singularity formation reads

$$[\tfrac{1}{4}(\partial W)^2 g^{\rho\sigma}-(\partial W)^{\sigma\alpha}(\partial W)^{\rho}{}_{\alpha}+\tfrac{i}{2}(\overline{\psi}\gamma^{\rho}\nabla^{\sigma}\psi-\nabla^{\sigma}\overline{\psi}\gamma^{\rho}\psi)+ \\ +M^2W^{\rho}W^{\sigma}-XW^{\sigma}S^{\rho}-\tfrac{1}{2}m\Phi g^{\rho\sigma}]u_{\rho}u_{\sigma}\geqslant 0 \tag{205}$$

and this is what we have to study.

In effective approximation, it becomes

$$\tfrac{i}{2}(\overline{\psi}\gamma^0\nabla_0\psi-\nabla_0\overline{\psi}\gamma^0\psi)-\tfrac{1}{2}m\Phi\geqslant 0 \tag{206}$$

and because (144) in the effective approximation is

$$i\gamma^0\nabla_0\psi+i\vec{\gamma}\cdot\vec{\nabla}\psi-\tfrac{X^2}{M^2}S_{\sigma}\gamma^{\sigma}\pi\psi-m\psi=0 \tag{207}$$

we may use this in the above to get

$$\tfrac{i}{2}(\vec{\nabla}\overline{\psi}\cdot\vec{\gamma}\psi-\overline{\psi}\vec{\gamma}\cdot\vec{\nabla}\psi)+\tfrac{X^2}{M^2}S_{\sigma}S^{\sigma}+\tfrac{1}{2}m\Phi\geqslant 0 \tag{208}$$

whose structure is similar to the condition of Kerlick but with the sign of the nonlinear interaction inverted. We may now follow Kerlick argument by neglecting the derivative term, and, by employing (102), we get that

$$-4\tfrac{X^2}{M^2}\phi^4+m\phi^2\cos\beta\geqslant 0 \tag{209}$$

which for for large densities are be violated, and quite easily too.

Therefore, because of the torsion–spin coupling, the energy condition is not verified and gravitational singularity formation is no longer a necessity [39].

We already said that torsion in effective approximation generates interactions which, without the spin-dependent part, are similar to what we would get by using the Higgs potential. Therefore, it is not surprising that singularity avoidance could be achieved also by the Higgs [40]. The difference is in the mass scale: the Higgs potential can only be used to avoid singularities in black holes, as it does not work before symmetry breaking, while torsion can be used to avoid singularities for black holes and the big bang, since torsion is always a massive field even prior to any mass generation mechanism.

Notice that this mechanism is proper to the Einsteinian gravitation. In fact, in order for this mechanism to work, one must have a theory in which gravitation can become repulsive if the energy density switches sign and in which the energy density is allowed to switch its overall sign. None of this would ever be possible in a theory of gravitation in which the source is not the energy but the mass, since the mass can never be negative.

5.2. Pauli Exclusion

The above-commented mechanism with which one may avoid the formation of singularity at a gravitational level is reminiscent of the degeneracy pressure encountered in the usual treatment of neutron stars. Consequently, the correlated Pauli exclusion principle comes to the mind. Such a principle stems from the fact that, in the construction of electronic levels, obtained by solving the non-relativistic matter field equations in a Coulomb potential, the solutions are given in terms of a quantum number n giving the energy level of the external shell, accounting for a total of n^2 electrons. However, the number of observed electrons $2n^2$ and hence there must be a two-fold degeneracy. This means that solutions of the matter field equation come in pairs of two, so that each electronic shell can be filled twice by the same state. The exclusion principle presented in this way is the original form by Pauli. Pauli's initial idea to assign a two-fold degeneracy was most straightforwardly that of introducing the concept of spin: the connection is very simple, based on the fact that irreducible representations of particles of spin s have exactly $d=2s+1$ independent components. For particles of spin $s=1/2$, this means $d=2/2+1=2$ components, so that it is possible to think that these two components be precisely the two states that account for the double state of multiplicity. Mathematically, this can be seen from the fact that the spinor field has, for each chiral part, two components. Indeed, recalling (105), we have that spinor fields can be written as

$$\psi=\phi e^{-i\alpha}e^{-\frac{i}{2}\beta\pi}\begin{pmatrix}1\\0\\1\\0\end{pmatrix} \quad \text{and} \quad \psi=\phi e^{-i\alpha}e^{-\frac{i}{2}\beta\pi}\begin{pmatrix}0\\1\\0\\1\end{pmatrix} \tag{210}$$

where the first is a spin-up (spin $+1/2$) eigenstate while the second is a spin-down (spin $-1/2$) eigenstate. These are the two opposite-spin eigenvalues of the same eigenspinor. As a consequence of this structure, superposition of two opposite-spin spinors is allowed and thus, if the two initial spinors are solutions, then also their superposition is another solution. This mechanism is indeed what happens in the hydrogen atom.

Nevertheless, the Pauli exclusion principle is not only this. Such a principle must also include a mechanism for which no more than two states can superpose. Quantum mechanics does not solve this problem. In quantum field theory, however, a solution is proposed, and the commonly accepted paradigm is described by the spin-statistic theorem: this theorem says that in a theory that is Lorentz covariance and causal, with positive norms and energies, half-integer spin particles cannot occupy more than one state at a time (while integer spin particles can). However, for this result to take place, the theorem must engage, and this is subject to the conditions granted by its hypotheses. In a classical theory of fields, Lorentz covariance and causality are ensured, but positive norms and energies are not. In fact, we have seen that negative energies are not only possible but also needed to ensure the mechanism to avoid the formation of singularities.

It so appears that the exclusion principle and singularity avoidance can not both be implemented in the same framework. In addition, usually, the common behavior is that of implementing the spin-statistic theorem and leaving the singularity formation unsolved. However, one can instead consider the complementary position of ensuring singularity avoidance and leave the Pauli exclusion open.

However, in a theory where spinors interact with torsion, we have seen that, in effective approximation, the torsionally-induced spin–contact interactions of the spinor give rise to self-interactions for the spinor field. These nonlinear contributions in the matter field equations are enough to ensure that no superposition of two identical solutions can also be a solution. This entails the exclusion principle.

Notice that, in case the two solutions are not identical, that is, if the two solutions correspond to opposite spins, their superposition is allowed by the double-valuedness that characterizes the spinorial fields in general cases.

6. Conditions on Energy

In this second section, we intend to deepen the investigation of the problem of the positive energies. We conclude with comments on the macroscopic approximation.

6.1. Positive Energy

In the development of field theories, it is not uncommon for some properties to be present in a given approximation but not in the full theory. Thus, particles behave in a certain manner in classical mechanics and very differently in quantum mechanics, and quantum particles behave in a given way in quantum mechanics and rather differently in the relativistic version of quantum mechanics.

Following a bottom-up approach in terms of successive generalizations, fewer and fewer properties will be found within the most general theory that is possible. There is, however, a property that does not appear to follow such a pattern, which is the energy.

From classical mechanics to quantum mechanics, to relativistic quantum mechanics, to relativistic quantum mechanics of spinning fields, the energy of a particle is always taken to be positive, either because it is proven positive, or because we force it to be positive by correcting the theory in an appropriate way.

Forcing the energy to be positive does have a number of consequences, not only for the interpretation, but also to obtain results like the spin-statistic theorem as discussed above. However, we have seen that the exclusion principle can also be entailed in a different manner, and there should be no surprise in finding a generalization of field the theory in which some energies happen to be negative after all.

Allowing negative energies has considerable advantages too, not only for the fact that the mathematics tells us that they are possible, but also to obtain results like the avoidance of singularities as we had discussed before.

Just the same, even assuming that energies can be negative, we have that they will have to turn out to be positive in those approximations in which we know they are.

To see this is in fact the case, let us consider the energy given as the right-hand side of (142), and that is

$$\begin{aligned}T^{\rho\sigma}=&\tfrac{1}{4}F^2g^{\rho\sigma}-F^{\rho\alpha}F^{\sigma}{}_{\alpha}+\\&+\tfrac{1}{4}(\partial W)^2g^{\rho\sigma}-(\partial W)^{\sigma\alpha}(\partial W)^{\rho}{}_{\alpha}+\\&+M^2(W^{\rho}W^{\sigma}-\tfrac{1}{2}W^2g^{\rho\sigma})+\\&+\tfrac{i}{4}(\overline{\psi}\gamma^{\rho}\nabla^{\sigma}\psi-\nabla^{\sigma}\overline{\psi}\gamma^{\rho}\psi+\overline{\psi}\gamma^{\sigma}\nabla^{\rho}\psi-\nabla^{\rho}\overline{\psi}\gamma^{\sigma}\psi)-\\&-\tfrac{1}{2}X(W^{\sigma}\overline{\psi}\gamma^{\rho}\pi\psi+W^{\rho}\overline{\psi}\gamma^{\sigma}\pi\psi)\end{aligned}\tag{211}$$

in general. In particular, as the electro-dynamic and torsional contributions are positive, we will consider only

$$\begin{aligned}E^{\rho\sigma}=&\tfrac{i}{4}(\overline{\psi}\gamma^{\rho}\nabla^{\sigma}\psi-\nabla^{\sigma}\overline{\psi}\gamma^{\rho}\psi+\\&+\overline{\psi}\gamma^{\sigma}\nabla^{\rho}\psi-\nabla^{\rho}\overline{\psi}\gamma^{\sigma}\psi)-\\&-\tfrac{1}{2}X(W^{\sigma}\overline{\psi}\gamma^{\rho}\pi\psi+W^{\rho}\overline{\psi}\gamma^{\sigma}\pi\psi)\end{aligned}\tag{212}$$

as pure spinorial contribution and which is not positive.

To better see this, we go in the frame where the spinor assumes the polar form (105) in which

$$\begin{aligned}E_{\rho\sigma}=&\phi^2\big[P_{\sigma}u_{\rho}+P_{\rho}u_{\sigma}+\\&+[\tfrac{1}{4}(\Omega^{ij}{}_{\sigma}\varepsilon_{\rho ijk}+\Omega^{ij}{}_{\rho}\varepsilon_{\sigma ijk})\eta^{ka}+\\&+\xi^{a}_{\rho}(\nabla\beta/2-XW)_{\sigma}+\xi^{a}_{\sigma}(\nabla\beta/2-XW)_{\rho}]s_a\big]\end{aligned}\tag{213}$$

whose time-time component is not positive defined as the straightforward check would immediately show.

In it, the momentum is given by (164) as

$$P^{\nu}=m\cos\beta u^{\nu}+Y_{\mu}u^{[\mu}s^{\nu]}+Z_{\mu}s_{\rho}u_{\sigma}\varepsilon^{\mu\rho\sigma\nu}\tag{214}$$

whose time component is also not positive defined.

However, if we could justify the assumption in terms of which we neglect all contributions coming from the spin, then the energy would reduce to

$$E_{00}=2\phi^2P_0u_0\tag{215}$$

for the time-time component.

The momentum becomes

$$P^0 = m\cos\beta u^0 \tag{216}$$

for the time component.

Therefore, if now the Yvon–Takabayashi angle vanishes, then the energy is ensured to be positive defined.

Summarizing, we can say that, if

$$\beta \to 0 \tag{217}$$

$$s_a \to 0 \tag{218}$$

then the energy of spinor fields is necessarily positive [41].

These two conditions together condense a very simple situation, as we are going to discuss in what follows.

6.2. Macroscopic Limit

In the previous part, we have discussed how the energy is positive if $\beta \to 0$ and $s_a \to 0$ happen to occur.

To understand the meaning of these conditions, let us consider again the field equations for the gravitational field and for electro-dynamics (142) and (143) and compute the divergences: they are respectively given by

$$\begin{aligned}\nabla_\rho[\tfrac{1}{4}F^2 g^{\rho\sigma} - F^{\rho\alpha}F^\sigma{}_\alpha + \\ +\tfrac{1}{4}(\partial W)^2 g^{\rho\sigma} - (\partial W)^{\sigma\alpha}(\partial W)^\rho{}_\alpha + \\ +M^2(W^\rho W^\sigma - \tfrac{1}{2}W^2 g^{\rho\sigma}) + \\ +\tfrac{i}{4}(\overline{\psi}\gamma^\rho\nabla^\sigma\psi - \nabla^\sigma\overline{\psi}\gamma^\rho\psi + \overline{\psi}\gamma^\sigma\nabla^\rho\psi - \nabla^\rho\overline{\psi}\gamma^\sigma\psi) - \\ -\tfrac{1}{2}X(W^\sigma\overline{\psi}\gamma^\rho\pi\psi + W^\rho\overline{\psi}\gamma^\sigma\pi\psi)] = 0\end{aligned} \tag{219}$$

and

$$\nabla_\mu(\overline{\psi}\gamma^\mu\psi) = 0 \tag{220}$$

identically, as we already know from the first chapter.

By substituting the polar form (105) and implementing the above conditions $\beta \to 0$ and $s_a \to 0$, we get

$$\nabla_\rho(\tfrac{1}{4}F^2 g^{\rho\sigma} - F^{\rho\alpha}F^\sigma{}_\alpha + 2m\phi^2 u^\rho u^\sigma) = 0 \tag{221}$$

and

$$\nabla_\mu(2\phi^2 u^\mu) = 0 \tag{222}$$

as it is straightforward to see.

Evaluating the divergence of the former and employing the latter, we obtain the expression

$$-2q\phi^2 u^\alpha F^\sigma{}_\alpha + 2m\phi^2 u^\rho\nabla_\rho u^\sigma = 0 \tag{223}$$

having used the Maxwell field equations.

After the necessary simplifications, we get

$$mu^\rho\nabla_\rho u^\sigma = qF^{\sigma\alpha}u_\alpha \tag{224}$$

which is just the Newton law in the presence of Lorentz force.

This is what we have in macroscopic approximation.

Thus, we can interpret $\beta \to 0$ and $s_a \to 0$ as the conditions that implement the known macroscopic approximation.

This is reasonable because vanishing the internal dynamics and all information about internal structures essentially means that we are considering situations where internal contributions are concealed within the spinorial field, which means we are in macroscopic approximation.

Spinor fields have energy density that can be negative as a consequence of all contributions of spin and internal dynamics, and it is only when these are hidden that the positivity of the energy density is also ensured for spinors.

6.3. Three: Special Models

This third and last chapter will be about the application of the above theory for phenomenological cases: we will in fact consider what the effects are of torsion for the two standard models of particles and cosmology.

7. Particles and Cosmology

In the first chapter, we have encountered the theorem of the polar form, which specified that, if both scalars Θ and Φ do not vanish identically, then we can always find special frames where the most general spinor is as in (105).

However, what if $\Theta = \Phi \equiv 0$ everywhere? The answer to this question has already been given in [15], and it is that we could still find a special frame in which the most general spinor can be written in some type of polar form.

Specifically, if $\Theta = \Phi \equiv 0$, then we can always find special frames where the most general spinor is given by

$$\psi = \frac{1}{\sqrt{2}}(\mathbb{I}\cos\tfrac{\alpha}{2} - \pi \sin\tfrac{\alpha}{2})\begin{pmatrix} 1 \\ 0 \\ 0 \\ 1 \end{pmatrix} \tag{225}$$

up to the reversal of the third axis.

Spinor fields undergoing these constraints are called flag-dipoles, and they contain two special cases: one with constraint $S^a = 0$ and called flagpoles, written as

$$\psi = \frac{1}{\sqrt{2}}\begin{pmatrix} 1 \\ 0 \\ 0 \\ 1 \end{pmatrix} \tag{226}$$

up to the reversal of the third axis and extinguishing the class of Majorana spinors; the other with a constraint given by $M^{ab} = 0$ and called dipoles, written as

$$\psi = \begin{pmatrix} 1 \\ 0 \\ 0 \\ 0 \end{pmatrix} \tag{227}$$

up to the reversal of the third axis and the switch between chiral parts and accounting for the Weyl spinors.

Therefore, as it can be seen quite clearly, Majorana as well as Weyl spinors can always be Lorentz transformed into a frame in which they remain with a fixed structure, and, consequently, they have no degree of freedom at all.

This may sound surprising, and thus we are going to give a direct proof of this statement for the Weyl spinors.

To do that, consider a general Weyl spinor, for instance left-handed, in the form

$$\psi = \begin{pmatrix} ae^{i\alpha} \\ be^{i\beta} \\ 0 \\ 0 \end{pmatrix}$$

where the two complex components have been written in polar form. Consider now as Lorentz transformation the rotation of angle θ around the second axis given by

$$\boldsymbol{\Lambda}_{R2} = \begin{pmatrix} \cos\theta/2 & -\sin\theta/2 & 0 & 0 \\ \sin\theta/2 & \cos\theta/2 & 0 & 0 \\ 0 & 0 & \cos\theta/2 & -\sin\theta/2 \\ 0 & 0 & \sin\theta/2 & \cos\theta/2 \end{pmatrix}$$

followed by the rotation of angle φ around the third axis

$$\boldsymbol{\Lambda}_{R3} = \begin{pmatrix} e^{i\varphi/2} & 0 & 0 & 0 \\ 0 & e^{-i\varphi/2} & 0 & 0 \\ 0 & 0 & e^{i\varphi/2} & 0 \\ 0 & 0 & 0 & e^{-i\varphi/2} \end{pmatrix}$$

applied to the spinor. The results are given by expression

$$\psi' = \begin{pmatrix} \cos\theta/2 & -\sin\theta/2 & 0 & 0 \\ \sin\theta/2 & \cos\theta/2 & 0 & 0 \\ 0 & 0 & \cos\theta/2 & -\sin\theta/2 \\ 0 & 0 & \sin\theta/2 & \cos\theta/2 \end{pmatrix} \cdot \\ \cdot \begin{pmatrix} e^{i\varphi/2} & 0 & 0 & 0 \\ 0 & e^{-i\varphi/2} & 0 & 0 \\ 0 & 0 & e^{i\varphi/2} & 0 \\ 0 & 0 & 0 & e^{-i\varphi/2} \end{pmatrix} \begin{pmatrix} ae^{i\alpha} \\ be^{i\beta} \\ 0 \\ 0 \end{pmatrix}$$

and that is

$$\psi' = \begin{pmatrix} a\cos\theta/2e^{i\varphi/2}e^{i\alpha} - b\sin\theta/2e^{-i\varphi/2}e^{i\beta} \\ a\sin\theta/2e^{i\varphi/2}e^{i\alpha} + b\cos\theta/2e^{-i\varphi/2}e^{i\beta} \\ 0 \\ 0 \end{pmatrix}$$

after multiplication. The spin-down component is zero if

$$a\sin\theta/2e^{i\varphi/2}e^{i\alpha} + b\cos\theta/2e^{-i\varphi/2}e^{i\beta} = 0$$

which can be worked out to be

$$\frac{a}{b}e^{i(\alpha-\beta)} = -e^{-i\varphi}\cot\theta/2$$

splitting into

$$\cot\theta/2 = -a/b$$
$$\varphi = \beta - \alpha$$

for the two angles. Thus, we can always find a combination of two rotations that brings the spin-down component to vanish identically. When this is done, we have

$$\psi' = \sqrt{a^2+b^2}e^{i(\beta+\alpha)/2}\begin{pmatrix}1\\0\\0\\0\end{pmatrix}$$

for spin-up Weyl spinors. With another rotation of angle $\zeta = \beta + \alpha$ around the third axis given as the above

$$\boldsymbol{\Lambda}_{R3} = \begin{pmatrix} e^{-i\zeta/2} & 0 & 0 & 0\\ 0 & e^{i\zeta/2} & 0 & 0\\ 0 & 0 & e^{-i\zeta/2} & 0\\ 0 & 0 & 0 & e^{i\zeta/2}\end{pmatrix}$$

the phase can also be vanished. Thus, we have

$$\psi'' = \sqrt{a^2+b^2}\begin{pmatrix}1\\0\\0\\0\end{pmatrix}$$

and employing a boost of rapidity $\eta = \ln|a^2+b^2|$ along the third axis given by

$$\boldsymbol{\Lambda}_{B3} = \begin{pmatrix} e^{-\eta/2} & 0 & 0 & 0\\ 0 & e^{\eta/2} & 0 & 0\\ 0 & 0 & e^{\eta/2} & 0\\ 0 & 0 & 0 & e^{-\eta/2}\end{pmatrix}$$

the module is also removed, and we get

$$\psi''' = \begin{pmatrix}1\\0\\0\\0\end{pmatrix}$$

for the final form of the Weyl spinor. Obviously, the same would be true if we intended to keep only the spin-down component. In addition, of course, the same remains true for the right-handed case. This result for Weyl spinors is general.

Although more calculations would be needed, it would still be straightforward to see that, by employing exactly the same method, we would obtain exactly the same result also if we were to consider the Majorana spinors.

Such a result may be surprising, but it is a mathematical consequence of the definition of Majorana and Weyl spinors alone and therefore it is true in full generality.

Thus, these spinors do not have degrees of freedom.

If we take this to conclude that these spinors cannot be physical, then we are bound to accept that such spinors cannot form the matter content of any theory, in particular, the standard model of particle physics as we know.

This leaves us with a remarkable consequence: if these spinors, and in particular Weyl spinors, cannot be used in physics, and in particular in the standard model of particle physics, then we cannot employ neutrinos as defined at the moment. Neutrinos need be right-handed too, and, after the symmetry breaking, they must get a Dirac mass.

Because the charge count of the standard model cannot change, neutrinos are sterile.

We will next try to see what happens when sterile neutrinos with a Dirac mass term are then included. Of course, the first application is neutrino oscillations.

We now try to see what the effects of torsion can be. However, in order to do so, we have first to make one little digression in order to generalize the theory.

Throughout the entire presentation, we have been considering single spinor fields, but clearly the treatment of two spinor fields, or even more spinor fields, is doable, and it is achieved by replicating the spinor field Lagrangian as many times as the number of independent spinor fields.

For instance, in the case of two spinor fields, we have

$$\begin{aligned}\mathscr{L}=&-\tfrac{1}{4}(\partial W)^2+\tfrac{1}{2}M^2W^2-\tfrac{1}{k}R-\tfrac{2}{k}\Lambda-\tfrac{1}{4}F^2+\\&+i\overline{\psi}_1\gamma^\mu\nabla_\mu\psi_1+i\overline{\psi}_2\gamma^\mu\nabla_\mu\psi_2+\\&-X_1\overline{\psi}_1\gamma^\mu\pi\psi_1W_\mu-X_2\overline{\psi}_2\gamma^\mu\pi\psi_2W_\mu+\\&-m_1\overline{\psi}_1\psi_1-m_2\overline{\psi}_2\psi_2\end{aligned}\tag{228}$$

as it is reasonable to expect.

Taking the variation with respect to torsion gives

$$\begin{aligned}\nabla_\nu(\partial W)^{\nu\mu}+M^2W^\mu=&X_1\overline{\psi}_1\gamma^\mu\pi\psi_1+\\&+X_2\overline{\psi}_2\gamma^\mu\pi\psi_2\end{aligned}\tag{229}$$

as the torsion field equations with two sources.

In effective approximation, we obtain expressions

$$M^2W^\mu\approx X_1\overline{\psi}_1\gamma^\mu\pi\psi_1+X_2\overline{\psi}_2\gamma^\mu\pi\psi_2\tag{230}$$

which can be plugged back into the Lagrangian giving

$$\begin{aligned}\mathscr{L}=&-\tfrac{1}{k}R-\tfrac{2}{k}\Lambda-\tfrac{1}{4}F^2+i\overline{\psi}_1\gamma^\mu\nabla_\mu\psi_1+i\overline{\psi}_2\gamma^\mu\nabla_\mu\psi_2+\\&+\tfrac{1}{2}\left|\tfrac{X_1}{M}\right|^2\overline{\psi}_1\gamma^\mu\psi_1\overline{\psi}_1\gamma_\mu\psi_1+\tfrac{1}{2}\left|\tfrac{X_2}{M}\right|^2\overline{\psi}_2\gamma^\mu\psi_2\overline{\psi}_2\gamma_\mu\psi_2-\\&-\tfrac{X_1}{M}\tfrac{X_2}{M}\overline{\psi}_1\gamma^\mu\pi\psi_1\overline{\psi}_2\gamma_\mu\pi\psi_2-m_1\overline{\psi}_1\psi_1-m_2\overline{\psi}_2\psi_2\end{aligned}\tag{231}$$

in which each spinor has self-interaction and between spinors there is mutual interaction.

The extension to three spinor fields, or n spinor fields, is similar: there are n self-interactions, always attractive, and $\frac{1}{2}n(n-1)$ mutual interactions, being either attractive or repulsive according to X_iX_j being positive or negative.

This extension is interesting for $n=3$ because this is the situation we have for neutrinos. By neglecting all the interactions apart from the effective interactions, and in them neglecting the self-interaction so to have only the mutual interactions, one may calculate the Hamiltonian

$$\mathscr{H}=\sum_{ij}\overline{\nu}_i(U_{ij}-X_iX_j\gamma^\mu\pi\nu_i\overline{\nu}_j\pi\gamma_\mu)\nu_j\tag{232}$$

where the Latin indices run over the three labels associated with the three different flavors of neutrinos. Hence, the matrix $U_{ij}-X_iX_j\gamma^\mu\pi\nu_i\overline{\nu}_j\pi\gamma_\mu$ is the combination of the constant matrix U_{ij} describing kinematic phases that arise from the mass terms, as usual, plus the field-dependent matrix $X_iX_j\gamma^\mu\pi\nu_i\overline{\nu}_j\pi\gamma_\mu$ describing the dynamical phases that arise from the torsionally-induced nonlinear potentials, those of the present theory.

Dealing with the nonlinear potentials is problematic, but, in reference [42], this problem is solved by taking neutrinos dense enough to make the torsion field background homogeneous and thus constant. The phase difference is

$$\Delta\Phi\approx\left(\frac{\Delta m^2}{2E}+\frac{1}{4}|W^0-W^3|\right)L\tag{233}$$

having assumed $W_1 = W_2 = 0$ and where L is the length of the oscillations. In [43], it was seen that (233) in the case in which the neutrino mass difference is small becomes

$$\Delta\Phi \approx \left(\Delta m^2 + m \frac{X_{\text{eff}}^2}{4M^2} |\bar{\nu}\gamma_\mu \nu \bar{\nu}\gamma^\mu \nu|^{\frac{1}{2}} \right) \frac{L}{2E} \tag{234}$$

where m is the value of the nearly-equal masses of neutrinos while X_{eff}^2 is a combination of the coupling constants and with the dependence L/E as the ratio between length and energy of the oscillations, as it is expected.

The phase difference due to the oscillation has the kinematic contribution, as a difference of the squared masses, plus a dynamic contribution, proportional to the neutrino mass density distribution. The novelty torsion introduced is that, even in the case in which neutrino masses were to be non-zero but with insufficient non-degeneracy in mass spectrum, we might still have oscillations, and therefore an ampler margin of freedom before having some tension. Notice also that both m and X_{eff}^2 depend on the masses and coupling constants of the two neutrinos involved so that they would be different for another pair of neutrinos, making it clear how the parameters of the oscillation depend on the specific pair of neutrinos, as it should be.

This is an immediate and clear effect that the neutrinos with Dirac mass term and interacting in terms of torsion give to us for some new physical insight beyond what is commonly expected from the standard model of particles.

What about the standard model of cosmology? To give an answer to this question, we move on to examine some consequences torsion may have for Dark Matter.

To begin with, we specify that, although we still do not exactly know what dark matter is, nevertheless it has to be a form of matter: albeit many models may fit galactic rotation curves, only dark matter as a real form of matter fits all galactic behaviors [44].

Hence, given dark matter as a form of matter, massive and weakly interacting, we will additionally take it to be described by $\frac{1}{2}$-spin spinor fields. This makes it possible to have the effects due to torsional interactions.

In reference [45], the torsional effects have been studied in a classical context to see how galactic dynamics could be modified by torsion, and, in [46], we have applied those results to the case in which torsion was coupled to spinors to see how galactic dynamics could be modified by torsion and how torsion could be sourced by dark matter.

Thus, here as before, torsion is not used as an alternative but as a correction over pre-existing physics. Having this in mind, we recall that, in [46], we showed how, if spinors are the source of torsion, the gravitational field in galaxies turns out to be increased: from (142), we see that, in the case of the effective approximation (165), we get

$$R^{\rho\sigma} - \frac{1}{2} R g^{\rho\sigma} - \Lambda g^{\rho\sigma} = \frac{k}{2} (E^{\rho\sigma} - \frac{1}{2} \frac{X^2}{M^2} S^\mu S_\mu g^{\rho\sigma}) \tag{235}$$

showing that the spinor field with the torsionally-induced nonlinear interactions has an effective energy, which is written as the usual term plus a nonlinear contribution.

For this contribution, we have to recall that we are not considering a single spinor field, as we have done when in particle physics, but collective states of spinor fields, as it is natural to assume in cosmology, with the consequence that it is not possible to employ the re-arrangements we used before and thus $S^\mu S_\mu$ cannot be reduced. Generally, we do not know how to compute it, but, as the square of a density, it may be positive.

In Ref. [46], we have been discussing precisely what would happen if the spin density square happened to be positive, and we have found that the contribution to the energy would change the gravitational field as to allow for a constant behavior of the rotation curves of galaxies, discussing the value of the torsion–spin coupling constant that is required to fit the galactic observations.

The details of the calculations were based on the fact that in this occurrence and within the approximations of slow rotational velocity and weak gravitational field, the acceleration felt by a point-particle was given by

$$\mathrm{div}\,\vec{a} \approx -m\rho - K^2\rho^2 \tag{236}$$

in which the Newton gravitational constant has been normalized and where K is the effective value of the torsional constant, with constant tangential velocity obtained for densities scaling down as r^{-2} in general. In the standard approach to dark matter, there are only Newtonian source contributions scaling down as r^{-1} and so a modification to the density distribution has to be devised, and it is the well known Navarro–Frenk–White profile. In the presence of torsional corrections, the Newtonian profile suffices because, even if the density drops as r^{-1}, it is squared in the torsional correction and the r^{-2} drop is obtained. These similarities suggest that the torsion correction might be what gives the Navarro–Frenk–White profile. After all, the NFW profile is obtained in n-body dynamics as those assumed here provided that the n spinors interact through torsion in terms of some axial-vector simplified model.

Nor is the idea of modeling dark matter, through the NFW profile, unexpected in terms of torsion, since this is precisely what a specific type of effective theories does.

In quite recent years, there has been a shift of approach in looking for physics beyond the standard model, and in particular dark matter. The new way of tackling the issue is based on the idea of studying all types of effective interactions that can be put in a Lagrangian, and, among all of them, there is the axial-vector spin-contact interaction.

However, in even more recent years, this approach has been generalized, shifting the attention from the effective interactions to the mediated interactions, known as simplified models [47]. However, the story does not change, since among all these there is the axial-vector mediated term

$$\Delta\mathscr{L} = -g\overline{\chi}\gamma^{\mu}\pi\chi B_{\mu} \tag{237}$$

where χ is the dark matter particle and B_{μ} is the axial-vector mediator, and where the structure of the interaction is that of the torsion–spin coupling, as it should be quite easily recognizable for the reader at this moment.

When the standard model has been acknowledged to need a complementation, we have been striving to have it placed within a more general model, which should have also contained some new physics, and in particular dark matter. It has been the constant failure in this project that prompted us to reverse the strategy, pushing us to look for simplified models, namely models that can immediately describe dark matter, or in general new physics, and leaving the task of including them, together with the standard model, into a more general model for later, and better, times. If we were, therefore, to see that the dark matter, or generally some new physics, were actually described by one of these simplified models, the following step would be to include it beside the standard model within a more general model, and at this point it should be clear what is our ultimate claim for this entire section.

Our claim is that, if such a simplified model is the one described by the axial-vector mediator, then we will need not look very far, as the general model would be torsion.

We next move to study a more direct effect about a cosmological situation [48,49].

The problem is quite simply the fact that the cosmological constant has a measured value that, in natural units, is about one hundred and twenty orders of magnitude off of the theoretically predicted one. Normally, this would have made physicists reject the theories in which its value is calculated, but those theories are quantum field theory and the standard model, being very successful otherwise.

Philosophers may argue that, in the face of a bad result disproving a theory, there can be no good result that can support it: the history of physics is loaded with examples of good agreements between observations and predictions that were based on theories later seen to be false. In addition, in this specific situation, the bad agreement is not only bad, but it is the worst in all of physics ever. Nowadays, the common behavior would be to claim that this is not really a bad agreement, since new physics might intervene to make the agreement acceptable. It does not take very experienced philosophers to see that this argument could always be invoked to push the problems under the carpet of an even higher energy frontier, and when this frontier will be unreachable, the predictivity of the theory will be annihilated. In this work, we try to embrace a philosophic approach, or merely be reasonable, admitting that such a discrepancy is lethal.

As a consequence, it follows that all theories predicting contributions to the cosmological constant must be dramatically re-adjusted. As we said above, these are the theory of quantum fields, with cosmological constant contributions due to zero-point energy, and the standard model, with cosmological constant contributions given by the same mechanism that gives mass to all fields and that is the spontaneous symmetry breaking.

As for the contribution coming from the general theory of quantum fields in terms of the zero-point energies, we have to recall that the zero-point energies are the result of quantization implemented with commutation relationships. In a normal-ordered quantum theory of fields, or simply in the classical theory of fields, zero-point energy does not appear, and thus no further contribution arises in the cosmological constant.

Leaving us without zero-point energy, it becomes necessary to find a way to compute the Casimir force without using any zero-point energy. It is worth remembering that Casimir forces derived from van der Waals forces was indeed the very first way to describe this phenomenon in the original paper by Casimir and Polder. A more recent account can be found in [50]. See also [51,52].

As for the contribution of the standard model in terms of the mechanism of symmetry breaking, recall that, after the break-down of the symmetry, we have the generation of the masses of all particles interacting with the Higgs field plus that of an effective cosmological constant $\frac{1}{2}\lambda^2 v^4$ with a value around 10^{120} in natural units. If this term is to disappear, we need to vanish either λ or v, but, as vanishing the former would imply no symmetry breaking, the only possibility is to vanish v so that symmetry breaking can still occur though not spontaneously. We may look for a dynamical symmetry breaking.

To begin our investigation, the very first thing we want to do is remark that, as the reader may have noticed, we never treated the scalar field. The reason was merely to keep an already heavy presentation from being heavier.

Still, it is now time to put in some scalar field. The scalar field complementing the Lagrangian (166) gives

$$\begin{aligned}\mathscr{L}=-\tfrac{1}{4}(\partial W)^2+\tfrac{1}{2}M^2W^2-\tfrac{1}{k}R-\tfrac{2}{k}\Lambda-\tfrac{1}{4}F^2+\\+i\overline{\psi}\gamma^\mu\nabla_\mu\psi+\nabla^\mu\phi^\dagger\nabla_\mu\phi-\\-X\overline{\psi}\gamma^\mu\boldsymbol{\pi}\psi W_\mu-\tfrac{1}{2}\Xi\phi^2W^2-Y\overline{\psi}\psi\phi-\\-m\overline{\psi}\psi+\mu^2\phi^2-\tfrac{1}{2}\lambda^2\phi^4\end{aligned} \tag{238}$$

where the X, Ξ, Y are the constants related to torsion with spinor and scalar interactions.

It is quite interesting to notice that, within this complementation, there is also the term ϕ^2W^2 coupling torsion to the scalar. This may look strange, since torsion is supposed to be sourced by the spin density, which is equal to zero for scalar fields. Therefore, we should expect to have torsion without a pure source of scalar fields, although we will have scalar contributions in torsion field equations.

In fact, upon variation of the Lagrangian, we obtain

$$\nabla_\alpha(\partial W)^{\alpha\nu}+(M^2-\Xi\phi^2)W^\nu=X\overline{\psi}\gamma^\nu\boldsymbol{\pi}\psi \tag{239}$$

in which there is indeed a scalar contribution, although in the form of an interaction giving an effective mass term.

There is, immediately, something rather striking about this expression: in a cosmic scenario, for a universe in an FLRW metric, we would have that the torsion, to respect the same symmetries of isotropy and homogeneity, would have to possess only the temporal component. However, in this case, the dynamical term would disappear leaving

$$(M^2-\Xi\phi^2)W^\nu = X\overline{\psi}\gamma^\nu\pi\psi \tag{240}$$

as the torsion field equations we would have had in the effective limit, though now the result is exact. The source would have to be the sum of the spin density of all spinors in the universe, and because the spin vector points in all directions, statistically the source vanishes too and

$$(M^2-\Xi\phi^2)W^\nu = 0 \tag{241}$$

which tells us that, if torsion is present, then

$$M^2 = \Xi\phi^2 \tag{242}$$

and, if Ξ is positive, the scalar acquires the value

$$\phi^2 = M^2/\Xi \tag{243}$$

which is constant throughout the universe.

A constant scalar all over the universe is the condition needed for slow-roll in inflationary scenarios, and in this case there arises an effective cosmological constant

$$\Lambda_{\text{effective}} = \Lambda + \frac{1}{2}\left|\frac{\lambda}{2}\left|\frac{M}{\Xi}\right|^2\right|^2 \tag{244}$$

in the Lagrangian (238), driving the scale factor of the FLRW metric and therefore driving the inflation itself. Inflation will last, so long as symmetry conditions hold, but as the universe expands and the density of sources decreases, local anisotropies are no longer swamped, and their presence will spoil the symmetries that engaged the above mechanism, bringing inflation to an end [53].

As the universe expands in a non-inflationary scenario, the torsional field equation would no longer lose the dynamic term due to the symmetries, but it might still lose it due to the fact that massive torsion can have an effective approximation. In this case, we would still have

$$(M^2-\Xi\phi^2)W^\nu \approx X\overline{\psi}\gamma^\nu\pi\psi \tag{245}$$

although only as an approximated form. We may plug it back into the initial Lagrangian (238) obtaining

$$\begin{aligned}\mathscr{L} = &-\tfrac{1}{k}R - \tfrac{2}{k}\Lambda - \tfrac{1}{4}F^2 + i\overline{\psi}\gamma^\mu\nabla_\mu\psi + \nabla^\mu\phi^\dagger\nabla_\mu\phi + \\ &+\tfrac{1}{2}X^2(M^2-\Xi\phi^2)^{-1}\overline{\psi}\gamma^\nu\psi\overline{\psi}\gamma_\nu\psi - \\ &-Y\overline{\psi}\psi\phi - m\overline{\psi}\psi + \mu^2\phi^2 - \tfrac{1}{2}\lambda^2\phi^4\end{aligned} \tag{246}$$

as the resulting effective Lagrangian. The presence of an effective interaction involving spinors and scalars, having a structure much richer than that of the Yukawa interaction, is obvious. In addition, we observe that, if for vanishingly small scalar this reduces to the above effective interaction for spinors, in the presence of larger values for the scalar, it can even become singular. We might speculate that such a value is the maximum allowed for the scalar as the one at which the above mechanism of inflation takes place.

Consider now the case $\mu=0$ in the above Lagrangian.

The scalar potential is minimized by $\phi^2 = v^2$ such that

$$\lambda^2 v^2 = \frac{1}{2}\Xi X^2 (M^2 - \Xi v^2)^{-2} |\overline{\psi}\gamma^\nu \psi \overline{\psi}\gamma_\nu \psi|_{\mathrm{v}} \tag{247}$$

linking the square of the Higgs vacuum to the square of the density of the spinor vacuum. Therefore, the dynamical symmetry breaking mechanism occurs eventually.

This break-down of symmetry is a dynamical one because the vacuum is not a constant, but it is the vacuum expectation value of the spinor distribution.

After dynamical symmetry breaking is implemented in the Lagrangian, the effective cosmological constant is still proportional to the Higgs vacuum, but the Higgs vacuum is now proportional to the spinor vacuum. Where material distributions tend to zero, as we would have in cosmology the vacuum for the spinor trivializes, the vacuum for the Higgs trivializes as well and the cosmological constant is no longer generated [54].

The picture that emerges is one for which symmetry breaking is no longer a mechanism that happens throughout the universe but only when spinors are present, with the consequence that, if spinors are not present, the effective cosmological constant is similarly not present. The cosmological constant due to spontaneous symmetry breaking in the standard model is thus avoidable.

No zero-point energy leaves no contribution apart from those due to phase transitions, which can be quenched by a symmetry breaking that is not spontaneous but dynamical, and no effective cosmological constant arises.

In this third chapter, we presented and discussed the possible torsional dynamics in the cosmology and particle physics standard models. Now, it is time to pull together all the loose ends in order to display the general overview.

We have seen and stated repeatedly that torsion can be thought as an axial-vector massive field coupling to the axial-vector bi-linear spinor field according to the term

$$\Delta\mathscr{L}^{\text{Q-spinor}}_{\text{interaction}} = -X\overline{\psi}\gamma^\mu \pi \psi W_\mu \tag{248}$$

of which we have one for every spinor. Effective approximations involving two, three, or even more spinor fields have been discussed, with a particular care for the case of neutrino oscillations, for which we have detailed in what way the results of [42] can be generalized in order to have

$$\Delta\Phi \approx \frac{L}{2E}\left(\Delta m^2 + m\frac{X^2_{\text{eff}}}{4M^2}|\overline{\nu}\gamma_\mu \nu \overline{\nu}\gamma^\mu \nu|^{\frac{1}{2}}\right) \tag{249}$$

describing the phase difference for almost degenerate neutrino masses, as consisting of the L/E dependence modulating the usual kinetic contribution, plus a new dynamic contribution, so that, even if the neutrino mass spectrum were to be degenerate, torsion would still induce an effective mechanism of oscillation. As these considerations have nothing special about neutrinos, and thus they may as well be extended to all leptons, we then proceeded in studying such extension. However, once the Lagrangian terms of the weak interaction after symmetry breaking and the torsion for an electron and a left-handed neutrino were taken in the effective approximation, we saw that, due to the cleanliness of the scattering and the precision of the measurements, the standard model correction induced by the torsion had to be very small, and, if this occurs because the torsion mass is large, then the effective approximation is no longer viable. We have then re-considered the case without effective approximations, allowing also for sterile right-handed neutrinos in order to maintain the feasibility of the dynamical neutrino oscillations discussed above, therefore reaching the general Lagrangian

$$\begin{aligned}\Delta\mathscr{L}^{\text{Q/weak-spinor}}_{\text{interaction}} = & -X_e\overline{e}\gamma^\mu \pi e W_\mu - X_\nu \overline{\nu}\gamma^\mu \pi \nu W_\mu + \\ & + \frac{g}{\sqrt{2}}\left(W^-_\mu \overline{\nu}\gamma^\mu e_L + W^+_\mu \overline{e}_L \gamma^\mu \nu\right) + \\ & + \frac{g}{\cos\theta} Z_\mu [\tfrac{1}{2}(\overline{\nu}\gamma^\mu \nu - \overline{e}_L \gamma^\mu e_L) + |\sin\theta|^2 \overline{e}\gamma^\mu e]\end{aligned} \tag{250}$$

showing that, while the sterile right-handed neutrino is by construction insensitive to weak interactions, it is sensitive to the universal torsion interaction, suggesting that, to see torsional interactions, we must pass for neutrino physics.

After having extensively wandered in the microscopic domain of particle physics, we move to see what type of effect torsion might have for a macroscopic application of a yet unseen particle, dark matter, and we have seen that, in the case of effective approximation, the spinor source in the gravitational field equations becomes of the form

$$R^{\rho\sigma}-\frac{1}{2}Rg^{\rho\sigma}-\Lambda g^{\rho\sigma}=\frac{k}{2}(E^{\rho\sigma}-\frac{1}{2}\frac{X^2}{M^2}S^{\mu}S_{\mu}g^{\rho\sigma}) \tag{251}$$

showing that, if the spin density square happens to be positive, the contribution to the energy would change the gravitational field as to allow for a constant behavior of the rotation curves of galaxies. We have discussed that this behavior comes from having a matter density scaling according to r^{-2} for large distances. Such behavior, usually, is granted by the Navarro–Frenk–White profile or, here, is due to the presence of torsion, suggesting that the NFW profile is just the manifestation of torsional interactions, and ultimately that dark matter may be described in terms of the axial-vector simplified model, sorting out one privileged type among all possible simplified models now in fashion in particle physics. Then, we proceeded to include into the picture also the scalar fields, getting

$$\begin{aligned}\mathscr{L}=&-\frac{1}{4}(\partial W)^2+\frac{1}{2}M^2W^2-\frac{1}{k}R-\frac{2}{k}\Lambda-\frac{1}{4}F^2+\\&+i\overline{\psi}\gamma^{\mu}\nabla_{\mu}\psi+\nabla^{\mu}\phi^{\dagger}\nabla_{\mu}\phi-\\&-X\overline{\psi}\gamma^{\mu}\pi\psi W_{\mu}-\frac{1}{2}\Xi\phi^2W^2-Y\overline{\psi}\psi\phi-\\&-m\overline{\psi}\psi+\mu^2\phi^2-\frac{1}{2}\lambda^2\phi^4\end{aligned} \tag{252}$$

showing that, in general, the torsion, besides its coupling to the spinor, may also couple to the scalar, with the scalar behaving as a sort of correction to the mass of torsion and a kind of re-normalization factor in the torsion–spinor effective interactions. We discussed how, within a homogeneous isotropic universe, the torsion field equations grant the condition $M^2=\Xi\phi^2$ so that, if Ξ were positive, then the scalar field would acquire a constant value, slow-roll will take place, and inflation could engage. After inflation has ended, torsion contributions to the scalar sector may induce a dynamical symmetry breaking. This may solve the cosmological constant problem in a new manner.

8. Conclusions

In this review, we have constructed the most general geometry with torsion as well as curvature, and, after having also introduced gauge fields in a similarly geometric manner, we have also built a genuinely geometric theory of spinor fields. We have seen how, under the assumption of being at the least-order derivative, the most general fully covariant system of field equations has been found for all physical fields in interaction. Separating torsion from all other fields and splitting spinor fields in their irreducible components allowed us to better see that torsion can be seen as an axial-vector massive field mediating the interaction between chiral projections. A formal integration of the spinorial degrees of freedom has also been discussed in some detail. Studying special situations, we have seen that the torsionally-induced spin-contact interactions are attractive. In addition, we have examined the conditions under which they can be removed with a choice of chiral gauge.

We have then seen that torsional effects for spinor fields can give rise to the conditions for which the gravitational singularities are no longer bound to form. We have hence established a parallel to the instance of the Pauli exclusion principle. We have discussed the problem of positive energies for spinors. In addition, we have determined the conditions under which positivity of the energy can be ensured.

Eventually, we have discussed problems inherent to the standard model of particles, specifically for neutrino oscillations and dark matter. In addition, we have commented on a possible solution to the cosmological constant problem.

In a geometry which, in its most general form, is naturally equipped with torsion, and for a physics which, for the most exhaustive form of coupling, has to couple the spin of matter, the fact that torsion couples to the spin of spinor material field distributions is just as well suited as a coupling can possibly be. In addition, its consequences about the stability of such field distributions are certainly worth receiving further attention.

In the standard model of particles, there are different facts to consider. Assuming the existence of right-handed sterile neutrinos, the torsion–spin coupling gives dynamic corrections to the oscillation pattern that can become important contributions if we were to see that the neutrino masses were too close to one another to fit the observed patterns. By assuming dark matter as constituted by a form of matter with spinorial structure, the torsion field, with its axial-vector massive character, might give rise to the known NFW profile. In cosmology, the most urgent of the problems is that of the cosmological constant, which can be solved, or at least quenched, by a theory in which spontaneous symmetry breaking is replaced by dynamical symmetry breaking, as is the case when torsion is allowed to interact through spinors with the scalar.

In the first two instances, that is, for the case of neutrino oscillations and dark matter, the new contributions are condensed in $\Delta\mathscr{L} = -X\overline{\psi}\gamma^{\mu}\pi\psi W_{\mu}$ as axial-vector coupling. In the last instance that is for the dynamical symmetry breaking, new physics are described in terms of the $\Delta\mathscr{L} = -X\overline{\psi}\gamma^{\mu}\pi\psi W_{\mu} - \frac{1}{2}\Xi\phi^{2}W^{2}$ contribution as what gives the torsion–spin and scalar interaction. These two potentials for torsion are the only two potentials that can be added into the Lagrangian within the restriction of renormalizability.

It is important to call attention to the fact that, as the general presentation goes, there are two ways to have torsion in differential geometry: the first is the one followed here, and it is the one more mathematical in essence, based on the general argument that torsion is present simply because there is no reason to set it to zero. The second one is more physical, based on the argument that torsion *must* be present since it necessarily arises after gauging translations much in the same way in which curvature is present as it arose after gauging rotations. In this approach, torsion and curvature are the Yang–Mills fields that inevitably emerge because we are considering the gauge theory of the full Poincaré group [55]. The usefulness of this approach has been remarkable in addressing problems related to supersymmetry, and especially supergravity. For more details, we refer the reader to [56,57], and in particular [58]. More recent papers that deal with torsion-gravity as a gauge theory are [59–61]. Still important is the work in [62].

This latter approach of gauging the full Poincaré group is based on the tetradic formalism, on which an overview by Tecchiolli can be found in this Special Issue [63].

Readers might not have failed to notice that there is a great absent in the presentation: field quantization, and there are reasons for it to be so. In spite of all successful predictions and precise measurements, it would not be a proper behavior to deny all mathematical problems the theory of quantum fields still has. From the fact that the equal-time commutator relations may not make sense at all [64] to the non-existence of interaction pictures [65], to mention only the most important of the problems, the rigorous mathematical treatment of quantum fields is yet to be achieved. What can torsion do for this? Honestly, it seems unlikely that any change in the field content can change things for the general structure of the theory from its roots. However, it may still be possible that, after all, torsion could address problems appearing later on in the development of the theory. For example, we have already discussed how torsion could be responsible for avoiding singularities in the case of spinor fields. This tells us that torsion may similarly be responsible for the fact that the elementary particles might not be point-like. If this were the case, then torsion would certainly have something to say about the problem of ultraviolet divergences. Torsion may be what gives a physical meaning to regularization and normalization, with the torsion mass giving the scale of the physical cut-off. There is still quite a way to go in fixing, or at least alleviating, the problems of the quantum field theoretical approach. It does not look unreasonable, however, that torsion might be there to help again.

Then, there are all the possible extensions. To begin, there is the fact that we wrote all field equations under the constraint of being at the least-order derivative possible. This requirement also coincides with renormalizability for all equations except those for gravity. If we wish to have renormalizability for all equations, then the gravitational field equations must be taken at the fourth-order derivative in the metric tensor [66]. This is certainly an opportunity for further research, especially for the effects torsion may have for the problems of singularity formation and positive energies. Another point needing some strengthening is related to the fact that torsion has always been taken completely antisymmetrically. Such a symmetry was duly justified in terms of fundamental arguments, and, for that matter, the completely antisymmetric torsion couples neatly with the completely antisymmetric spin of the Dirac field. However, if higher-spin fields were to be found or more general geometric backgrounds were to be needed, more general torsion would make their appearance in the theory. Studying what may be the role of a trace torsion or of the remaining irreducible torsion component is also an important task.

Regarding the mass generation of spinors through spin–torsion interactions, it is necessary to direct the attention toward [67–69], and recently [70].

Again, a recent work is that of Diether and Christian later in this Special Issue [71].

In particle physics more in general, high-energy experimental constraints on torsion have been placed, especially in [72–76]. Going to cosmology, dark matter has also been studied in the presence of torsion: after the already-mentioned [45], the reader may find it interesting to also have a look at [77–81]. As for the problem of singularity formation during the Big Bang, the following references may be of help [82–86].

More mathematical extensions have been addressed along the years in various manners. In fact, all possible alternatives and extensions of Einstein gravity can also be generalized for the torsional case: for instance, conformal gravity with torsion has been established in [87], while $f(R)$-types of torsion-gravity have been studied in [88]. As for the latter, the reader may also find the problem of junction conditions interesting [89].

For a review of such a problem, we invite the reader to the paper by Vignolo that can also be found later in this Special Issue [90].

From a purely mathematical, general point of view, interesting features of the torsional background in the presence of spinors have been investigated in [91,92].

The dynamics of the torsion field may also in principle allow the propagation of parity violating modes, although many constraints have been placed recently [93,94].

Anomalies and constraints on torsion were studied in [95,96].

The possibilities introduced by not neglecting torsion in gravity for Dirac fields can also be more mathematical in essence. Above all, it is paramount to mention all the exact solutions for the coupled system of field equations that may be found. In [97–99], we found exact solutions for the Dirac field in its own gravitational field. Including torsion in gravity and allowing the coupling to the spinor can only increase the interesting features that exact solutions could have. Of course, finding exact solutions for a system of interacting fields is a very difficult enterprise and so we must expect a slow evolution.

It is clearly impossible to draw a complete list of references. Nevertheless, those presented here, and their own references, might be taken as a fair list to help the reader.

We wish to conclude this exposition with one personal note on aesthetics. It is very often stated in philosophical debates that a theory is considered to be beautiful when it has some sense of inevitability built into itself. That is, a sense for which there is nothing that can be modified, or removed, from the theory without looking like a form of unnecessary assumption. We see torsion gravity precisely like that. Such a theory is formed by requiring that some very general principles of symmetry be respected for the four-dimensional continuum space–time. If these hypotheses are put in, they determine the development of the theory without anything else to be postulated. It is at this point therefore that a theory of gravity with no torsion, where we would remove an object that would otherwise be naturally present, has to be regarded as something arbitrary.

The most beautiful, in the sense of the most necessary, theory of gravity is the one in which torsion is allowed to occupy the place that is its own a priori.

Funding: This research received no external funding.

Institutional Review Board Statement: Not applicable.

Informed Consent Statement: Not applicable.

Data Availability Statement: Not applicable.

Acknowledgments: In the last sixteen years, the single constant source of material (and more importantly immaterial) support has been Marie-Hélène Genest, my wife. It is to her that I give my continuous and everlasting thanks for everything.

Conflicts of Interest: The author declares no conflict of interest.

References

1. Cartan, E. Sur les variétés à connexion affine et la théorie de la relativité généralisée (première partie). *Ann. Sci. Ecole Norm. Sup.* **1923**, *40*, 325. [CrossRef]
2. Cartan, E. Sur les variétés à connexion affine et la théorie de la relativité généralisée (première partie) (Suite). *Ann. Sci. Ecole Norm. Sup.* **1924**, *41*, 1. [CrossRef]
3. Cartan, E. Sur les variétés à connexion affine et la théorie de la relativité généralisée (deuxième partie). *Ann. Sci. Ecole Norm. Sup.* **1925**, *42*, 17. [CrossRef]
4. Cartan, E. Sur une généralisation de la notion de courbure de Riemann et les espaces à torsion. *Compt. Rend. Acad. Sci.* **1922**, *174*, 593.
5. Hehl, F.W.; Obukhov, Y.N. Elie Cartan's torsion in geometry and in field theory, an essay. *Ann. Fond. Broglie* **2007**, *32*, 157.
6. Sciama, D.W. On the analogy between charge and spin in general relativity. In *Recent Developments in General Relativity*; Springer: London, UK, 1962.
7. Kibble, T.W.B. Lorentz invariance and the gravitational field. *J. Math. Phys.* **1961**, *2*, 212. [CrossRef]
8. Hehl, F.W.; Heyde, P.V.D.; Kerlick, G.D.; Nester, J.M. General relativity with spin and torsion: Foundations and prospects. *Rev. Mod. Phys.* **1976**, *48*, 393. [CrossRef]
9. Fabbri, L. Higher-Order Theories of Gravitation. *arXiv* **2008**, arXiv:0806.2610.
10. Fabbri, L. On a completely antisymmetric Cartan torsion tensor. *Ann. Fond. Broglie* **2007**, *32*, 215.
11. Hayashi, K. Restrictions on gauge theory of gravitation. *Phys. Lett. B* **1976**, *65*, 437. [CrossRef]
12. Yu, X. The Ω-field theory of gravitation and cosmology. *Astrophys. Space Sci.* **1989**, *154*, 321. [CrossRef]
13. Audretsch, J.; Lämmerzahl, C. Constructive axiomatic approach to spacetime torsion. *Class. Quant. Grav.* **1988**, *5*, 1285. [CrossRef]
14. Macias, A.; Lämmerzahl, C. On the dimensionality of space–time. *J. Math. Phys.* **1993**, *34*, 4540.
15. Fabbri, L. A generally-relativistic gauge classification of the Dirac fields. *Int. J. Geom. Meth. Mod. Phys.* **2016**, *13*, 1650078. [CrossRef]
16. Lounesto, P. *Clifford Algebras and Spinors*; Cambridge University Press: London, UK, 2001.
17. Cavalcanti, R.T. Classification of Singular Spinor Fields and Other Mass Dimension One Fermions. *Int. J. Mod. Phys. D* **2014**, *23*, 1444002. [CrossRef]
18. Fabbri, L. On the problem of Unicity in Einstein-Sciama–Kibble Theory. *Ann. Fond. Broglie* **2008**, *33*, 365.
19. Sabbata, V.D.; Sivaram, C. *Spin and Torsion in Gravitation*; World Scientific: Singapore, 1994.
20. Sivaram, C.; Sinha, K.P. F-Gravity and Spinors in General Relativity. *Lett. Nuovo Cim.* **1975**, *13*, 357. [CrossRef]
21. Fabbri, L. On Geometrically Unified Fields and Universal Constants. *Gen. Relativ. Gravit.* **2013**, *45*, 1285. [CrossRef]
22. Velo, G.; Zwanziger, D. Propagation and quantization of Rarita-Schwinger waves in an external electromagnetic potential. *Phys. Rev.* **1969**, *186*, 1337. [CrossRef]
23. Velo, G.; Zwanziger, D. Noncausality and other defects of interaction lagrangians for particles with spin one and higher. *Phys. Rev.* **1969**, *188*, 2218. [CrossRef]
24. Fabbri, L. Least-Order Torsion-Gravity for Chiral-Spinor Fields, induced Self-Interacting Potentials and Parity Conservation. *Gen. Relativ. Gravit.* **2014**, *46*, 1663. [CrossRef]
25. Fabbri, L. A discussion on the most general torsion-gravity with electrodynamics for Dirac spinor matter fields. *Int. J. Geom. Meth. Mod. Phys.* **2015**, *12*, 1550099. [CrossRef]
26. Belyaev, A.S.; Shapiro, I.L. The action for the propagating torsion and the limits on the torsion parameters from present experimental data. *Phys. Lett. B* **1998**, *425*, 246. [CrossRef]
27. Fabbri, L. Torsion Gravity for Dirac Fields. *Int. J. Geom. Meth. Mod. Phys.* **2017**, *14*, 1750037. [CrossRef]
28. Steane, A.M. An introduction to spinors. *arXiv* **2013**, arXiv:1312.3824.
29. Barut, A.O.; Ziino, G. On parity conservation and the question of the 'missing' (right-handed) neutrino. *Mod. Phys. Lett. A* **1993**, *8*, 1011. [CrossRef]

30. Fabbri, L. On a purely geometric approach to the Dirac matter field and its quantum properties. *Int. J. Theor. Phys.* **2014**, *53*, 1896. [CrossRef]
31. Bahamonde, S.; Valcarcel, J.G. Observational constraints in metric-affine gravity. *Eur. Phys. J. C* **2021**, *81*, 495. [CrossRef]
32. Nambu, Y.; Jona-Lasinio, G. Dynamical Model of Elementary Particles Based on an Analogy with Superconductivity. 1. *Phys. Rev.* **1961**, *122*, 345. [CrossRef]
33. Nambu, Y.; Jona-Lasinio, G. Dynamical Model of Elementary Particles Based on an Analogy with Superconductivity. 2. *Phys. Rev.* **1961**, *124*, 246. [CrossRef]
34. Fabbri, L. The Spin-Torsion coupling and Causality for the Standard Model. *Mod. Phys. Lett. A* **2011**, *26*, 2091. [CrossRef]
35. Fabbri, L.; Vignolo, S. A modified theory of gravity with torsion and its applications to cosmology and particle physics. *Int. J. Theor. Phys.* **2012**, *51*, 3186. [CrossRef]
36. Salesi, G.; Recami, E. About the kinematics of spinning particles. *Adv. Appl. Clifford Algebr.* **1997**, *7*, S253.
37. Recami, E.; Salesi, G. Kinematics and hydrodynamics of spinning particles. *Phys. Rev. A* **1998**, *57*, 98. [CrossRef]
38. Kerlick, G.D. Cosmology and Particle Pair Production via Gravitational Spin Spin Interaction in the Einstein-Cartan-Sciama–Kibble Theory of Gravity. *Phys. Rev. D* **1975**, *12*, 3004. [CrossRef]
39. Fabbri, L. Singularity-free spinors in gravity with propagating torsion. *Mod. Phys. Lett. A* **2017**, *32*, 1750221. [CrossRef]
40. Fabbri, L. Black Hole singularity avoidance by the Higgs scalar field. *Eur. Phys. J. C* **2018**, *78*, 1028. [CrossRef]
41. Fabbri, L. A geometrical assessment of spinorial energy conditions. *Eur. Phys. J. Plus* **2017**, *132*, 156. [CrossRef]
42. Adak, M.; Dereli, T.; Ryder, L.H. Neutrino oscillations induced by space–time torsion. *Class. Quant. Grav.* **2001**, *18*, 1503. [CrossRef]
43. Fabbri, L.; Vignolo, S. A torsional completion of gravity for Dirac matter fields and its applications to neutrino oscillations. *Mod. Phys. Lett. A* **2016**, *31*, 1650014. [CrossRef]
44. Clowe, D.; Bradač, M.; Gonzalez, A.H.; Markevitch, M.; Randall, S.W.; Jones, C.; Zaritsky, D. A direct empirical proof of the existence of dark matter. *Astrophys. J. Lett.* **2006**, *648*, L109. [CrossRef]
45. Tilquin, A.; Schucker, T. Torsion, an alternative to dark matter? *Gen. Relativ. Gravit.* **2011**, *43*, 2965. [CrossRef]
46. Fabbri, L. Weakly-Interacting Massive Particles in Torsionally-Gravitating Dirac Theory. *Int. J. Mod. Phys. D* **2013**, *22*, 1350071. [CrossRef]
47. Abdallah, J.; Araujo, H.; Arbey, A.; Ashkenazi, A.; Belyaev, A.; Berger, J.; Boehm, C.; Boveia, A.; Brennan, A.; Brooke, J.; et al. Simplified Models for Dark Matter Searches at the LHC. *Phys. Dark Univ.* **2015**, *9*, 8. [CrossRef]
48. Rugh, S.E.; Zinkernagel, H. The quantum vacuum and the cosmological constant problem. *Stud. Hist. Philos. Mod. Phys.* **2002**, *33*, 663. [CrossRef]
49. Martin, J. Everything you always wanted to know about the cosmological constant problem (but were afraid to ask). *Comptes Rendus Phys.* **2012**, *13*, 566. [CrossRef]
50. Schwinger, J. Casimir effect in source theory. *Lett. Math. Phys.* **1975**, *1*, 43. [CrossRef]
51. Jaffe, R.L. The Casimir effect and the quantum vacuum. *Phys. Rev. D* **2005**, *72*, 021301. [CrossRef]
52. Nikolic, H. Proof that Casimir force does not originate from vacuum energy. *Phys. Lett. B* **2016**, *761*, 197. [CrossRef]
53. Fabbri, L. A simple assessment on inflation. *Int. J. Theor. Phys.* **2017**, *56*, 2635. [CrossRef]
54. Fabbri, L. A simple assessment on the hierarchy problem. *Int. J. Geom. Meth. Mod. Phys.* **2016**, *13*, 1650068. [CrossRef]
55. Utiyama, R. Invariant theoretical interpretation of interaction. *Phys. Rev.* **1956**, *101*, 1597. [CrossRef]
56. MacDowell, S.W.; Mansouri, F. Unified Geometric Theory of Gravity and Supergravity. *Phys. Rev. Lett.* **1977**, *38*, 739. [CrossRef]
57. Stelle, K.; West, P. Minimal Auxiliary Fields for Supergravity. *Phys. Lett. B* **1978**, *74*, 330. [CrossRef]
58. Chamseddine, A.; West, P. Supergravity as a Gauge Theory of Supersymmetry. *Nucl. Phys.* **1977**, *B129*, 39. [CrossRef]
59. Nieh, H.T. Torsion in Gauge Theory. *Phys. Rev. D* **2018**, *97*, 044027. [CrossRef]
60. Carlevaro, N.; Lecian, O.M.; Montani, G. Lorentz Gauge Theory and Spinor Interaction. *Int. J. Mod. Phys. A* **2008**, *23*, 1282. [CrossRef]
61. Lecian, O.M.; Montani, G.; Carlevaro, N. Novel Analysis of Spinor Interactions and non-Riemannian Geometry. *Eur. Phys. J. Plus* **2013**, *128*, 19. [CrossRef]
62. Edelen, D.G.B. Direct gauging of the Poincaré group. *Int. J. Theor. Phys.* **1985**, *24*, 1173. [CrossRef]
63. Tecchiolli, M. On the Mathematics of Coframe Formalism and Einstein-Cartan Theory—A Brief Review. *Universe* **2019**, *5*, 206. [CrossRef]
64. Streater, R.F.; Wightman, A.S. *PCT, Spin and Statistics, and All That*; Princeton University Press: Princeton, NJ, USA, 2000.
65. Haag, R. On quantum field theories. *Mat. Fys. Med.* **1955**, *29*, 1.
66. Stelle, K.S. Renormalization of Higher Derivative Quantum Gravity. *Phys. Rev. D* **1977**, *16*, 953. [CrossRef]
67. Zubkov, M.A. Gauge theory of Lorentz group as a source of the dynamical electroweak symmetry breaking. *J. High Energy Phys.* **2013**, *9*, 044. [CrossRef]
68. Zubkov, M.A. Torsion instead of Technicolor. *Mod. Phys. Lett. A* **2010**, *25*, 2885. [CrossRef]
69. Zubkov, M.A. Dynamical torsion as the microscopic origin of the neutrino seesaw. *Mod. Phys. Lett. A* **2014**, *29*, 1450111. [CrossRef]
70. Castillo-Felisola, O.; Corral, C.; Villavicencio, C.; Zerwekh, A.R. Fermion Masses Through Condensation in Spacetimes with Torsion. *Phys. Rev. D* **2013**, *88*, 124022. [CrossRef]

71. Diether, C.F., III; Christian, J. On the Role of Einstein-Cartan Gravity in Fundamental Particle Physics. *Universe* **2020**, *6*, 112. [CrossRef]
72. Freidel, L.; Minic, D.; Takeuchi, T. Quantum gravity, torsion, parity violation and all that. *Phys. Rev. D* **2005**, *72*, 104002. [CrossRef]
73. Obukhov, Y.N.; Silenko, A.J.; Teryaev, O.V. Spin-torsion coupling and gravitational moments of Dirac fermions: Theory and experimental bounds. *Phys. Rev. D* **2014**, *90*, 124068. [CrossRef]
74. De Almeida, F.M.L.; De Andrade, F.R.; do Vale, M.A.B.; Nepomuceno, A.A. Torsion limits from $t\bar{t}$ production at the LHC. *Phys. Rev. D* **2018**, *97*, 075036. [CrossRef]
75. Castillo-Felisola, O.; Corral, C.; Schmidt, I.; Zerwekh, A.R. Updated limits on extra dimensions through torsion and LHC data. *Mod. Phys. Lett. A* **2014**, *29*, 1450081. [CrossRef]
76. Castillo-Felisola, O.; Corral, C.; Kovalenko, S.; Schmidt, I. Torsion in Extra Dimensions and One-Loop Observables. *Phys. Rev. D* **2014**, *90*, 024005. [CrossRef]
77. Belyaev, A.S.; Thomas, M.C.; Shapiro, I.L. Torsion as a Dark Matter Candidate from the Higgs Portal. *Phys. Rev. D* **2017**, *95*, 095033. [CrossRef]
78. Alvarez-Castillo, D.; Cirilo-Lombardo, D.J.; Zamora-Saa, J. Dark matter candidates, helicity effects and new affine gravity with torsion. *J. High Energy Astrophys* **2017**, *13*, 10. [CrossRef]
79. Castillo-Felisola, O.; Corral, C.; Kovalenko, S.; Schmidt, I.; Lyubovitskij, V.E. Axions in gravity with torsion. *Phys. Rev. D* **2015**, *91*, 085017. [CrossRef]
80. Grensing, G. Dark matter and torsion. *Gen. Relativ. Gravit.* **2021**, *53*, 49. [CrossRef]
81. Shaposhnikov, M.; Shkerin, A.; Timiryasov, I.; Zell, S. Einstein-Cartan Portal to Dark Matter. *Phys. Rev. Lett.* **2021**, *126*, 161301. [CrossRef] [PubMed]
82. Magueijo, J.; Zlosnik, T.G.; Kibble, T.W.B. Cosmology with a spin. *Phys. Rev. D* **2013**, *87*, 063504. [CrossRef]
83. Magueijo, J.; Złośnik, T. Parity violating Friedmann Universes. *Phys. Rev. D* **2019**, *100*, 084036. [CrossRef]
84. Alexander, S.; Bambi, C.; Marciano, A.; Modesto, L. Fermi-bounce Cosmology and scale invariant power-spectrum. *Phys. Rev. D* **2014**, *90*, 123510. [CrossRef]
85. Unger, G.; Poplawski, N. Big bounce and closed universe from spin and torsion. *Astrophys. J.* **2019**, *870*, 78. [CrossRef]
86. Farnsworth, S.; Lehners, J.-L.; Qiu, T. Spinor driven cosmic bounces and their cosmological perturbations. *Phys. Rev. D* **2017**, *96*, 083530. [CrossRef]
87. Fabbri, L. Metric-Torsional Conformal Gravity. *Phys. Lett. B* **2012**, *707*, 415. [CrossRef]
88. Fabbri, L.; Vignolo, S. Dirac fields in $f(R)$-gravity with torsion. *Class. Quant. Grav.* **2011**, *28*, 125002. [CrossRef]
89. Vignolo, S.; Cianci, R.; Carloni, S. On the junction conditions in $f(R)$-gravity with torsion. *Class. Quant. Grav.* **2018**, *35*, 095014. [CrossRef]
90. Vignolo, S. Some Mathematical Aspects of f(R)-Gravity with Torsion: Cauchy Problem and Junction Conditions. *Universe* **2019**, *5*, 224. [CrossRef]
91. Peeters, K.; Waldron, A. Spinors on manifolds with boundary: APS index theorems with torsion. *J. High Energy Phys.* **1999**, *2*, 24. [CrossRef]
92. Randono, A. Do Spinors Frame-Drag? *Phys. Rev. D* **2010**, *81*, 024027. [CrossRef]
93. Kostelecky, V.A.; Russell, N.; Tasson, J. New Constraints on Torsion from Lorentz Violation. *Phys. Rev. Lett.* **2008**, *100*, 111102. [CrossRef]
94. Fabbri, L. Non-causal Propagation for Higher-Order Interactions of Torsion with Spinor Fields. *Int. J. Theor. Phys.* **2018**, *57*, 1683. [CrossRef]
95. Yajima, S.; Kimura, T. Anomalies in Four-dimensional Curved Space With Torsion. *Prog. Theor. Phys.* **1985**, *74*, 866. [CrossRef]
96. Fabbri, L. M.Tecchiolli, Restrictions on torsion–spinor field theory. *Mod. Phys. Lett. A* **2019**, *34*, 1950311. [CrossRef]
97. Cianci, R.; Fabbri, L.; Vignolo, S. Exact solutions for Weyl fermions with gravity. *Eur. Phys. J. C* **2015**, *75*, 478. [CrossRef]
98. Cianci, R.; Fabbri, L.; Vignolo, S. Critical exact solutions for self-gravitating Dirac fields. *Eur. Phys. J. C* **2016**, *76*, 595. [CrossRef]
99. Cianci, R.; Fabbri, L.; Vignolo, S. Axially-Symmetric Exact Solutions for Flagpole Fermions with Gravity. *Eur. Phys. J. Plus* **2020**, *135*, 131. [CrossRef]

MDPI

Review

Received: 22 October 2019; Accepted: 3 December 2019; Published: 6 December 2019

1. Introduction

Several open problems in modern physics at both ultraviolet and infrared scales seem to justify the need to enlarge or revise General Relativity (GR). For example, at astrophysical and cosmological scales, in order for observations to agree with the theoretical predictions of GR, it is necessary to assume the existence of the so-called dark matter and dark energy. But, up to now, at a fundamental level, no experimental evidence has been found to prove the existence of such unknown forms of matter and energy. This fact, together with other shortcomings of GR, represents the signal of a possible breakdown in our understanding of gravity; the possibility of developing extended or alternative theories of gravity is then to be seriously taken into account.

In the last thirty years [1,2], many extensions of GR have been actually proposed; among these, $f(R)$-gravity certainly remains one of the most direct and simplest [3–7]: it relies on the idea that the gravitational Lagrangian may depend on the Ricci scalar R in a more general way than the linear one as it happens in the Einstein-Hilbert action. Recently, $f(R)$-gravity has received great interest in view of its successes in accounting for both cosmic speed-up and missing matter at cosmological and astrophysical scales, respectively (see, for example, [8–10]).

At the same time, including the torsion tensor among the geometrical attributes of space-time is another way to extend GR. Cartan was the first to introduce torsion in the geometrical background; after him, Sciama and Kibble embodied it within the framework of Einstein gravity implementing the idea that spin can be source of torsion as energy does for curvature [11–13]. The resulting theory, known as Einstein–Cartan–Sciama–Kibble (ECSK) theory, has been the first generalization of GR trying to take the spin of elementary fields into account, and it still remains one of the most serious attempts in this direction [14–16].

Following this paradigm, $f(R)$-gravity with torsion consists in one of the simplest extensions of the ECSK theory, just as purely metric $f(R)$-gravity is with respect to GR. The key idea is again that of replacing the Einstein-Hilbert Lagrangian with a non-linear function of the scalar curvature. A remarkable consequence of the non-linearity of the gravitational Lagrangian is that torsion can be non-zero even without the presence of spin, as long as the trace of the matter stress–energy tensor is not constant [17–21]. This is a noticeable difference with respect to ECSK theory, where instead torsion can exist only coupled to spin. It is known that torsion may give rise to singularity-free and accelerated cosmological models [22], and a torsion arising from the non-linearity of the gravitational Lagrangian function could amplify this effects and make them possible even in the absence of spin. This is a feature that makes $f(R)$-gravity with torsion interesting enough to be studied in depth.

Of course, in order for any physical theory to be viable, it has to possess an associated initial value problem correctly formulated in such a way that the dynamical evolution is uniquely determined and consistent with causality requirements. More specifically, the following properties have to hold: (i) small perturbations of the initial data have to generate small perturbations in the subsequent dynamics; (ii) changes of the initial data have to preserve the causal structure of the theory. The initial value problem of the theory is well-posed if both these requests are satisfied.

It is well known that GR has a well-posed initial value problem, so resulting in a stable theory with a robust causal structure [23–26]. In order to be considered as a viable extension of the Einstein theory, $f(R)$-gravity should also have such a feature.

About this, by taking advantage of the dynamical equivalence with O'Hanlon theories [27], it is easily seen that purely metric $f(R)$-gravity possesses a well-posed Cauchy problem [28] regardless of the explicit form of the function $f(R)$.

As far as the theory with torsion is concerned, the issue is quite simple whenever the trace of the stress–energy tensor is constant: in this circumstance and in the absence of matter spin sources, in fact, the theory is equivalent to GR with or without a cosmological constant, depending on the explicit expression of the function $f(R)$. For instance, this is what happens in vacuo and in the case of coupling to electromagnetic or Yang–Mills fields. Instead, the coupling to other kinds of matter sources must be discussed carefully case by case. Here, we face the Cauchy problem in the presence of a perfect fluid or a Klein–Gordon scalar field. Making use of some different techniques, such as conformal transformations and dynamic equivalence with scalar-tensor theories, we formulate sufficient conditions to ensure that the related Cauchy problem is well-posed, also showing that there exist $f(R)$ functions that actually satisfy these requirements. The so-stated conditions can be adopted as a selection rule for viable $f(R)$-models with torsion.

Another important mathematical aspect concerning every theory of gravitation is related to the problem of matching different spacetimes like, for instance, joining together the interior with the exterior region of a relativistic stars. The requirements which have to be fulfilled to solder two different spacetimes are commonly known as junction conditions.

In GR, junction conditions have been investigated by different authors, including Lichnerowicz [29,30], Taub [31], Choquet–Bruhat [32] and Israel [33], and the solution of the problem is now very well known. In [34], the reader can find a very clear discussion about the topic.

On the contrary, at least in the authors' knowledge, very few works deal with junction conditions in ECSK theory: an analysis has been performed by Arkuszewski et al. [35], by means of the formalism of tensor–valued differential forms [36–38], while the same topic has been indirectly addressed by Bressange [39] following the same approach as in [34]. Concerning $f(R)$-gravity in purely metric formulation, a discussion of junction conditions has been proposed by Deruelle et al. [40] and Senovilla [41].

In this paper, we address the topic within the theory with torsion, analyzing the junction conditions for $f(R)$-gravity with torsion. Borrowing arguments and notations from [34], after formulating the junction conditions, we discuss their explicit form in the case of coupling to a Dirac field and a spin fluid. As we shall see, the resulting junction conditions are very similar to those existing in ECSK theory. However, this close similarity is only formal. Indeed, due to the contributions that the non linearity of the gravitational Lagrangian function $f(R)$ gives to the contortion tensor, the obtained junction conditions are seen to involve also the trace of the stress–energy tensor and its first derivatives evaluated on the separation hypersurface. This is a remarkable difference with respect to the ECSK theory, which translates into conditions also concerning the function $f(R)$. Therefore, as in the case of the Cauchy problem, the study of the junction conditions can help to distinguish viable from nonviable $f(R)$-models with torsion.

The layout of the paper is the following: In Section 2, we illustrate some generalities about $f(R)$-gravity with torsion. In Section 3, we address the Cauchy problem. In Section 4, we discuss the junction conditions. Finally, we devote Section 5 to conclusions. Throughout the paper, we use natural units ($\hbar = c = 8\pi G = 1$).

2. $f(R)$-Gravity with Torsion

In $f(R)$-gravity with torsion, the (gravitational) dynamical fields are given by a pseudo-Riemannian metric g and a metric compatible linear connection Γ, defined on the space-time manifold M. The covariant derivative induced by connection Γ is given by:

$$\nabla_{\partial_i}\partial_j = \Gamma_{ij}{}^{h}\,\partial_h. \tag{1}$$

The torsion and Riemann curvature tensors, induced by the dynamical connection Γ, are expressed as:

$$T_{ij}{}^{h} = \Gamma_{ij}{}^{h} - \Gamma_{ji}{}^{h}, \tag{2a}$$

$$R^{h}{}_{kij} = \partial_i \Gamma_{jk}{}^{h} - \partial_j \Gamma_{ik}{}^{h} + \Gamma_{ip}{}^{h} \Gamma_{jk}{}^{p} - \Gamma_{jp}{}^{h} \Gamma_{ik}{}^{p}. \tag{2b}$$

In view of the metric compatibility, the linear connection Γ can be decomposed as [14,15]:

$$\Gamma_{ij}{}^{h} = \tilde{\Gamma}_{ij}{}^{h} - K_{ij}{}^{h}, \tag{3}$$

where:

$$K_{ij}{}^{h} := \frac{1}{2}\left(-T_{ij}{}^{h} + T_{j}{}^{h}{}_{i} - T^{h}{}_{ij}\right) \tag{4}$$

is the so-called contorsion tensor, and $\tilde{\Gamma}_{ij}{}^{h}$ is the Levi–Civita connection induced by the metric g.

The field equations are obtained by varying an action functional of the form:

$$\mathcal{A}(g,\Gamma) = \int \left(\sqrt{|g|} f(R) + \mathcal{L}_m\right) ds, \tag{5}$$

where $R(g,\Gamma) = g^{ij} R_{ij}$ (with $R_{ij} := R^{h}{}_{ihj}$) denotes the scalar curvature associated with the connection Γ. The field equations result in [18–20]:

$$f'(R) R_{ij} - \frac{1}{2} f(R) g_{ij} = \mathcal{T}_{ij} \tag{6a}$$

and:

$$T_{ij}{}^{h} = \frac{1}{2f'}\left(\frac{\partial f'}{\partial x^p} + \mathcal{S}_{pq}{}^{q}\right)\left(\delta^p_j \delta^h_i - \delta^p_i \delta^h_j\right) + \frac{1}{f'} \mathcal{S}_{ij}{}^{h}, \tag{6b}$$

where $\mathcal{T}_{ij}$ and $\mathcal{S}_{ij}{}^{h}$ denote the stress–energy and the spin density tensors, respectively. In Equation (6a), attention must be paid to the order of the indexes, because the Ricci and stress–energy tensors R_{ij} and $\mathcal{T}_{ij}$ are not symmetric, in general.

It is worth noticing that, due to the independence between the metric tensor g_{ij} and the dynamical linear connection $\Gamma_{ij}{}^{h}$, the variation of the action functional (5) with respect to the metric tensor does not generate in Equation (6a) any term containing covariant derivatives of the scalar $f'(R)$ (for details, see [18]); this is a remarkable difference with respect to the purely metric formulation of $f(R)$-gravity [2], and it has important consequences: for instance, the theory with torsion is not of fourth derivative order, as is the purely metric $f(R)$-theory. Taking the trace of Equation (6a) into account, we get relation:

$$f'(R) R - 2f(R) = \mathcal{T} \tag{7}$$

between the curvature scalar R and the trace $\mathcal{T}$ of the stress–energy tensor.

From Equation (7), it is seen that if the trace $\mathcal{T}$ is constant, so R is. Of course, the same conclusion holds when $\mathcal{T}_{ij} = 0$. In such circumstances, the field equations of $f(R)$-gravity with torsion are seen to amount to the ones of Einstein–Cartan theory with (or without) cosmological constant if spin is present, or the ones of Einstein theory with (or without) cosmological constant in the absence of spin. This holds in general, with the exception of the particular case $\mathcal{T} = 0$ and $f(R) = \alpha R^2$. In such a case, indeed, Equation (7) is a trivial identity, and it does not impose any restriction on the scalar curvature R.

Therefore, from now on, we shall systematically suppose that $\mathcal{T}_{ij}$ is not zero and $\mathcal{T}$ is not constant, as well as that the relation (7) is invertible. In this way, the curvature scalar R can be thought as a suitable function of $\mathcal{T}$, namely:

$$R = R(\mathcal{T}). \tag{8}$$

The relation (8) plays a crucial role for the formulation of $f(R)$-gravity with torsion presented in this paper, as well as in our previous works. About this, it is worth noticing that the trace equation (7) gives rise to an algebraic or transcendental relation between the curvature scalar and the stress–energy tensor trace, but it is not a differential relation (unlike what happens in the purely metric formulation of $f(R)$-gravity). Therefore, the Dini theorem is generally applicable, and the relation (8) can be (almost always) supposed to locally exist. This allows us to express the torsion as a function of the matter fields and, therefore, to separate purely metric contributions from torsional ones within the Einstein-like equations, exactly as it happens in ECSK theory.

Defining the scalar field:

$$\varphi(\mathcal{T}) := f'(R(\mathcal{T})), \tag{9}$$

we can rewrite Equation (6a) in the equivalent form:

$$R_{ij} - \frac{1}{2}Rg_{ij} = \frac{1}{\varphi}\mathcal{T}_{ij} + \frac{1}{2\varphi}\left(f(R(\mathcal{T})) - f'(R(\mathcal{T}))R(\mathcal{T})\right)g_{ij}, \tag{10a}$$

$$T_{ij}{}^{h} = \frac{1}{2\varphi}\left(\frac{\partial\varphi}{\partial x^p} + \mathcal{S}_{pq}{}^{q}\right)\left(\delta^p_j\delta^h_i - \delta^p_i\delta^h_j\right) + \frac{1}{\varphi}\mathcal{S}_{ij}{}^{h}, \tag{10b}$$

which will be used in the following discussion. Making use of Equations (3), (4) and (10b), we can express the contorsion tensor as:

$$K_{ij}{}^{h} = \hat{K}_{ij}{}^{h} + \hat{S}_{ij}{}^{h}, \tag{11a}$$

$$\hat{S}_{ij}{}^{h} := \frac{1}{2\varphi}\left(-\mathcal{S}_{ij}{}^{h} + \mathcal{S}_{j}{}^{h}{}_{i} - \mathcal{S}^{h}{}_{ij}\right), \tag{11b}$$

$$\hat{K}_{ij}{}^{h} := -\hat{T}_j\delta^h_i + \hat{T}_p g^{ph} g_{ij}, \tag{11c}$$

$$\hat{T}_j := \frac{1}{2\varphi}\left(\frac{\partial\varphi}{\partial x^j} + \mathcal{S}_{jk}{}^{k}\right). \tag{11d}$$

Introducing the so-called torsion vector $T_i := T_{ij}{}^{j}$, we also mention the conservation laws [42]:

$$\nabla_a\mathcal{T}^{ai} + T_a\mathcal{T}^{ai} - \mathcal{T}_{ca}T^{ica} - \frac{1}{2}\mathcal{S}_{sta}R^{stai} = 0, \tag{12a}$$

$$\nabla_h\mathcal{S}^{ijh} + T_h\mathcal{S}^{ijh} + \mathcal{T}^{ij} - \mathcal{T}^{ji} = 0, \tag{12b}$$

which have to be satisfied by the stress–energy and spin density tensors of the matter fields. In particular, we recall that Equation (12b) amount to the antisymmetric part of the Einstein-like Equation (10a).

In the case that the spin density tensor is zero, separating the Levi–Civita terms from the torsional ones, we can rewrite the Einstein-like field Equation (10a) in the form [18]:

$$\begin{aligned}\tilde{R}_{ij} - \frac{1}{2}\tilde{R}g_{ij} = \frac{1}{\varphi}\mathcal{T}_{ij} + \frac{1}{\varphi^2}\bigg(&-\frac{3}{2}\frac{\partial\varphi}{\partial x^i}\frac{\partial\varphi}{\partial x^j} + \varphi\tilde{\nabla}_j\frac{\partial\varphi}{\partial x^i} + \frac{3}{4}\frac{\partial\varphi}{\partial x^h}\frac{\partial\varphi}{\partial x^k}g^{hk}g_{ij} \\ &-\varphi\tilde{\nabla}^h\frac{\partial\varphi}{\partial x^h}g_{ij} - V(\varphi)g_{ij},\bigg),\end{aligned} \tag{13}$$

where the effective potential for the scalar field φ:

$$V(\varphi) := \frac{1}{4}\left[\varphi F^{-1}((f')^{-1}(\varphi)) + \varphi^2(f')^{-1}(\varphi)\right], \tag{14}$$

has been introduced. In Equation (13), $\tilde{R}_{ij}$, $\tilde{R}$, and $\tilde{\nabla}$ denote, respectively, the Ricci tensor, the scalar curvature, and the covariant derivative associated with the Levi–Civita connection of the dynamical metric g_{ij}.

The Einstein-like Equation (13) (together with Equation (9)) are deducible from a scalar-tensor theory with Brans–Dicke parameter $\omega_0 = -3/2$. This can be seen by recalling the action functional of a (purely metric) scalar-tensor theory:

$$\mathcal{A}(g,\varphi) = \int \left[\sqrt{|g|}\left(\varphi\tilde{R} - \frac{\omega_0}{\varphi}\varphi_i\varphi^i - U(\varphi)\right) + \mathcal{L}_m\right] ds, \tag{15}$$

where φ is the scalar field, $\varphi_i := \dfrac{\partial\varphi}{\partial x^i}$ and $U(\varphi)$ is the potential of φ, $\mathcal{L}_m(g_{ij},\psi)$ is the matter Lagrangian, function of the metric and some other matter fields ψ, and ω_0 is the so called Brans–Dicke parameter. By varying (15) with respect to the metric tensor and the scalar field, one gets the field equations:

$$\tilde{R}_{ij} - \frac{1}{2}\tilde{R}g_{ij} = \frac{1}{\varphi}\mathcal{T}_{ij} + \frac{\omega_0}{\varphi^2}\left(\varphi_i\varphi_j - \frac{1}{2}\varphi_h\varphi^h g_{ij}\right) + \frac{1}{\varphi}\left(\tilde{\nabla}_j\varphi_i - \tilde{\nabla}_h\varphi^h g_{ij}\right) - \frac{U}{2\varphi}g_{ij}, \tag{16}$$

and:

$$\frac{2\omega_0}{\varphi}\tilde{\nabla}_h\varphi^h + \tilde{R} - \frac{\omega_0}{\varphi^2}\varphi_h\varphi^h - U' = 0, \tag{17}$$

where $\mathcal{T}_{ij} := -\dfrac{1}{\sqrt{|g|}}\dfrac{\delta\mathcal{L}_m}{\delta g^{ij}}$ and $U' := \dfrac{dU}{d\varphi}$. By inserting the trace of Equation (16) into Equation (17), one gets the equation:

$$(2\omega_0 + 3)\,\tilde{\nabla}_h\varphi^h = \mathcal{T} + \varphi U' - 2U. \tag{18}$$

A direct comparison immediately shows that, for $\omega_0 := -\dfrac{3}{2}$ and $U(\varphi) := \dfrac{2}{\varphi}V(\varphi)$ (where $V(\varphi)$ is defined in Equation (14)), Equation (16) becomes formally identical to the Einstein-like Equation (13) for $f(R)$-gravity with torsion. Moreover, in such a circumstance, Equation (18) reduces to the algebraic equation:

$$\mathcal{T} + 2V'(\varphi) - \frac{6}{\varphi}V(\varphi) = 0, \tag{19}$$

relating the matter trace $\mathcal{T}$ to the scalar field φ. In particular, it is easily seen that, under the condition $f'' \neq 0$, Equation (19) represents exactly the inverse relation of (9), namely:

$$\mathcal{T} + 2V'(\varphi) - \frac{6}{\varphi}V(\varphi) = 0 \quad\Longleftrightarrow\quad \mathcal{T} = F^{-1}((f')^{-1}(\varphi)), \tag{20}$$

being $F^{-1}(X) = f'(X)X - 2f(X)$. The conclusion follows that, when the matter Lagrangian does not depend on the dynamical connection (the dynamical connection does not couple with matter), $f(R)$-gravity with torsion is dynamically equivalent to a scalar-tensor theory with a Brans–Dicke parameter $\omega_0 = -\dfrac{3}{2}$.

For later use, we also notice that field equations (13) can be simplified by rewriting them the Einstein frame. In fact, performing the conformal transformation:

$$\bar{g}_{ij} = \varphi g_{ij}. \tag{21}$$

Equation (13) assumes the simpler form (see for example [18,43]):

$$\bar{R}_{ij} - \frac{1}{2}\bar{R}\bar{g}_{ij} = \frac{1}{\varphi}\mathcal{T}_{ij} - \frac{1}{\varphi^3}V(\varphi)\bar{g}_{ij}, \tag{22}$$

where $\bar{R}_{ij}$ and $\bar{R}$ are, respectively, the Ricci tensor and the curvature scalar induced by the conformal metric $\bar{g}_{ij}$.

The relationships between the the conservation laws existing in the Jordan and those holding in the Einstein frame are clarified by the following results [44,45]:

Proposition 1. *Equations* (13), (14), (19) *imply the standard conservation laws* $\tilde{\nabla}^j \mathcal{T}_{ij} = 0$.

Proposition 2. *The condition* $\tilde{\nabla}^j \mathcal{T}_{ij} = 0$ *is equivalent to the condition* $\bar{\nabla}^j \bar{\mathcal{T}}_{ij} = 0$, *where* $\bar{\mathcal{T}}_{ij} := \frac{1}{\varphi}\mathcal{T}_{ij} - \frac{1}{\varphi^3}V(\varphi)\bar{g}_{ij}$ *and* $\bar{\nabla}$ *denotes the covariant derivative associated to the conformal metric* $\bar{g}_{ij}$.

3. The Cauchy Problem

First of all, we notice that, if the trace of the stress–energy tensor $\mathcal{T}$ is constant, $f(R)$-gravity with torsion reduces to GR with (or without) cosmological constant. Therefore, when this is the case, the Cauchy problem is well-formulated and well-posed [23,26,46]. For instance, this happens in vacuo and in the presence of electromagnetic (or also Yang–Mills) fields (if $f(R) \neq \alpha R^2$).

The situation is more complicated if $\mathcal{T} \neq$ const.: in such a circumstance, the theory no longer amounts to GR, so the classical Bruhat results [46] do not apply, and the well-formulation and well-posedness of the Cauchy problem is not automatically assured.

To overcome this issue, the Cauchy problem could be addressed by exploiting the dynamical equivalence with scalar-tensor theories with Brans–Dicke parameter $\omega_0 = -\frac{3}{2}$. Unfortunately, here a difficulty occurs: The d'Alembertian $g^{pq}\nabla_p\nabla_q\varphi$ disappears from Equation (18), and we no longer have the possibility of deriving the expression of the d'Alembertian as a function of the dynamical variables and their derivatives up to the first-order. In other words, we cannot eliminate the second-order derivatives of the scalar field φ from the Einstein-like Equation (16).

An alternative idea is to pass from the Jordan to the Einstein frame, making use of the conformal transformation technique. Following this approach, we can derive sufficient conditions for the well-posedness of the Cauchy problem for $f(R)$-gravity with torsion in the presence of a perfect fluid [44,47] or a Klein–Gordon scalar field [48]. Such conditions result in suitable requirements imposed on the function $f(R)$, so they can be assumed as a sort of selection rule for viable $f(R)$ models. In addition, we show that the function $f(R) = R + \alpha R^2$ satisfies the above mentioned conditions, in such a way that the set of viable $f(R)$ models is not empty.

3.1. The Cauchy Problem in Presence of a Perfect Fluid

We discuss the Cauchy problem in presence of a perfect fluid. As mentioned above, implementing a conformal transformation (21), we show that the initial value problem can be analyzed by applying the same results stated in [23,24,46] for GR.

To start with, given the metric g_{ij} of signature $(-+++)$ in the Jordan frame, let us consider a perfect fluid having stress–energy tensor of the form:

$$\mathcal{T}_{ij} = (\rho + p)\, U_i U_j + p\, g_{ij} \tag{23a}$$

and undergoing the usual conservation laws:

$$\tilde{\nabla}_j \mathcal{T}^{ij} = 0. \tag{23b}$$

In Equation (23a), ρ and p are the matter–energy density and the pressure of the fluid, respectively, while U_i denotes the components of the four velocity of the fluid (with $g^{ij}U_iU_j = -1$). Performing the conformal transformation (21), we rewrite the field equations in the Einstein frame as:

$$\bar{R}_{ij} - \frac{1}{2}\bar{R}\bar{g}_{ij} = \bar{\mathcal{T}}_{ij}, \tag{24a}$$

and:

$$\bar{\nabla}_j \bar{\mathcal{T}}^{ij} = 0, \tag{24b}$$

where:

$$\bar{\mathcal{T}}_{ij} = \frac{1}{\varphi}(\rho + p)\, U_i U_j + \left(\frac{p}{\varphi^2} - \frac{V(\varphi)}{\varphi^3}\right)\bar{g}_{ij}, \tag{25}$$

plays the role of the effective stress–energy tensor.

In view of Proposition 2, Equation (24b) is equivalent to Equation (23b). This is a crucial aspect for our discussion, allowing us to apply to the present case the results stated in [23,24,46]. To see this point, we first suppose that the scalar field φ is positive, that is $\varphi > 0$. The opposite case $\varphi < 0$ differs from the former only for some technical aspects, and it will be briefly outlined after. Of course, it is implicitly assumed that $\varphi \neq 0$ at least in a neighborhood of the initial space-like surface.

Under the assumed conditions, the stress–energy tensor (25) can be rewritten in the form:

$$\bar{\mathcal{T}}_{ij} = \frac{1}{\varphi^2}(\rho + p)\, \bar{U}_i \bar{U}_j + \left(\frac{p}{\varphi^2} - \frac{V(\varphi)}{\varphi^3}\right)\bar{g}_{ij}, \tag{26}$$

where $\bar{U}_i = \sqrt{\varphi}U_i$ is the four velocity of the fluid in the Einstein frame. Introducing the effective mass-energy density:

$$\bar{\rho} := \frac{\rho}{\varphi^2} + \frac{V(\varphi)}{\varphi^3}, \tag{27a}$$

and the effective pressure:

$$\bar{p} := \frac{p}{\varphi^2} - \frac{V(\varphi)}{\varphi^3}, \tag{27b}$$

we can express the stress–energy tensor (26) in the standard form:

$$\bar{\mathcal{T}}_{ij} = (\bar{\rho} + \bar{p})\, \bar{U}_i \bar{U}_j + \bar{p}\, \bar{g}_{ij}. \tag{28}$$

We notice that, given an equation of state of the form $\rho = \rho(p)$ and assuming that the relation (27b) is invertible ($p = p(\bar{p})$), by composition with Equation (27a), we obtain an effective equation of state $\bar{\rho} = \bar{\rho}(\bar{p})$. Moreover, we recall that the explicit expressions of the scalar field φ and the potential $V(\varphi)$ depend on the specific form of the function $f(R)$. As a consequence, the request that the relation (27b) is invertible together with the condition $\varphi > 0$ (or, equivalently, $\varphi < 0$) can be assumed as criteria for the viability of the functions $f(R)$, providing us with suitable selection rules for admissible gravitational Lagrangian function $f(R)$ (see also [9]).

After that, in order to discuss the Cauchy problem, we can follow step-by-step the Bruhat's arguments [23,24,46]. In particular, we recall that the Cauchy problem for the system of differential equations (24), with stress–energy tensor given by Equation (28) and equation of state $\bar{\rho} = \bar{\rho}(\bar{p})$, is well-posed if the condition:

$$\frac{d\bar{\rho}}{d\bar{p}} \geq 1, \tag{29}$$

is satisfied. The requirement (29) is easily verified by means of the relation:

$$\frac{d\bar{\rho}}{d\bar{p}} = \frac{d\bar{\rho}/dp}{d\bar{p}/dp} \geq 1, \tag{30}$$

together with expressions (27) and the equation of state $\rho = \rho(p)$. Once again, condition (30) depends on the expressions of φ and $V(\varphi)$; thus, it is strictly related to the form of the function $f(R)$. Therefore, condition (30) represents a further criterion for the admissibility of $f(R)$-models.

For the sake of completeness, we conclude by outlining the case $\varphi < 0$. Supposing again that the signature of the metric in the Jordan frame is $(-+++)$, the signature of the conformal metric is now $(+---)$, and the components of the four velocity of the fluid in the Einstein frame are $\bar{U}_i = \sqrt{-\varphi}U_i$. The effective stress–energy tensor is expressed as:

$$\bar{\mathcal{T}}_{ij} = -\frac{1}{\varphi^2}(\rho + p)\,\bar{U}_i\bar{U}_j + \left(\frac{p}{\varphi^2} - \frac{V(\varphi)}{\varphi^3}\right)\bar{g}_{ij} = (\bar{\rho} + \bar{p})\,\bar{U}_i\bar{U}_j - \bar{p}\,\bar{g}_{ij}, \tag{31}$$

where, as above, the quantities:

$$\bar{\rho} := -\frac{\rho}{\varphi^2} - \frac{V(\varphi)}{\varphi^3}, \tag{32a}$$

and:

$$\bar{p} := -\frac{p}{\varphi^2} + \frac{V(\varphi)}{\varphi^3}, \tag{32b}$$

represent the effective mass-energy and the effective pressure.

After that, everything proceeds again as in [23,24,46], with the exception of a technical aspect: if ρ and p are positive, the quantity $r := \bar{\rho} + \bar{p} = -\frac{\rho + p}{\varphi^2}$ is now negative. About this, the reader can easily verify that, with the choice $\log(-f^{-2}r)$ instead of $\log(f^{-2}r)$ as in [23,24,46], the Bruhat's arguments apply equally well.

As a simple example, we consider the model $f(R) = R + \alpha R^2$ coupled with dust. In the Jordan frame, the matter stress–energy tensor is given by $\mathcal{T}_{ij} = \rho U_i U_j$, and the trace of the Einstein-like Equation (6a) yields the relation:

$$(1 + 2\alpha R)R - 2R - 2\alpha R^2 = -\rho \qquad \Longleftrightarrow \qquad R = \rho. \tag{33}$$

The scalar field (9) assumes the form:

$$\varphi(\rho) = f'(R(\rho)) = 1 + 2\alpha\rho. \tag{34}$$

Taking into account small values of the density $\rho << 1$ (for instance, the present cosmological baryonic matter density) and choosing values of $|\alpha|$ not comparable with $1/\rho$, we can reasonably suppose $\varphi > 0$, independently of the sign of the parameter α. We have to calculate the potential (14):

$$V(\varphi) = \frac{1}{4}\left[\varphi F^{-1}((f')^{-1}(\varphi)) + \varphi^2 (f')^{-1}(\varphi)\right]. \tag{35}$$

To this end, since $(f')^{-1}(\varphi) = \rho$, from Equation (34) we get the relation:

$$\frac{1}{4}\varphi^2 (f')^{-1}(\varphi) = \frac{1}{4}(1 + 2\alpha\rho)^2\rho, \tag{36}$$

and considering that $F^{-1}(Y) = f'(Y)Y - 2f(Y)$, we have the identities:

$$\frac{1}{4}F^{-1}((f')^{-1}(\varphi)) = \frac{1}{4}F^{-1}(\rho) = -\rho \tag{37}$$

and:

$$\frac{1}{4}\varphi F^{-1}((f')^{-1}(\varphi)) = -\frac{(1 + 2\alpha\rho)\rho}{4}. \tag{38}$$

We conclude that:

$$V(\varphi(\rho)) = \frac{\alpha\rho^2(1+2\alpha\rho)}{2}. \tag{39}$$

In the Einstein frame, the stress–energy tensor (26) is expressed as:

$$\bar{\mathcal{T}}_{ij} = \frac{\rho}{\varphi^2}\bar{U}_i\bar{U}_j - \frac{V(\varphi)}{\varphi^3}\bar{g}_{ij}. \tag{40}$$

Tensor (40) can be considered as the stress–energy tensor of a perfect fluid with density and pressure given, respectively, by:

$$\bar{\rho} := \frac{\rho}{\varphi^2} + \frac{V(\varphi)}{\varphi^3} = \frac{2\rho + \alpha\rho^2}{2(1+2\alpha\rho)^2} \tag{41a}$$

and:

$$\bar{p} := -\frac{V(\varphi)}{\varphi^3} = -\frac{\alpha\rho^2}{2(1+2\alpha\rho)^2}. \tag{41b}$$

It is an easy matter to verify that the function (41b) is invertible. Indeed, for $\rho > 0$, one has:

$$\frac{d\bar{p}}{d\rho} = -\frac{4\alpha\rho}{4(1+2\alpha\rho)^3} \neq 0. \tag{42}$$

In addition, we have:

$$\frac{d\bar{\rho}}{d\rho} = \frac{4-4\alpha\rho}{4(1+2\alpha\rho)^3}, \tag{43}$$

so that:

$$\frac{d\bar{\rho}}{d\bar{p}} = \frac{d\bar{\rho}/dp}{d\bar{p}/dp} = \frac{-1+\alpha\rho}{\alpha\rho} \geq 1 \quad \Longleftrightarrow \quad \alpha < 0 \tag{44}$$

With condition (29) satisfied, it is then proved that the model $f(R) = R + \alpha R^2$, with $\alpha < 0$, possesses a well-posed Cauchy problem when coupled with dust.

3.2. The Cauchy Problem in the Presence of a Scalar Field

We take the coupling with a Klein–Gordon scalar field into account. Again, we give sufficient conditions for the well-posedness of the related Cauchy problem. To this end, let us denote by ψ a Klein–Gordon scalar field with self-interacting potential $U(\psi) = \frac{1}{2}m^2\psi^2$. The corresponding stress–energy tensor is given by:

$$\mathcal{T}_{ij} = \frac{\partial\psi}{\partial x^i}\frac{\partial\psi}{\partial x^j} - \frac{1}{2}g^{ij}\left(\frac{\partial\psi}{\partial x^p}\frac{\partial\psi}{\partial x^q}g^{pq} + m^2\psi^2\right). \tag{45}$$

The associated Klein–Gordon equation is expressed as:

$$\tilde{\nabla}_j\frac{\partial\psi}{\partial x^i}g^{ij} = m^2\psi, \tag{46}$$

where $\tilde{\nabla}$ denotes the Levi–Civita covariant derivative induced by the metric g_{ij}. The trace of tensor (45) is:

$$\mathcal{T} := \mathcal{T}_{ij}g^{ij} = -\frac{\partial\psi}{\partial x^p}\frac{\partial\psi}{\partial x^q}g^{pq} - 2m^2\psi^2. \tag{47}$$

The trace (47) depends explicitly on the metric tensor g_{ij}. Because of this, the conformal transformation cannot be applied directly to the field equations (13), with the scalar field φ defined by (9). Indeed, if we proceed in this way, both the metric g_{ij} and $\bar{g}_{ij}$ would appear in the conformally transformed equations (22). This difficulty can be overcome making use of the already mentioned dynamical equivalence with $\omega_0 = -\frac{3}{2}$ Brans–Dicke gravity. The idea is then to discuss the Cauchy problem for a $\omega_0 = -\frac{3}{2}$ Brans–Dicke theory coupled with the given Klein–Gordon field ψ. The field equations of such a theory are the Einstein-like Equation (13), the Equation (19), and the Klein–Gordon Equation (46), where the scalar field φ is a dynamical variable related to the trace $\mathcal{T}$ through Equation (19). After implementing the conformal transformation (21), the Einstein-like Equation (13) assumes the simpler form (22). At the same time, recalling the relation:

$$\bar{\Gamma}_{ij}{}^{h} = \tilde{\Gamma}_{ij}{}^{h} + \frac{1}{2\varphi}\frac{\partial\varphi}{\partial x^j}\delta^h_i - \frac{1}{2\varphi}\frac{\partial\varphi}{\partial x^p}g^{ph}g_{ij} + \frac{1}{2\varphi}\frac{\partial\varphi}{\partial x^i}\delta^h_j, \tag{48}$$

linking the Levi–Civita connection $\tilde{\Gamma}_{ij}{}^{h}$ associated with the metric g_{ij} to the Levi–Civita connection $\bar{\Gamma}_{ij}{}^{h}$ induced by the conformal metric $\bar{g}_{ij}$, we can write the Klein–Gordon equation in terms of the conformal metric $\bar{g}_{ij}$ as:

$$-\frac{\partial\psi}{\partial x^i}\bar{g}^{ij}\frac{\partial\varphi}{\partial x^j} + \varphi\bar{\nabla}_j\frac{\partial\psi}{\partial x^i}\bar{g}^{ij} = m^2\psi, \tag{49}$$

where $\bar{\nabla}_j$ indicates the covariant derivative associated with the conformal metric $\bar{g}_{ij}$. Analogously, we can express the trace $\mathcal{T}$ as function of $\bar{g}_{ij}$, that is:

$$\mathcal{T} = -\frac{\partial\psi}{\partial x^p}\frac{\partial\psi}{\partial x^q}\varphi\bar{g}^{pq} - 2m^2\psi^2\,. \tag{50}$$

The relation corresponding to (19) now links the scalar field φ to the Klein–Gordon field ψ, its partial derivatives $\frac{\partial\psi}{\partial x^i}$, and the conformal metric $\bar{g}_{ij}$. Moreover, as it has been already pointed out, the quantity:

$$\bar{\mathcal{T}}_{ij} := \frac{1}{\varphi}\Sigma_{ij} - \frac{1}{\varphi^3}V(\varphi)\bar{g}_{ij}, \tag{51}$$

represents an effective stress–energy tensor. On the other hand, the Klein–Gordon equation (46) implies the conservation laws $\tilde{\nabla}^j\mathcal{T}_{ij} = 0$, thus also identifying $\bar{\nabla}^j\bar{\mathcal{T}}_{ij} = 0$ (see Proposition 2). This is a key point, allowing us to making use of harmonic coordinates and then to apply similar arguments as in [23,24,26].

More specifically, after rewriting the Einstein-like equations (22) in the equivalent form:

$$\bar{R}_{ij} = \bar{\mathcal{T}}_{ij} - \frac{1}{2}\bar{\mathcal{T}}\bar{g}_{ij}, \tag{52}$$

we adopt harmonic coordinates obeying the condition:

$$\bar{\nabla}_p\bar{\nabla}^p x^i = -\bar{g}^{pq}\bar{\Gamma}^i_{pq} = 0\,, \tag{53}$$

in such a way that equations (52) can be expresed as (see, for example, [23,26]):

$$\bar{g}^{pq}\frac{\partial^2\bar{g}_{ij}}{\partial x^p\partial x^q} = f_{ij}(\bar{g},\partial\bar{g},\psi,\partial\psi), \tag{54}$$

where f_{ij} are suitable functions depending only on the metric $\bar{g}$, the scalar field ψ, and their first order derivatives.

In addition to this, we suppose that Equation (19) is solvable with respect to the variable φ, and then to derive from Equation (19) itself a function of the form:

$$\varphi = \varphi(\bar{g},\psi,\frac{\partial\psi}{\partial x^p}\frac{\partial\psi}{\partial x^q}\bar{g}^{pq}), \tag{55}$$

expressing the scalar field φ as a suitable function of the metric $\bar{g}$, the Klein–Gordon field ψ, and its first order derivatives. We notice that, in view of Equation (50), the dependence of φ on the derivatives of ψ is necessarily of the form indicated in Equation (55). Once again, the solvability with respect the scalar field φ to about Equation (19) depends on the explicit form of the potential $V(\varphi)$ which is defined in terms of the function $f(R)$ via the relation (14). Therefore, the possibility of solving Equation (19) with respect to φ can be taken as a rule to select viable $f(R)$-models. Moreover, from Equation (55), we obtain the identity:

$$\frac{\partial\varphi}{\partial x^i} = \frac{\partial\varphi}{\partial\left(\frac{\partial\psi}{\partial x^s}\frac{\partial\psi}{\partial x^t}\bar{g}^{st}\right)} 2\frac{\partial\psi}{\partial x^q}\bar{g}^{pq}\frac{\partial^2\psi}{\partial x^i\partial x^p} + f_i(\bar{g},\partial\bar{g},\psi,\partial\psi). \tag{56}$$

Inserting Equation (56) in Equation (49) and taking Equation (53) into account, we get the final form of the Klein–Gordon equation expressed as:

$$\left(\bar{g}^{ip} - \frac{2}{\varphi}\frac{\partial\varphi}{\partial\left(\frac{\partial\psi}{\partial x^s}\frac{\partial\psi}{\partial x^t}\bar{g}^{st}\right)}\frac{\partial\psi}{\partial x^j}\bar{g}^{ji}\frac{\partial\psi}{\partial x^q}\bar{g}^{pq}\right)\frac{\partial^2\psi}{\partial x^i\partial x^p} = f(\bar{g},\partial\bar{g},\psi,\partial\psi). \tag{57}$$

In Equations (56) and (57), f_i and f denote suitable functions of $\bar{g}_{ij}$, ψ, and their first order derivatives only.

Now, Equations (54) and (57) form a second order quasi-diagonal system of partial differential equations for the unknowns $\bar{g}_{ij}$ and ψ. The matrix of the principal parts of such a system is diagonal, and its elements are the differential operators:

$$\bar{g}^{pq}\frac{\partial^2}{\partial x^p\partial x^q}, \tag{58a}$$

and:

$$\left(\bar{g}^{ip} - \frac{2}{\varphi}\frac{\partial\varphi}{\partial\left(\frac{\partial\psi}{\partial x^s}\frac{\partial\psi}{\partial x^t}\bar{g}^{st}\right)}\frac{\partial\psi}{\partial x^j}\bar{g}^{ji}\frac{\partial\psi}{\partial x^q}\bar{g}^{pq}\right)\frac{\partial^2}{\partial x^i\partial x^p}. \tag{58b}$$

The operator (58a) is the wave operator associated with the metric $\bar{g}_{ij}$, while the operator (58b) is very similar to the sound wave operator involved in the analysis of the Cauchy problem for GR coupled with an irrotational perfect fluid [23,46]. It follows that the Cauchy problem associated with the system of Equations (54) and (57) can be discussed borrowing arguments and results from [23,46]. More in detail, we recall that if the quadratic form associated with (58b) is of Lorentzian signature and, if the characteristic cone of the operator (58b) is exterior to the metric cone, the system (54) and (57) is causal and Leray hyperbolic [49,50]. Under these conditions, the associated Cauchy problem is well-posed in suitable Sobolev spaces. Still borrowing from [23,46], if the signature of $\bar{g}_{ij}$ is $(+---)$, the above mentioned conditions are satisfied whenever the vector $\frac{\partial\psi}{\partial x^j}\bar{g}^{ij}$ is timelike and the inequality:

$$-\frac{2}{\varphi}\frac{\partial\varphi}{\partial\left(\frac{\partial\psi}{\partial x^s}\frac{\partial\psi}{\partial x^t}\bar{g}^{st}\right)} \geq 0, \tag{59}$$

holds. Of course, when the signature of the metric $\bar{g}_{ij}$ is $(-+++)$, the sign of inequality (59) has to be inverted. As it has been already remarked, the function (55) depends on the potential (14), which is determined by the explicit form of the function $f(R)$. Therefore, we can adopt requirement (59) as a criterion to single out viable $f(R)$-models with torsion.

As an illustrative example, we consider again the model $f(R) = R + \alpha R^2$. From the relation $F^{-1}(X) = f'(X)X - 2f(X) = -X$, the identity $(f')^{-1}(\varphi) = \frac{\varphi - 1}{2\alpha}$, and the expression (14), we easily obtain the effective potential:

$$V(\varphi) = \frac{1}{8\alpha}(\varphi - 1)^2\varphi. \tag{60}$$

Equation (60), together with Equations (19) and (50), yields:

$$\varphi = \frac{\left(\frac{1}{2\alpha} + 2m^2\psi^2\right)}{\left(\frac{1}{2\alpha} - \frac{\partial\psi}{\partial x^s}\frac{\partial\psi}{\partial x^t}\bar{g}^{st}\right)}, \tag{61}$$

which describes the scalar field φ as a function of the metric $\bar{g}_{ij}$, the Klein–Gordon field ψ, and its first order derivatives. By deriving (61), we have:

$$\frac{\partial\varphi}{\partial\left(\frac{\partial\psi}{\partial x^s}\frac{\partial\psi}{\partial x^t}\bar{g}^{st}\right)} = \frac{\varphi}{\left(\frac{1}{2\alpha} - \frac{\partial\psi}{\partial x^s}\frac{\partial\psi}{\partial x^t}\bar{g}^{st}\right)}. \tag{62}$$

If the metric $\bar{g}_{ij}$ has signature $(+ - - -)$, we see that the requirement (59) is fulfilled if $\alpha < 0$ and if $\bar{g}^{pq}\frac{\partial\psi}{\partial x^q}$ is a time-like vector field. On the contrary, when the signature of $\bar{g}_{ij}$ is $(- + + +)$, α has to be positive.

4. The Junction Conditions

In this section, we address the junction conditions issue within the framework of $f(R)$-gravity with torsion. As mentioned in the Introduction, the junction condition problem is crucial for any theory of gravitation; for instance, in order to join together the interior with the exterior region of a relativistic star, we need to know how matching different solutions of the field equations of the theory at a given hypersurface. After deriving general junction conditions, in order to highlight the main differences with respect to ECSK theory, we give two illustrative examples. For reasons of greater clarity and better readability, the proposed examples are presented in two separate subsections.

Let us consider a hypersurface Σ which separates two different regions $\mathcal{M}^+$ and $\mathcal{M}^-$ of spacetime. To begin with, let us deal with the case in which the hypersurface Σ is either timelike or spacelike; the case of null hypersurface will be discussed later. Let us denote by $\left(g^+_{ij}, \Gamma^+_{ij}{}^h\right)$ and $\left(g^-_{ij}, \Gamma^-_{ij}{}^h\right)$ two solutions of the field equations (10), defined in $\mathcal{M}^+$ and $\mathcal{M}^-$, respectively. We want to discuss how to solder together at Σ the two given Einstein–Cartan geometries, in order to obtain a unique solution of the field equations on the whole spacetime.

To this end, we refer Σ to local coordinates y^A $(A = 1, \ldots, 3)$, and we adopt a coordinate system x^i, locally overlapping both $\mathcal{M}^+$ and $\mathcal{M}^-$ in an open set containing Σ. After that, considering the arc length s between any point $p \in \mathcal{M}$ and Σ along the geodesic normal to Σ (with respect to one of the two given metric tensors) and passing through p itself, we define a function s which, without loss of generality, can be set negative in $\mathcal{M}^-$, positive in $\mathcal{M}^+$, and equal to zero at Σ. Indicating by n^i the unit normal (with respect to the chosen metric tensor) outgoing from Σ, one has the relations:

$$dx^i = n^i ds, \qquad n_i = \epsilon\partial_i s \qquad \text{and} \qquad n^i n_i = \epsilon, \tag{63}$$

where $\epsilon = 1$ if Σ is spacelike, and $\epsilon = -1$ if Σ is timelike. Moreover, given any geometric quantity W defined on both sides of the hypersurface Σ, we denote by:

$$[W] := W\left(\mathcal{M}^+\right)_{|\Sigma} - W\left(\mathcal{M}^-\right)_{|\Sigma} \tag{64}$$

the jump of W across Σ. The issue of matching different geometries at a given hypersurface Σ is usually discussed in the framework of distribution-valued tensors [29,30,32,51,52]. In this regard, denoting by $\Theta(s)$ (with $\Theta(0) := 1$) the Heaviside distribution, we introduce the following geometrical objects:

$$g_{ij} = \Theta(s) g^+_{ij} + (1-\Theta(s))\, g^-_{ij}, \tag{65a}$$

$$\Gamma_{ij}{}^{h} = \Theta(s)\Gamma^{+}_{ij}{}^{h} + (1-\Theta(s))\,\Gamma^{-}_{ij}{}^{h}, \tag{65b}$$

with the requirement that the quantities (65) define a solution of the field equations (10) in the distributional sense. To satisfy this request, the quantities (65) and all the the geometric quantities induced by them have to be well defined as distributions. In particular, this must apply to the Riemann and the Einstein tensors. Moreover, consistency between (65), (3) implies the identity:

$$\Gamma_{ij}{}^{h} = \Theta(s)\left(\tilde\Gamma^{+}_{ij}{}^{h} - K^{+}_{ij}{}^{h}\right) + [1-\Theta(s)]\left(\tilde\Gamma^{-}_{ij}{}^{h} - K^{-}_{ij}{}^{h}\right), \tag{66}$$

where $\tilde\Gamma_{ij}{}^{h}$ are the Christoffel coefficients associated with the metric (65a). By differentiating (65), we get the relations:

$$\partial_k g_{ij} = \Theta(s)\partial_k g^+_{ij} + (1-\Theta(s))\,\partial_k g^-_{ij} + \epsilon\delta(s)\left[g_{ij}\right] n_k, \tag{67a}$$

$$\partial_k \Gamma_{ij}{}^{h} = \Theta(s)\partial_k\Gamma^{+}_{ij}{}^{h} + (1-\Theta(s))\,\partial_k\Gamma^{-}_{ij}{}^{h} + \epsilon\delta(s)\left[\Gamma_{ij}{}^{h}\right] n_k, \tag{67b}$$

where, referring the reader to [31,51,52] and references therein for the definition of the Dirac δ-function with support on the submanifold $\Sigma : s = 0$, we have used the identities $\frac{\partial s}{\partial x^i} = \epsilon n_i$ and $\frac{d\Theta(s)}{ds} = \delta(s)$.

Making use of Equation (67), as well as of the identities $\Theta^2(s) = \Theta(s)$ and $\Theta(s)\,(1-\Theta(s)) = 0$, it is easily seen that the Levi–Civita contribution to the connection $\Gamma_{ij}{}^{h}$ contains a singular term having expression:

$$\frac{1}{2} g^{+hk}_{|\Sigma}\left(\left[g_{ik}\right] n_j + \left[g_{jk}\right] n_i - \left[g_{ij}\right] n_k\right)\epsilon\delta(s). \tag{68}$$

Requirement (65b) implies then the vanishing of the term (68); thus,

$$\left[g_{ij}\right] = 0, \tag{69}$$

amounting to the fact that the two metrics have to coincide on the hypersurface Σ. In addition, from Equation (67b), we get the expression of the Riemann tensor of the the connection (65b):

$$R^{p}{}_{qij} = \Theta(s) R^{+p}{}_{qij} + (1-\Theta(s))\, R^{-p}{}_{qij} + \delta(s) A^{p}{}_{qij}, \tag{70}$$

where we have denoted by:

$$A^{p}{}_{qij} := \epsilon\left(\left[\Gamma_{jq}{}^{p}\right] n_i - \left[\Gamma_{iq}{}^{p}\right] n_j\right) \tag{71}$$

the tensor connected with the presence of the δ-function term in the Riemann tensor (70). Once again, decomposition (3) can be used, so that we can rewrite the tensor (71) as the sum:

$$A^{p}{}_{qij} = \tilde A^{p}{}_{qij} + \bar A^{p}{}_{qij}, \tag{72}$$

where:

$$\tilde A^{p}{}_{qij} = \epsilon\left(\left[\tilde\Gamma_{jq}{}^{p}\right] n_i - \left[\tilde\Gamma_{iq}{}^{p}\right] n_j\right) \tag{73}$$

and:

$$\bar A^{p}{}_{qij} = \epsilon\left(\left[-K_{jq}{}^{p}\right] n_i + \left[K_{iq}{}^{p}\right] n_j\right) \tag{74}$$

are quantities related to the Levi–Civita and contortion, respectively.

The continuity of the metric tensor across the hypersurface Σ implies that its derivatives may have discontinuities only along the normal direction. Then, there exists a tensor field on Σ:

$$k_{ij} := \epsilon \left[\partial_h g_{ij}\right] n^h, \tag{75}$$

such that:

$$\left[\partial_h g_{ij}\right] = k_{ij} n_h. \tag{76}$$

From Equation (76), we get the expressions:

$$\left[\tilde{\Gamma}_{ij}{}^{h}\right] = \frac{1}{2}\left(k^h{}_j n_i + k^h{}_i n_j - k_{ij} n^h\right), \tag{77}$$

which, inserted into Equation (73), yield the explicit representation:

$$\tilde{A}^p{}_{qij} = \frac{\epsilon}{2}\left(k^p{}_j n_q n_i - k^p{}_i n_q n_j - k_{qj} n^p n_i + k_{qi} n^p n_j\right). \tag{78}$$

By contraction of Equation (78), we have:

$$\tilde{A}_{qj} := \tilde{A}^p{}_{qpj} = \frac{\epsilon}{2}\left(k^p{}_j n_q n_p - k n_q n_j - k_{qj}\epsilon + k_{qp} n^p n_j\right) \tag{79}$$

and:

$$\tilde{A} := \tilde{A}^q{}_q = \epsilon\left(k_{pq} n^p n^q - \epsilon k\right), \tag{80}$$

with $k := k_{ij} g^{ij}$. Making use of Equations (56), (80), we introduce the tensor:

$$\tilde{H}_{qj} = \tilde{A}_{qj} - \frac{1}{2}\tilde{A} g_{qj} = \frac{\epsilon}{2}\left(k^p{}_j n_q n_p - k n_q n_j - k_{qj}\epsilon + k_{qp} n^p n_j\right) - \frac{\epsilon}{2}\left(k_{st} n^s n^t - \epsilon k\right) g_{qj}, \tag{81}$$

which represents the δ-function part of the Einstein tensor, generated by Levi–Civita connection. Tensor (81) is symmetric and tangent to the hypersurface Σ. In fact, it is a straightforward matter to verify that $\tilde{H}_{qj} n^j = 0$. If we denote by $E^i_A := \frac{\partial x^i}{\partial y^A}$, the tensor $\tilde{H}_{qj}$ can be expressed as $\tilde{H}^{qj} = \tilde{H}^{AB} E^q_A E^j_B$, with [34]:

$$\tilde{H}_{AB} := \tilde{H}_{qj} E^q_A E^j_B = -\frac{1}{2} k_{qj} E^q_A E^j_B + \frac{1}{2} k_{pq} h^{pq} h_{AB}, \tag{82}$$

where $h^{pq} := g^{pq} - \epsilon n^p n^q$ and $h_{AB} := g_{ij} E^i_A E^j_B$ are the projection operator and the induced metric on the hypersurface Σ, respectively.

Analogously, we can single out the contributions given by contortion to the δ-function part of the Einstein tensor. By contraction, from Equation (74), we in fact:

$$\bar{A}_{qj} := \bar{A}^p{}_{qpj} = \epsilon\left(\left[-K_{jq}{}^{p}\right] n_p + \left[K_{pq}{}^{p}\right] n_j\right) \tag{83}$$

and:

$$\bar{A} := \bar{A}^q{}_q = 2\epsilon\left[K_{pq}{}^{p}\right] n^q. \tag{84}$$

By means of expressions (83), (84), we define the tensor:

$$\bar{H}_{qj} := \bar{A}_{qj} - \frac{1}{2}\bar{A} g_{qj} = \epsilon\left(\left[-K_{jq}{}^{p}\right] n_p + \left[K_{pq}{}^{p}\right] n_j\right) - \epsilon\left[K_{st}{}^{s}\right] n^t g_{qj}, \tag{85}$$

which, in general, is neither symmetric nor tangent to the hypersurface Σ. All the obtained results allow us to express the effective stress–energy tensor appearing on the right-hand side of Equation (10a) in the form:

$$\hat{\mathcal{T}}_{qj} = \Theta(s)\left[\frac{1}{\varphi}\mathcal{T}_{qj} + \frac{1}{\varphi}U(\mathcal{T})g_{qj}\right]^{+} + (1-\Theta(s))\left[\frac{1}{\varphi}\mathcal{T}_{qj} + \frac{1}{\varphi}U(\mathcal{T})g_{qj}\right]^{-} - \delta(s)H_{qj}, \tag{86}$$

where:

$$H_{qj} = \tilde{H}_{qj} + \bar{H}_{qj}, \tag{87}$$

and where, for simplicity, we have denoted by $U(\mathcal{T}) := \frac{1}{2}\left[f(R(\mathcal{T})) - f'(R(\mathcal{T}))R(\mathcal{T}))\right]$.

From Equation (86), it follows that the request that the Einstein-like equations (10a) have a smooth transition across the hypersurface Σ is then equivalent to require that the tensor H_{qj} vanishes at Σ. Therefore, the remaining junction conditions can be obtained by imposing the vanishing of all projections of the tensor H_{qj} on Σ. About this, we have:

- the completely orthogonal projection of H_{qj} on Σ is automatically zero:

$$H_{qj}n^q n^j = \bar{H}_{qj}n^q n^j = -\epsilon\left[K_j{}^{qp}\right]n_p n_q n^j = 0 \tag{88}$$

  because $\tilde{H}_{qj}$ is tangent to Σ and the contorsion is antisymmetric in the last two indexes;
- the tangent–orthogonal projection of H_{qj} is:

$$H_{qj}E^q_A n^j = \bar{H}_{qj}E^q_A n^j = -\epsilon\left[K_{jq}{}^{p}\right]n_p E^q_A n^j + \left[K_{pq}{}^{p}\right]E^q_A. \tag{89}$$

  According to [35], the quantity in Equation (89) results in the jump of trace of the projection on Σ of the contorsion tensor. In fact, it is easily seen that the identity:

$$\left[K_{jq}{}^{p}h^j_i h^i_p E^q_A\right] = \left[K_{jq}{}^{p}\left(\delta^j_p - \epsilon n_p n^j\right)E^q_A\right] = -\epsilon\left[K_{jq}{}^{p}\right]n_p n^j E^q_A + \left[K_{pq}{}^{p}\right]E^q_A, \tag{90}$$

  holds.
- the orthogonal–tangent projection of H_{qj} is zero:

$$H_{qj}E^j_A n^q = \bar{H}_{qj}E^j_A n^q = \epsilon\left(-\left[K_{jq}{}^{p}\right]n_p n^q E^j_A + \left[K_{pq}{}^{p}\right]n_j E^j_A n^q\right) = 0, \tag{91}$$

  in view of the antisymmetry properties of the contorsion tensor and the orthogonality between the vectors n^i and E^i_A;
- the totally tangent projection of H_{qj} is given by:

$$H_{qj}E^q_A E^j_B = \tilde{H}_{qj}E^q_A E^j_B + \bar{H}_{qj}E^q_A E^j_B = \tilde{H}_{AB} + \epsilon\left(-\left[K_{jq}{}^{p}\right]n_p E^q_A E^j_B + \left[K_q{}^{qp}\right]n_p h_{AB}\right). \tag{92}$$

Summarizing everything, it is seen that the vanishing of the tensor H_{qj} needs the quantities (89), (92) to be zero at Σ. In particular, as it happens in GR, it can be shown that the condition $H_{qj}E^q_A E^j_B = 0$ is connected with the vanishing of the jump of the extrinsic curvature across Σ. To clarify this point, let us introduce the quantity:

$$Q_{AB} := \left(\nabla_i n_j\right)E^j_A E^i_B, \tag{93}$$

which generalizes the notion of extrinsic curvature for an arbitrary linear connection (3). From Equation (93) together with Equations (3) and (77), we get the relation:

$$[Q_{AB}] = \left[\nabla_i n_j E^j_A E^i_B\right] = \left[\nabla_i n_j\right]E^j_A E^i_B = \frac{\epsilon}{2}k_{ij}E^i_A E^j_B + \left[K_{ji}{}^{h}\right]n_h E^i_A E^j_B. \tag{94}$$

Comparing Equations (82), (92) and (94), it is then an easy matter to prove the identity:

$$H_{AB} := H_{kj} E^k_A E^j_B = -\epsilon \left([Q_{AB}] - [Q] \, h_{AB} \right), \tag{95}$$

where $[Q] := [Q_{AB}] \, h^{AB}$. It follows that the requirements $H_{AB} = 0$ and $[Q_{AB}] = 0$ at Σ are equivalent.

The request of vanishing of the quantities (89), (92) involves the Levi–Civita connection and the spin tensor (via the contorsion tensor) but also the trace of the energy–impulse tensor and its first derivatives. This is because of the torsional contributions given by the non-linearity of the function $f(R)$ and it represents a significant difference from the ECSK theory. In order to better clarify this last aspect, in the next subsections, we illustrate two examples dealing with the spin fluid and the Dirac field.

Before doing this, for the sake of completeness, we briefly outline also the case of null hypersurface. Then, let Σ be a null hypersurface described by an equation $\Phi(x^i) = 0$, where Φ is a smooth function. We suppose that $\mathcal{M}^+$ and $\mathcal{M}^-$ correspond to the domains where Φ is positive and negative, respectively. Again, we discuss the matching on Σ of two solutions of the field equations in the form:

$$g_{ij} = \Theta(\Phi) g^+_{ij} + (1 - \Theta(\Phi)) \, g^-_{ij}, \tag{96a}$$

$$\Gamma_{ij}{}^h = \Theta(\Phi) \Gamma^+_{ij}{}^h + (1 - \Theta(\Phi)) \, \Gamma^-_{ij}{}^h. \tag{96b}$$

The null normal vector is defined as $n_i = \alpha^{-1} \partial_i \Phi$, where α is a suitable non–zero function on Σ. By means of analogous arguments to those given above, it is immediately seen that the metric tensor (96a) has to be continuous across the hypersurface Σ, namely $[g_{ij}] = 0$. Following a usual procedure, let us then introduce a transverse vector field N^i satisfying the requirements $N^i n_i = 1$ and $N^i N_i = 0$. We have the relations $[n^i] = [N^i] = 0$. We also introduce the transverse metric:

$$h_{ij} = g_{ij} - n_i N_j - n_j N_i. \tag{97}$$

Due to the continuity of the metric tensor across Σ, its derivatives may have discontinuities only along the transverse direction. This implies the existence of a tensor field γ_{ij} on Σ, such that:

$$\gamma_{ij} = [\partial_s g_{ij}] N^s \qquad \Longleftrightarrow \qquad [\partial_s g_{ij}] = \gamma_{ij} n_s. \tag{98}$$

By Equation (98), we can express the jump of the Christoffel symbols as:

$$\left[\tilde{\Gamma}_{ij}{}^h \right] = \frac{1}{2} \left(\gamma^h{}_j n_i + \gamma^h{}_i n_j - \gamma_{ij} n^h \right). \tag{99}$$

Making use of Equation (99) and following the identical procedure illustrated above, it is easily seen that the δ-function part of the Einsein tensor is now given by the sum:

$$H^{ij} = \tilde{H}^{ij} + \bar{H}^{ij}, \tag{100}$$

where:

$$\tilde{H}^{ij} := \frac{\alpha}{2} \left(\gamma^i_h n^h n^j + \gamma^j_h n^h n^i - \gamma^h_h n^i n^j - \gamma_{hk} n^h n^k g^{ij} \right) \tag{101}$$

represents the contribution due to Levi–Civita terms, and:

$$\bar{H}^{ij} := \alpha \left(- \left[K^{jih} \right] n_h + \left[K_h{}^{ih} \right] n^j + \left[K_h{}^{hk} \right] n_k g^{ij} \right) \tag{102}$$

represents the contribution given by contorsion terms. As in the case of spacelike or timelike hypersurfaces, smooth transition across the hull hypersurface Σ at the level of Einstein-like equations requires the vanishing of the tensor (100).

4.1. The Coupling to a Spin Fluid

Let us consider a Weyssenhoff spin fluid with stress–energy and the spin tensors, respectively, given by [15,53,54]:

$$\mathcal{T}^{ij} = U^i P^j + p\left(U^i U^j - g^{ij}\right), \tag{103a}$$

and:

$$\mathcal{S}_{ij}{}^{h} = \mathcal{S}_{ij} U^h, \tag{103b}$$

where U^i is the 4-velocity, P^j denotes the 4-density of energy–momentum, $\mathcal{S}_{ij} = -\mathcal{S}_{ji}$ is the spin density, and p is the pressure of the fluid. By means of the conservation laws for the spin (12b), which are equivalent to the antisymmetric part of Einstein-like equations (10a), we can express the stress–energy tensor (103a) as [54]:

$$\mathcal{T}_{ij} = (\rho + p)\, U_i U_j - p g_{ij} - U_i \hat{T}_h \mathcal{S}^h{}_j - U_i \tilde{\nabla}_h \left(\mathcal{S}_{kj} U^h\right) U^k, \tag{104}$$

where $\rho := U^i P_i$ and $\tilde{\nabla}_h$ is the covariant derivative with respect to the Levi–Civita connection induced by the metric g_{ij}. In view of the usual convective condition $\mathcal{S}_{ij} U^j = 0$ [53,55] and the representation (11), it is easily seen that the vanishing at Σ of the quantities (89), (92) yields the explicit equations:

$$-\epsilon\left[K_{jq}{}^{p}\right] n_p E^q_A n^j + \left[K_{pq}{}^{p}\right] E^q_A = -\epsilon\left[\frac{1}{\varphi}\mathcal{S}_{qj} U_p\right] n^p n^j E^q_A - \left[\frac{1}{\varphi}\frac{\partial \varphi}{\partial x^q}\right] E^q_A = 0, \tag{105a}$$

$$\begin{aligned} &\tilde{H}_{AB} + \epsilon\left(-\left[K_{jq}{}^{p}\right] n_p E^q_A E^j_B + \left[K_q{}^{qp}\right] n_p h_{AB}\right) = \\ &\tilde{H}_{AB} + \epsilon\left(\left[\frac{1}{2\varphi}\left(\mathcal{S}_{jq} U^p + \mathcal{S}^p{}_q U_j + \mathcal{S}^p{}_j U_q\right)\right] n_p E^q_A E^j_B + \left[\frac{1}{\varphi}\frac{\partial \varphi}{\partial x^p}\right] n^p h_{AB}\right) = 0. \end{aligned} \tag{105b}$$

Equation (105b) can be decomposed into its symmetric and antisymmetric parts, thus giving rise to the further conditions:

$$\left[\frac{1}{2\varphi}\mathcal{S}_{jq} U^p\right] n_p E^q_A E^j_B = 0, \tag{106a}$$

$$\tilde{H}_{AB} + \epsilon\left(\left[\frac{1}{2\varphi}\left(\mathcal{S}^p{}_q U_j + \mathcal{S}^p{}_j U_q\right)\right] n_p E^q_A E^j_B + \left[\frac{1}{\varphi}\frac{\partial \varphi}{\partial x^p}\right] n^p h_{AB}\right) = 0. \tag{106b}$$

In order to illustrate a specific case, we imagine having to join together two static and spherically symmetric metrics:

$$ds^2_{\pm} = e^{\nu^{\pm}}\, dt^2 - e^{\lambda^{\pm}}\, dr^2 - r^2\left(d\theta^2 + \sin^2\theta\, d\phi^2\right), \tag{107}$$

solutions of Equation (10) coupled to a spin fluid. It is convenient to rename the spherical coordinates as $x^0 := t, x^1 := r, x^2 := \theta, x^3 := \phi$ in such a way that the 4-velocity of the fluid (supposed to be at rest in the chosen frame) is described by $U^i = U^0 \delta^i_0$, with $U^0 = e^{-\frac{\nu}{2}}$, and the unit normal to the hypersurface $\Sigma : x^1 = \text{const.}$ is given by $n^i = n^1 \delta^i_1$ with $n^1 = e^{-\frac{\lambda}{2}}$. The functions ν and λ, as well as all the involved matter fields, depend only on the radial variable r.

According to the convective condition $\mathcal{S}_{ij} U^j = 0$ and the stated spherical symmetry, we suppose that the spins of the particles composing the fluid are all aligned in the r direction; this means that only the components $\mathcal{S}_{23} = -\mathcal{S}_{32}$ of the spin density are non-zero [55]. Under these conditions, the stress–energy tensor of the spin fluid assumes the usual form:

$$\mathcal{T}_{ij} = (\rho + p)\, U_i U_j - p g_{ij}. \tag{108}$$

Using the above assumptions, it is easily seen that the constraints (105a), (106a) are automatically satisfied, while Equation (106b) reduces to:

$$\tilde{H}_{AB} + \epsilon \left[\frac{1}{\varphi} \frac{\partial \varphi}{\partial x^p} \right] n^p h_{AB} = 0. \tag{109}$$

Recalling the identity $\tilde{H}_{AB} = -\epsilon \left([\tilde{Q}_{AB}] - [\tilde{Q}]\, h_{AB} \right)$ [34], it is seen that Equation (109) relates the quantity $\left[\frac{1}{\varphi} \frac{\partial \varphi}{\partial x^p} \right]$ to the jump across Σ of the extrinsic curvature $\tilde{Q}_{AB}$ associated with the Levi–Civita connection of the metric (107). We note that, in the case of ECSK theory, condition (109) becomes $\tilde{H}_{AB} = 0$, which is the same condition holding in General Relativity [34].

Because of Equations (8) and (9), in general, the condition (109) involves the derivatives of matter fields. To see this point more in detail, we again take the model $f(R) = R + \alpha R^2$ into account. Due to Equation (108), from the trace equation (7) and the definition (9), we have the relations:

$$-R = \mathcal{T} = \rho - 3p, \tag{110}$$

and:

$$\varphi = 1 + 2\alpha\,(3p - \rho)\,. \tag{111}$$

Moreover, it is easy to verify that:

$$\tilde{Q}_{00} = \frac{1}{2} \frac{\partial \nu}{\partial r} e^{\nu - \frac{\lambda}{2}}{}_{|r=r_0} \tag{112}$$

is the only non–vanishing component of the extrinsic curvature $\tilde{Q}_{AB}$ induced by the metric (107) on the hypersurface $\Sigma: \quad r = r_0 \quad$ const.. In view of this, requirement (109) is seen to reduce to the following two conditions:

$$\left[\frac{2\alpha \left(3\frac{\partial p}{\partial r} - \frac{\partial \rho}{\partial r} \right)}{1 + 2\alpha\,(3p - \rho)} \right] = 0, \tag{113a}$$

and:

$$\left[\frac{\partial \nu}{\partial r} \right] = 0. \tag{113b}$$

As an even more specific example, we suppose to have to joining together the interior spacetime $\mathcal{M}^-$ of a star with spin properties, with the exterior region $\mathcal{M}^+$ assumed empty. In such a circumstance, we have $\mathcal{T}^+_{ij} = 0$ and $\mathcal{S}^{+\;h}_{ij} = 0$, and in $\mathcal{M}^+$ the field equations (10) are identical to the Einstein equations (without cosmological constant) in vacuo; their unique solution $\left(g^+_{ij}, \Gamma^{+\;h}_{ij} \right)$ is then given by the Schwartzchild metric:

$$g^+_{ij}\, dx^i dx^j = \left(1 - \frac{2M}{r} \right) dt^2 - \left(1 - \frac{2M}{r} \right)^{-1} dr^2 - r^2 \left(d\theta^2 + \sin^2\theta\, d\phi^2 \right), \tag{114}$$

together with its Levi–Civita connection $\Gamma^{+\;h}_{ij} = \tilde{\Gamma}^{+\;h}_{ij}$. Consequently, the junction conditions (69), (113) assume the explicit form:

$$e^{\nu^-(r_0)} = \left(1 - \frac{2M}{r_0} \right), \qquad e^{\lambda^-(r_0)} = \left(1 - \frac{2M}{r_0} \right)^{-1}, \tag{115a}$$

$$\left(\frac{\partial \nu^-}{\partial r} \right)_{|r=r_0} = \frac{2M}{r_0\,(r_0 - 2M)}, \qquad \left(3\frac{\partial p^-}{\partial r} - \frac{\partial \rho^-}{\partial r} \right)_{|r=r_0} = 0. \tag{115b}$$

On Equation (115), some comments are in order. Due to the second equation (115b), at Σ the spin fluid must behave like a sort of radiation, having a barotropic factor of the form $w = \left(\frac{\partial p^-}{\partial \rho^-} \right)_{|r=r_0} = 1/3$

at the boundary $r = r_0$. This fact is quite general: for all static and spherically symmetric solutions (107) of $f(R)$-gravity with torsion, the condition $\left(-3\frac{\partial p^-}{\partial r} + \frac{\partial \rho^-}{\partial r}\right)_{|r=r_0} = \frac{\partial T^-}{\partial r}_{|r=r_0} = 0$ is always sufficient (together with (113b)) to fulfill the requirement (109), and it becomes necessary also whenever $\frac{\partial \varphi^-}{\partial T}_{|r=r_0} \neq 0$ (like in the case $f(R) = R + \alpha R^2$, where $\frac{\partial \varphi^-}{\partial T}_{|r=r_0} = -2\alpha$). On the other hand, whenever the condition $\frac{\partial \varphi^-}{\partial T}_{|r=r_0} = 0$ is imposed, it yields a relation between density and pressure at the separation hypersurface, which constraints the equation of state [56].

4.2. The Coupling to a Dirac Field

Let ψ be a Dirac field with Lagrangian function given by:

$$\mathcal{L}_m = \left[\frac{i}{2}\left(\bar{\psi}\gamma^i D_i\psi - D_i\bar{\psi}\gamma^i\psi\right) - m\bar{\psi}\psi\right], \tag{116}$$

where $D_i\psi = \frac{\partial \psi}{\partial x^i} + \omega_i{}^{\mu\nu}\sigma_{\mu\nu}\psi$ and $D_i\bar{\psi} = \frac{\partial \bar{\psi}}{\partial x^i} - \bar{\psi}\omega_i{}^{\mu\nu}\sigma_{\mu\nu}$ are the covariant derivatives of the Dirac fields, $\sigma_{\mu\nu} = \frac{1}{8}\left[\gamma_\mu, \gamma_\nu\right]$, $\gamma^i = \gamma^\mu e^i_\mu$ with γ^μ denoting Dirac matrices and where m is the mass of the Dirac field. In what follows, the notation for which:

$$\gamma^\mu\gamma^\nu\gamma^\lambda = \gamma^\mu\eta^{\nu\lambda} - \gamma^\nu\eta^{\mu\lambda} + \gamma^\lambda\eta^{\mu\nu} + i\epsilon^{\mu\nu\lambda\tau}\gamma_5\gamma_\tau, \tag{117}$$

is used. From (116), we derive the Dirac equations:

$$i\gamma^h D_h\psi + \frac{i}{2}T_h\gamma^h\psi - m\psi = 0, \tag{118}$$

where, due to to the fact that torsion is no longer totally antisymmetric, the torsion vector $T_h := T_{hj}{}^j$ is present. The stress–energy and the spin density tensors are given by [15,42]:

$$\mathcal{T}_{ij} = \frac{i}{4}\left(\bar{\psi}\gamma_i D_j\psi - D_j\bar{\psi}\gamma_i\psi\right), \tag{119}$$

and:

$$\mathcal{S}_{ij}{}^h = -\frac{1}{4}\eta^{\mu\sigma}\epsilon_{\sigma\nu\lambda\tau}\left(\bar{\psi}\gamma_5\gamma^\tau\psi\right)e^h_\mu e^\nu_i e^\lambda_j. \tag{120}$$

In what follows, we can systematically assume that $\bar{\psi}\psi \neq 0$. Indeed, if $\bar{\psi}\psi = 0$, the trace of the stress–energy tensor would be constantly zero and the theory would amount to an ECSK-like theory for which the solution of the junction conditions problem is already known [35]. Therefore, without loss of generality, we can limit ourselves to dealing with spinor fields of type-1 and type-2 according to the Lounesto classification [57–59].

Making use of representation (11), it is seen that in this case the vanishing at Σ of the quantities (89), (92) yields the conditions:

$$-\epsilon\left[K_{jq}{}^p\right]n_p E^q_A n^j + \left[K_{pq}{}^p\right]E^q_A = -\epsilon\left[\hat{K}_{jq}{}^p\right]n_p E^q_A n^j + \left[\hat{K}_{pq}{}^p\right]E^q_A = -\left[\frac{1}{\varphi}\frac{\partial \varphi}{\partial x^q}\right]E^q_A = 0, \tag{121a}$$

$$\begin{aligned} \tilde{H}_{AB} + \epsilon\left(-\left[K_{jq}{}^p\right]n_p E^q_A E^j_B + \left[K_q{}^{qp}\right]n_p h_{AB}\right) &= \\ \tilde{H}_{AB} + \epsilon\left(-\left[\hat{K}_{jq}{}^p\right]n_p E^q_A E^j_B + \left[\hat{K}_q{}^{qp}\right]n_p h_{AB} - \left[\hat{S}_{jq}{}^p\right]n_p E^q_A E^j_B\right) &= \\ \tilde{H}_{AB} + \epsilon\left(\left[\frac{1}{\varphi}S_{jq}{}^p\right]n_p E^q_A E^j_B + \left[\frac{1}{\varphi}\frac{\partial \varphi}{\partial x^p}\right]n^p h_{AB}\right) &= 0. \end{aligned} \tag{121b}$$

Splitting Equation (121b) in its symmetric and antisymmetric parts, we obtain the equations:

$$\tilde{H}_{AB} + \epsilon \left[\frac{1}{\varphi} \frac{\partial \varphi}{\partial x^p} \right] n^p h_{AB} = 0, \tag{122a}$$

$$\left[\frac{1}{\varphi} \mathcal{S}_{jq}{}^{p} \right] n_p E^q_A E^j_B = 0. \tag{122b}$$

As an illustrative example, we suppose joining two axially symmetric spacetimes, solutions of the field equations resulting again from the model $f(R) = R + \alpha R^2$. More in detail, we assume that the metric tensors in both the regions $\mathcal{M}^-$ and $\mathcal{M}^+$ are of Lewis–Papapetrou kind, expressed in spherical coordinates as:

$$g^{\pm}_{ij}\, dx^i dx^j = -B^2_{\pm}(r^2\, d\theta^2 + dr^2) - A^2_{\pm}(-W_{\pm}\, dt + d\phi)^2 + C^2_{\pm}\, dt^2, \tag{123}$$

where all functions $A_{\pm}(r,\theta)$, $B_{\pm}(r,\theta)$, $C_{\pm}(r,\theta)$, and $W_{\pm}(r,\theta)$ depend on the r and θ variables only. We assume that $\mathcal{M}^+$ is empty, while $\mathcal{M}^-$ is filled with a Dirac field. We also suppose that in $\mathcal{M}^+$ the metric is the Kerr one. This is consistent with the fact that $R + \alpha R^2$ gravity with torsion in vacuo is equivalent to GR and, therefore, admits the same solutions. In the Lewis–Papapetrou form (123), the coefficients of the Kerr metric are expressed as:

$$A^2_+(r,\theta) = \left[a^2 + \frac{\left(-a^2+m^2+2mr+r^2\right)^2}{4r^2} \right] \sin^2\theta + \frac{ma^2\left(-a^2+m^2+2mr+r^2\right)\sin^4\theta}{r\left(\frac{\left(-a^2+m^2+2mr+r^2\right)^2}{4r^2} + a^2\cos^2\theta\right)}, \tag{124a}$$

$$B^2_+(r,\theta) = \frac{a^2\cos^2\theta}{r^2} + \frac{1}{4} + \frac{m}{r} + \frac{3m^2-a^2}{2r^2} + \frac{m^3-a^2m}{r^3} + \frac{a^4-2a^2m^2+m^4}{4r^4}, \tag{124b}$$

$$C^2_+(r,\theta) = \frac{m^2a^2\left(-a^2+m^2+2mr+r^2\right)^2\sin^4\theta}{\left(\left(a^2+\frac{\left(-a^2+m^2+2mr+r^2\right)^2}{4r^2}\right)\sin^2\theta + \frac{ma^2\left(-a^2+m^2+2mr+r^2\right)\sin^4\theta}{r\left(\frac{\left(-a^2+m^2+2mr+r^2\right)^2}{4r^2}+a^2\cos^2\theta\right)}\right)} \times$$
$$\frac{1}{r^2\left(\frac{\left(-a^2+m^2+2mr+r^2\right)^2}{4r^2} + a^2\cos^2\theta\right)^2} + 1 - \frac{m\left(-a^2+m^2+2mr+r^2\right)}{r\left(\frac{\left(-a^2+m^2+2mr+r^2\right)^2}{4r^2} + a^2\cos^2\theta\right)}, \tag{124c}$$

$$W_+(r,\theta) = \frac{ma\left(-a^2+m^2+2mr+r^2\right)\sin^2\theta}{\left(a^2+\frac{\left(-a^2+m^2+2mr+r^2\right)^2}{4r^2}\right)\sin^2\theta + \frac{ma^2\left(-a^2+m^2+2mr+r^2\right)\sin^4\theta}{r\left(\frac{\left(-a^2+m^2+2mr+r^2\right)^2}{4r^2}+a^2\cos^2\theta\right)}} \times$$
$$\frac{1}{r\left(\frac{\left(-a^2+m^2+2mr+r^2\right)^2}{4r^2} + a^2\cos^2\theta\right)}, \tag{124d}$$

where a and m are the parameters entering the Kerr metric. We want to analyze the junction conditions at the hypersurface $\Sigma: \quad r = r_0$ const. To this end, by using Equations (118) and (119), we preliminarily notice that in the regions $\mathcal{M}^-$ and $\mathcal{M}^+$ we have, respectively:

$$\varphi^- = 1 + 2\alpha R = 1 - 2\alpha\mathcal{T} = 1 - \alpha m\bar{\psi}\psi, \tag{125a}$$

and:

$$\varphi^+ = 1. \tag{125b}$$

In view of Equation (125), the constraint (121a) implies that the scalar $\bar{\psi}\psi$ is forced to be constant on the hypersurface Σ. Moreover, it is easily seen that the requirement (122b) is equivalent to the conditions:

$$\bar{\psi}\gamma_5\gamma^0\psi_{|\Sigma} = 0, \qquad \bar{\psi}\gamma_5\gamma^2\psi_{|\Sigma} = 0, \qquad \bar{\psi}\gamma_5\gamma^3\psi_{|\Sigma} = 0, \tag{126}$$

which have to be satisfied at Σ by the spinor field ψ. The remaining condition (122a) can be discussed by rewriting it in the equivalent form:

$$\left[\tilde{Q}_{AB}\right] = -\frac{1}{2}\left[\frac{1}{\varphi}\frac{\partial\varphi}{\partial x^h}\right] n^h h_{AB}, \tag{127}$$

where $\left[\tilde{Q}_{AB}\right]$ indicates the jump across Σ of the extrinsic curvatures induced by the metrics (123). Denoting by $\tilde{A} := A^+(r_0,\theta) = A^-(r_0,\theta)$, $\tilde{B} := B^+(r_0,\theta) = B^-(r_0,\theta)$, $\tilde{C} := C^+(r_0,\theta) = C^-(r_0,\theta)$, and $\tilde{W} := W^+(r_0,\theta) = W^-(r_0,\theta)$ for simplicity, we have that the non-zero components of $\left[\tilde{Q}_{AB}\right]$ are:

$$\left[\tilde{Q}_{\theta\theta}\right] = -r_0^2\left[\partial_r B\right], \tag{128a}$$

$$\left[\tilde{Q}_{\phi\phi}\right] = -\frac{\tilde{A}}{\tilde{B}}\left[\partial_r A\right], \tag{128b}$$

$$\left[\tilde{Q}_{t\phi}\right] = \frac{\tilde{A}\left(2\tilde{W}\left[\partial_r A\right] + \tilde{A}\left[\partial_r W\right]\right)}{2\tilde{B}}, \tag{128c}$$

$$\left[\tilde{Q}_{tt}\right] = \frac{\tilde{C}\left[\partial_r C\right] - \tilde{A}\tilde{W}^2\left[\partial_r A\right] - \tilde{A}^2\tilde{W}\left[\partial_r W\right]}{\tilde{B}}. \tag{128d}$$

Due to Equations (125) and (128), the non-trivial equations of (127) result to have explicit expression:

$$\frac{\left[\partial_r B\right]}{\tilde{B}} = -\frac{\alpha m}{2(1-\alpha m\bar{\psi}\psi_{|\Sigma})}\partial_r\left(\bar{\psi}\psi\right)_{|\Sigma}, \tag{129a}$$

$$\frac{\left[\partial_r A\right]}{\tilde{A}} = -\frac{\alpha m}{2(1-\alpha m\bar{\psi}\psi_{|\Sigma})}\partial_r\left(\bar{\psi}\psi\right)_{|\Sigma}, \tag{129b}$$

$$\frac{2\left[\partial_r A\right]}{\tilde{A}} + \frac{\left[\partial_r W\right]}{\tilde{W}} = -\frac{\alpha m}{(1-\alpha m\bar{\psi}\psi_{|\Sigma})}\partial_r\left(\bar{\psi}\psi\right)_{|\Sigma}, \tag{129c}$$

$$\frac{\tilde{C}\left[\partial_r C\right] - \tilde{A}\tilde{W}^2\left[\partial_r A\right] - \tilde{A}^2\tilde{W}\left[\partial_r W\right]}{\tilde{C}^2 - \tilde{A}^2\tilde{W}^2} = -\frac{\alpha m}{2(1-\alpha m\bar{\psi}\psi_{|\Sigma})}\partial_r\left(\bar{\psi}\psi\right)_{|\Sigma}. \tag{129d}$$

From Equation (129), it is seen that the jumps of the r-derivatives of quantities $A^\pm$, $B^\pm$, and $C^\pm$ have to satisfy the relations:

$$\frac{\left[\partial_r A\right]}{\tilde{A}} = \frac{\left[\partial_r B\right]}{\tilde{B}} = \frac{\left[\partial_r C\right]}{\tilde{C}} = -\frac{\alpha m}{2(1-\alpha m\bar{\psi}\psi_{|\Sigma})}\partial_r\left(\bar{\psi}\psi\right)_{|\Sigma} = \frac{1}{2\varphi}\frac{\partial\varphi}{\partial(\bar{\psi}\psi)}\partial_r\left(\bar{\psi}\psi\right)_{|\Sigma}, \tag{130}$$

while the function $W(r,\theta)$ has to be of class C^1.

In conclusion, it is shown that, in the non–linear case $f(R) \neq R + \lambda$, the scalar field $\bar{\psi}\psi$ is also involved in the characterization of the junction conditions. In particular, the derivatives of the metric components with respect to the coordinate r can have some jumps at the hypersurface Σ, connected with the r–derivative of the scalar quantity $\bar{\psi}\psi$. This is a difference from the linear case $f(R) = R + \lambda$ (ECSK theory), where, instead, the metric has to be at least of class C^1.

5. Conclusions

The well-posedness of the Cauchy problem, as well as the well-formulation of the junction conditions, are crucial aspects of any theory of gravity. In fact, a well-posed initial value problem ensures uniqueness, continuity, and causality of solutions from initial data; at the same time, well-formulated junction conditions allow us to understand if and how two different space-times can be soldered at a given hypersurface, with obvious applications and consequences, for example, on an astrophysical level.

In this paper, we have discussed the Cauchy problem and the junction conditions within the framework of $f(R)$-gravity with torsion.

For what concerns the Cauchy problem, we have seen that the problem is always well-posed in vacuo and, in the absence of spin, every time the trace of the matter stress–energy tensor is constant; indeed, in such a circumstance, the theory amounts to an Einstein-like theory for which the well-posedness of the initial value problem is well-established: for instance, this is what happens in the case of coupling to an electromagnetic field or a Yang–Mills field. On the contrary, when the stress–energy tensor trace is not constant, the problem needs to be discussed case-by-case.

Here, we have faced the coupling to a perfect fluid and a Klein–Gordon scalar field. In both cases, we have derived sufficient conditions ensuring the well-posedness of the initial value problem. We have also proved that there exist $f(R)$ models with torsion, which actually satisfy the stated conditions: the model $f(R) = R + \alpha R^2$ does it. The key idea to achieve these results has been implementing a conformal transformation from the Jordan to the Einstein frame, proving that the conservation laws are formally preserved under such a transformation; this has allowed us to apply well-known Bruhat's results, holding in GR.

On the junction conditions, we have deduced the general requirements needed to solder at a given hypersurface two different solutions of $f(R)$-gravity with torsion. Despite a formal resemblance, junction conditions for $f(R)$-gravity with torsion differ from those holding in the ECSK theory because they involve the trace of the matter stress–energy tensor and its first derivatives; this is due to the contributions that the non-linearity of the gravitational function $f(R)$ gives to the contorsion tensor and, in general, it results in specific conditions that the matter fields have to satisfy at the separation hypersurface. In order to better clarify this aspect, we have given two illustrative examples, considering the model $f(R) = R + \alpha R^2$ coupled to a spin fluid and a Dirac field, respectively.

Finally, we have shown that the study of the initial value problem, as well as the junction conditions in the context of $f(R)$-gravity with torsion, singles out suitable conditions on the gravitational Lagrangian function $f(R)$ itself, which may be used as selection criteria for viable $f(R)$ models.

Conflicts of Interest: The author declares no conflict of interest.

References

1. Nojiri, S.; Odintsov, S.D. Introduction to modified gravity and gravitational alternative for dark energy. *Int. J. Geom. Meth. Mod. Phys.* **2007**, *4*, 115. [CrossRef]
2. Capozziello, S.; Faraoni, V. *Beyond Einstein Gravity*; Fundamental Theories of Physics *170*; Springer: Dordrecht, The Netherlands, 2011.
3. Capozziello, S. Curvature quintessence. *Int. J. Mod. Phys.* **2002**, *D11*, 483. [CrossRef]
4. Sotiriou, T.P.; Faraoni, V. Modified gravity with R–matter couplings and (non-) geodesic motion. *Class. Quant. Grav.* **2008**, *25*, 205002. [CrossRef]
5. Capozziello, S.; Francaviglia, M. Extended theories of gravity and their cosmological and astrophysical applications. *Gen. Rel. Grav.* **2008**, *40*, 357. [CrossRef]
6. Sotiriou, T.P.; Faraoni, V. $f(R)$ theories of gravity. *Rev. Mod. Phys.* **2010**, *82*, 451. [CrossRef]
7. De Felice, A.; Tsujikawa, S. $f(R)$ theories. *Living Rev. Rel.* **2010**, *13*, 3. [CrossRef]
8. Nojiri, S.; Odintsov, S.D. Unified cosmic history in modified gravity: From $F(R)$ theory to Lorentz non-invariant models. *Phys. Rep.* **2011**, *505*, 59. [CrossRef]
9. Olmo, G.J. Palatini approach to modified gravity: $f(R)$ theories and beyond. *Int. J. Mod. Phys.* **2011**, *D20*, 413. [CrossRef]
10. Mahato, P. Torsion, Scalar Field and $f(R)$ Gravity. *Ann. Fond. Broglie* **2007**, *32*, 297.

11. Sciama, D.W. On a non-symmetric theory of the pure gravitational field. *Math. Proc. Cambridge Philos. Soc.* **1958**, *54*, 72. [CrossRef]
12. Sciama, D.W. The Physical Structure of General Relativity. *Rev. Mod. Phys.* **1964**, *36*, 463. [CrossRef]
13. Kibble, T.W.B. Lorentz invariance and the gravitational field. *J. Math. Phys.* **1961**, *2*, 212. [CrossRef]
14. Hehl, F.W.; von der Heyde, P.; Kerlick, G.D. General relativity with spin and torsion and its deviations from Einstein's theory. *Phys. Rev. D* **1974**, *10*, 1066. [CrossRef]
15. Hehl, F.W.; von der Heyde, P.; Kerlick, G.D.; Nester, J.M. General relativity with spin and torsion: Foundations and prospects. *Rev. Mod. Phys.* **1976**, *48*, 393. [CrossRef]
16. Mahato, P. Torsion, Dirac field, dark matter and dark radiation. *Int. J. Mod. Phys.* **2007**, *A22*, 835. [CrossRef]
17. Rubilar, G. On the universality of Einstein-Cartan field equations in the presence of matter fields. *Class. Quantum Grav.* **1998**, *15*, 239. [CrossRef]
18. Capozziello, S.; Cianci, R.; Stornaiolo, C.; Vignolo, S. $f(R)$ gravity with torsion: The metric-affine approach. *Class. Quantum Grav.* **2007**, *24*, 6417. [CrossRef]
19. Capozziello, S.; Cianci, R.; Stornaiolo, C.; Vignolo, S. $f(R)$ gravity with Torsion: A geometric approach within the-bundles framework. *Int. J. Geom. Meth. Mod. Phys.* **2008**, *5*, 765. [CrossRef]
20. Capozziello, S.; Cianci, R.; Stornaiolo, C.; Vignolo, S. $f(R)$ cosmology with torsion. *Phys. Scripta* **2008**, *78*, 065010. [CrossRef]
21. Capozziello, S.; Vignolo, S. Metric-affine $f(R)$-gravity with torsion: An overview. *Annalen. Phys.* **2010**, *19*, 238. [CrossRef]
22. De Sabbata, V.; Sivaram, C. Torsion and inflation. *Astr. Space Sci.* **1991**, *176*, 141. [CrossRef]
23. Fourés–Bruhat, Y. Théorèmes d'existence en mécanique desfluides relativiste. *Bull. Soc. Math. Fr.* **1958**, *86*, 155. [CrossRef]
24. Choquet–Bruhat, Y. *Cauchy Problem, in Gravitation: An Introduction to Current Research*; Witten, L., ed.; Wiley: New York, NY, USA, 1962.
25. Synge, J.L. *Relativity: The General Theory*; North–Holland Publishing Company: Amsterdam, The Netherlands, 1971.
26. Wald, R.M. *General Relativity*; Chicago University Press: Chicago, IL, USA, 1984.
27. O'Hanlon, J. Intermediate-range gravity: A generally covariant model. *Phys. Rev. Lett.* **1972**, *29*, 137. [CrossRef]
28. Capozziello, S.; Vignolo, S. The Cauchy problem for $f(R)$-gravity: An Overview. *Int. J. Geom. Methods Mod. Phys.* **2012**, *9*, 1250006. [CrossRef]
29. Lichnerowicz, A. Sur les ondes de choc gravitationnelles. *C.R. Acad. Sci.* **1971**, *273*, 528.
30. Lichnerowicz, A. Ondes de choc gravitationnelles et electromagnetiques. *Inst. Naz. Alta Math. Symp. Math.* **1973**, *12*, 93.
31. Taub, A.H. Space–times with distribution valued curvature tensors. *J. Math. Phys.* **1980**, *21*, 1423. [CrossRef]
32. Choquet–Bruhat, Y.;. DeWitt–Morette, C. *Analysis, Manifolds and Physics*; (revised version); North–Holland: Amsterdam, The Netherlands, 1982.
33. Israel, W. Singular hypersurfaces and thin shells in general relativity. *Nuovo Cimento* **1996**, *44*, 1. [CrossRef]
34. Poisson, E. *A Relativist's Toolkit. The Mathematics of Black-Hole Mechanics*; Cambridge University Press: Cambridge, UK, 2004.
35. Arkuszewski, W.; Kopczynski, W.; Ponomariev, V.N. Matching conditions in the Einstein-Cartan theory of gravitation. *Commun. Math. Phys* **1975**, *45*, 183. [CrossRef]
36. Trautman, A. On the Einstein–Cartan Equations I. *Bull. Pol. Acad. Sci.* **1972**, *20*, 185.
37. Trautman, A On the Einstein–Cartan Equations II. *Bull. Pol. Acad. Sci.* **1972**, *20*, 503.
38. Trautman, A. On the Structure of the Einstein—Cartan Equations. *Inst. Naz. Alta Math. Symp. Math.* **1973**, *12*, 139.
39. Bressange, G.F. On the extension of the concept of thin shells to the Einstein-Cartan theory. *Class. Quantum Grav.* **2000**, *17*, 2509. [CrossRef]
40. Deruelle, N.; Sasaki, M.; Sendouda, Y. Junction Conditions in $f(R)$ Theories of Gravity. *Prog. Theor. Phys.* **2008**, *119*, 237. [CrossRef]
41. J. M. M. Senovilla,Junction conditions for $F(R)$ gravity and their consequences. *Phys. Rev. D* **2013**, *88*, 064015. [CrossRef]
42. Fabbri, L.; Vignolo, S. Dirac fields in $f(R)$-gravity with torsion. *Class. Quantum Grav.* **2011**, *28*, 125002. [CrossRef]

43. Olmo, G.J. Post-Newtonian constraints on $f(R)$ cosmologies in metric and Palatini formalism. *Phys. Rev. D* **2005**, *72*, 083505. [CrossRef]
44. Capozziello, S.; Vignolo, S. The Cauchy problem for metric-affine $f(R)$-gravity in the presence of perfect-fluid matter. *Class. Quantum Grav.* **2009**, *26*, 175013. [CrossRef]
45. Capozziello, S.; Vignolo, S. On the well-formulation of the initial value problem of metric-affine $f(R)$-gravity. *Int. J. Geom. Methods Mod. Phys.* **2009**, *6*, 985. [CrossRef]
46. Choquet–Bruhat, Y. *General Relativity and the Einstein Equations*; Oxford University Press Inc.: New York, NY, USA, 2009.
47. Capozziello, S.; Vignolo, S. A comment on 'The Cauchy problem of $f(R)$-gravity'. *Class. Quantum Grav.* **2009**, *26*, 168001. [CrossRef]
48. Capozziello, S.; Vignolo, S. The cauchy problem for metric-affine $f(R)$-gravity in presence of a Klein–Gordon scalar field. *Int. J. Geom. Methods Mod. Phys.* **2011**, *8*, 167. [CrossRef]
49. Leray, J. *Hyperbolic Differential Equations*; Institute for Advanced Study Pub.: Princeton, NJ, USA, 1953.
50. Choquet–Bruhat, Y. Isenberg, J.; Pollack, D. Applications of theorems of Jean Leray to the Einstein-scalar field equations. *J. Fied Point Theor. Appl.* **2006**, *1*, 31–46.
51. Dray, T. Tensor distributions in the presence of degenerate metrics. *Int. J. Mod. Phys. D* **1997**, *6*, 717. [CrossRef]
52. Hartley, D.; Tucker, R.W.; Tuckey, P.A.; Dray, T. Tensor distributions on signature-changing space-times. *Gen. Rel. Grav.* **2000**, *32*, 491. [CrossRef]
53. Obukhov, Y.N.; Korotky, V.A. The Weyssenhoff fluid in Einstein-Cartan theory. *Class. Quantum Grav.* **1987**, *4*, 1633. [CrossRef]
54. Vignolo, S.; Fabbri, L. Spin fluids in Bianchi-I $f(R)$-cosmology with torsion. *Int. J. Geom. Methods Mod. Phys.* **2012**, *9*, 1250054. [CrossRef]
55. Prasanna, A.R. Static fluid spheres in Einstein-Cartan theory. *Phys. Rev. D* **1975**, *11*, 2076. [CrossRef]
56. Vignolo, S.; Cianci, R.; Carloni, S. On the junction conditions in $f(R)$-gravity with torsion. *Class. Quantum Grav.* **2018**, *35*, 095014. [CrossRef]
57. Lounesto, P. *Clifford Algebras and Spinors*, 2nd ed.; Cambridge University Press: Cambridge, UK, 2002.
58. da Silva, J.M.H.; da Rocha, R. Unfolding physics from the algebraic classification of spinor fields. *Phys. Lett. B* **2013**, *718*, 1519. [CrossRef]
59. da Rocha, R.; Fabbri, L.; da Silva, J.M.H.; Cavalcanti, R.T.; Silva-Neto, J.A. Flag-dipole spinor fields in ESK gravities. *J. Math. Phys.* **2013**, *54*, 102505. [CrossRef]

Review

On the Mathematics of Coframe Formalism and Einstein–Cartan Theory—A Brief Review

Manuel Tecchiolli

Institute for Theoretical Physics, ETH Zürich, Wolfgang-Pauli-Str. 27, 8093 Zürich, Switzerland; manuelte@student.ethz.ch

Received: 3 September 2019; Accepted: 24 September 2019; Published: 27 September 2019

Abstract: This article is a review of what could be considered the basic mathematics of Einstein–Cartan theory. We discuss the formalism of principal bundles, principal connections, curvature forms, gauge fields, torsion form, and Bianchi identities, and eventually, we will end up with Einstein–Cartan–Sciama–Kibble field equations and conservation laws in their implicit formulation.

Keywords: general relativity; torsion-gravity; mathematical physics

1. Introduction

The formulation of torsion gravity and the consequent coupling with spin rely on a different formulation compared to the one of original works on General Relativity. This formulation regards geometrical objects called principal bundles. In this context, we can formulate General Relativity (or Einstein–Cartan–Sciama–Kibble (ECSK) theory in the presence of torsion) with a principal connection, which can be pulled back to the base manifold in a canonical way and further restricted to the only antisymmetric component, giving birth to the well-known spin connection. This process shows the possibility of formulating General Relativity as a proper gauge theory rather than using the affine formulation and Christoffel symbols Γ. What permits the equivalence of the two formulations is a bundle isomorphism called *tetrads* or *vierbein*, which is supposed to respect certain compatibility conditions. Then, we can define the associated torsion form and postulate the Palatini–Cartan action as a functional of such tetrads and spin connection. This leads to ECSK field equations.

We will first set up all the abstract tools of principal bundles, tetrads, and principal connection; secondly, we will derive the Einstein–Cartan–Sciama–Kibble theory in its implicit version; and finally, we will discuss conservation laws coming from local $SO(3,1)$ and diffeomorphism invariance of ECSK theory.

Throughout the article, we will give theorems and definitions. However, we would like to stress that hypotheses for such theorems will often be slightly redundant: we will take spaces and functions to be differentiable manifolds and smooth, even though weaker statements would suffice. This is because we prefer displaying the setup for formalizing the theory rather than presenting theorems and definitions with weaker hypotheses that we will never use for the theory. Nontheless, we will specify where such hypotheses are strengthened. In spite of this, the discussion will be rather general, probably more general than what is usually required in formulating ECSK (Einstein–Cartan–Sciama–Kibble) theory.

2. Bundle Structure

The introduction of a metric g and an orthogonality relation via a minkowskian metric η are two fundamental ingredients for building up a fiber bundle where we want the orthogonal group to act freely and transitively on the fibers. This will allow us to have a principal connection and to see the perfect analogy with an ordinary gauge theory ([1] chapter III).

Such a construction underlies the concept of *principal bundle*, and tetrads will be an isomorphism from the tangent bundle[1] TM to an associated bundle $\mathcal{V}$.

2.1. G-Principal Bundle

We give some definitions[2].

Definition 1 (G-principal bundle[3]). *Let M be a differentiable manifold and G be a Lie group.*

A G-principal bundle P is a fiber bundle $\pi : P \to M$ together with a smooth (at least continuous) right action $\mathfrak{P} : G \times P \to P$ such that $\mathfrak{P}$ acts freely and transitively on the fibers[4] of P and such that $\pi(\mathfrak{P}_g(p)) = \pi(p)$ for all $g \in G$ and $p \in P$.

We need to introduce a fundamental feature of fiber bundles.

Definition 2 (Local trivialization of a fiber bundle). *Let E be a fiber bundle over M, a differentiable manifold, with fiber projection $\pi : E \to M$, and let F be a space[5].*

A local trivialization (U, φ_U) of E, is a neighborhood $U \subset M$ of $u \in M$ together with a local diffeomorphism.

$$\varphi_U : U \times F \to \pi^{-1}(U) \tag{1}$$

such that $\pi(\varphi_U(u, f)) = u \in U$ for all $u \in U$ and $f \in F$.

This definition implies $\pi^{-1}(u) \simeq F \; \forall u \in U$.

Definition 3 (Local trivialization of a G-principal bundle). *Let P be a G-principal bundle.*

A local trivialization (U, φ_U) of P is a neighborhood $U \subset M$ of $u \in M$ together with a local diffeomorphism.

$$\varphi_U : U \times G \to \pi^{-1}(U) \tag{2}$$

such that $\pi(\varphi_U(u, g)) = u \in U$ for all $u \in U$ and $g \in G$ and such that

$$\varphi_U^{-1}(\mathfrak{P}_g(p)) = \varphi_U^{-1}(p)g = (u, g')g = (u, g'g). \tag{3}$$

1 Disjoint union of tangent spaces: $TM = \cup_{x \in M}\{x\} \times T_xM$
2 References [2–4] are recommended for further details.
3 We give the definition based on our purposes; in general, we can release some hypotheses. In particular, G needs to be only a locally compact topological group and M needs to be a topological Hausdorff space. This definition is a version with a stronger hypothesis than the one contained in Reference [5].
4 Fibers are $\pi^{-1}(x) \; \forall x \in M$.
5 In the present case, F will be a differentiable manifold, a vector space, a topological space, or a topological group. Furthermore, if we write "space", we mean one among these.

Observation 1: A fiber bundle is said to be *locally trivial* in the sense that it admits a local trivialization for all $x \in M$, namely there exists an open cover $\{U_i\}$ of M and a set of diffeomorphisms φ_i such that every $\{U_i, \varphi_i\}$ is a local trivialization[6].

Here, we recall the similarity with a differentiable manifold. For a manifold when we change charts, we have an induced diffeomorphism between the neighborhoods of the two charts, given by the composition of the two maps.

Thus, having two charts (U_i, ϕ_i) and (U_j, ϕ_j), we define the following:

$$\phi_j \circ \phi_i^{-1} : \phi_i(U_i \cap U_j) \to \phi_j(U_i \cap U_j). \tag{4}$$

At a level up, we have an analogous thing when we change trivialization. Of course, here, we have one more element: the element of fiber.

Taking two local trivializations (U_i, φ_i) and (U_j, φ_j) and given a smooth left action $\mathcal{T} : G \to \mathrm{Diffeo}(F)$ of G on F, we then have

$$(\varphi_j^{-1} \circ \varphi_i)(x, f) = (x, \mathcal{T}(g_{ij}(x))(f)) \qquad \forall x \in U_i \cap U_j, f \in F. \tag{5}$$

where the maps $g_{ij} : U_i \cap U_j \to G$ are called the *transition functions* for this change of trivialization and G is called the structure group.

Such functions obey the following transition functions conditions for all $x \in U_i \cap U_j$:

- $g_{ii}(x) = id$
- $g_{ij}(x) = (g_{ji}(x))^{-1}$
- $g_{ij}(x) = g_{ik}(x) g_{kj}(x)$ for all $x \in U_i \cap U_k \cap U_j$.

The last condition is called the *cocycle condition.*

Theorem 1 (Fiber bundle construction theorem). *Let M be a differentiable manifold, F be a space, and G be a Lie group with faithful smooth left action $\mathcal{T} : G \to \mathrm{Diffeo}(F)$ of G on F.*

Given an open cover $\{U_i\}$ of M and a set of smooth maps,

$$t_{ij} : U_i \cap U_j \to G \tag{6}$$

defined on each nonempty overlap, satisfying the transition function conditions.

Then, there exists a fiber bundle $\pi : E \to M$ such that

- *$\pi^{-1}(x) \simeq F$ for all $x \in M$*
- *its structure group is G, and*
- *it is trivializable over $\{U_i\}$ with transition functions given by t_{ij}.*

A proof of the theorem can be found in Reference [6] (Chapter 1).

2.2. *Coframe Bundle and Minkowski Bundle*

It is clear now that having E as a fiber bundle over M with fibers isomorphic to F and F' as a space equipped with the smooth action $\mathcal{T}'$ of G, implies the possibility of building a bundle E' *associated* to

[6] The bundle is said to be *trivial* if U = M.

E, which shares the same structure group and the same transition functions g_{ij}. By the fiber bundle construction theorem, we have a new bundle E' over M with fibers isomorphic to F'.

This bundle is called the *associated bundle* to E.

Depending on the nature of the associated bundle[7], we have the following two definitions:

Definition 4 (Associated G-principal bundle). *Let $\pi : E \to M$ be a fiber bundle over a differentiable manifold M, G be a Lie group, F' be a topological space, and $\mathfrak{P}$ be a smooth right action of G on F'. Let also E' be the associated bundle to E with fibers isomorphic to F'.*

If F' is the principal homogeneous space[8] for $\mathfrak{P}$, namely $\mathfrak{P}$ acts freely and transitively on F', then E' is called the G-principal bundle associated to E.

Definition 5 (Associated bundle to a G-principal bundle). *Let P be a G-principal bundle over M, F' be a space, and ρ: $G\to$ Diffeo(F') be a smooth effective left action of the group G on F'.*

We then have an induced right action of the group G over $P \times F'$ given by

$$(p, f') * g = (\mathfrak{P}_g(p), \rho(g^{-1})(f')). \tag{7}$$

We define the associated bundle E to the principal bundle P, as an equivalence relation:

$$E := P \times_\rho F' = \frac{P \times F'}{\sim}, \tag{8}$$

where $(p, f') \sim (\mathfrak{P}_g(p), \rho(g^{-1})(f'))$, $p \in P$, and $f' \in F'$ with projection $\pi_\rho : E \to M$ s.t. $\pi_\rho([p, f']) = \pi(p) = x \in M$.

Therefore $\pi_\rho : E \to M$ is a fiber bundle over M with $\pi_\rho^{-1}(x) \simeq F'$ for all $x \in M$.

Observation 2: The new bundle, given by the latter definition, is what we expected from a general associated bundle: a bundle with the same base space, different fibers, and the same structure group.

Idea: We take a G-principal bundle P as an associated bundle to TM, and we build a vector bundle associated to P with a fiber-wise metric η. We shall call this associated bundle $\mathcal{V}$.

First of all, we display the G-principal bundle as the G-principal bundle associated to TM.

Definition 6 (Orthonormal coframe). *Let (M, g) be a pseudo-riemannian n-dimensional differentiable manifold and (V, η) be an n-dimensional vector space with minkowskian metric η.*
A coframe at $x \in M$ is the linear isometry.

$$_xe := \{ {_xe} : T_xM \to V |\ {_xe}^*\eta := \eta_{ab}\, {_xe}^a {_xe}^b = g \}, \tag{9}$$

equivalently $_xe^a$ forms an ordered orthonormal basis in T_x^*M.

An orthonormal frame is defined as the dual of a coframe.

[7] We will be dealing with two particular types of associated bundles: a principal bundle associated to a vector bundle and a vector bundle associated to a principal bundle.

[8] The space where the orbits of G span all the space.

Observation 3: Locally, coframes can be identified with local covector fields. A necessary and sufficient condition for identifying them with global covector fields (namely a coframe for each point of the manifold) is to have a *parallelizable* manifold, namely a trivial tangent bundle.

Definition 7 (Orthonormal coframe bundle). *Let (M, g) be a differentiable n-dimensional manifold with pseudo-riemannian metric g and T^*M be its cotangent bundle (real vector bundle of rank n).*

We call the coframe bundle $F^*_O(M)$ the G-principal bundle where the fiber at $x \in M$ is the set of all orthonormal coframes at x and where the group $G = \mathrm{O}(n-1,1)$ acts freely and transitively on them.

The dual bundle of this is the orthonormal frame bundle, and it is denoted by $F_O(M)$, made up of orthonormal frames (dual of orthonormal coframes).

Observations 4:

i. The orthonormal frame bundle is an associated G-principal bundle to TM.
ii. We can consider the *Minkowski bundle* $\mathcal{V}$ the vector bundle over M with fibers V. It is clear that such a bundle and $F_O(M)$ are one of the associated bundles of the other via action of the orthogonal group $\mathrm{O}(n-1,1)$. Therefore, $\mathcal{V} := F_O(M) \times_\rho V$, where ρ is taken to be the fundamental representation of $O(n-1,1)$.
iii. We stress that this bundle $\mathcal{V}$ is not canonically isomorphic to TM; in general, there is no canonical choice of a representative of ${}_xe$ of the equivalence class $[{}_xe, v] \in \mathcal{V}$, of which the inverse ${}_xe^{-1}(v)$ gives rise to a canonical identification of a vector in T_xM. Namely, fixed a $v \in V$, not all choices of ${}_xe$ give rise to a fixed vector $X \in T_xM$. As a matter of fact, the reference metric fixed on V does not allow in general the existence of a canonical soldering (Section 7). In Reference [7], it is shown how to define the Minkowski bundle without deriving it from $F_O(M)$; the authors refer to that as *fake tangent bundles*.
iv. If the manifold is parallelizable, we have the bundle isomorphism $e : TM \to \mathcal{V}$, which is given by the identity map over M and ${}_xe : T_xM \to V \ \forall x \in M$. It can be regarded as a $\mathcal{V}$-valued 1-form $e \in \Omega^1(M, \mathcal{V})$. We can identify e with an element of $\Omega^1(M, V)$, thus with global sections of the cotangent bundle such that, at each point in M, the corresponding covectors ${}_xe^a$ obey $\eta_{ab}\, {}_xe^a\, {}_xe^b = g$.

We are now ready to define tetrads.

Definition 8 (Tetrads). *Let $\rho : O(3,1) \to \mathrm{Aut}(V)$ be the fundamental representation.*

Tetrads are the bundle isomorphisms $e : TM \to \mathcal{V}$. They are identifiable with elements $e \in \Omega^1(M, \mathcal{V})$, and if M is parallelizable, tetrads can be identified with $\Omega^1(M, V) \ni e^a v_a$ such that $\{v_a\}$ is an orthonormal basis of V, $e^a \in \Omega^1(M)$, and $\eta_{ab} e^a e^b = g$.

3. Principal Connection

Is there any difference?

In the ordinary formulation of General Relativity (as in the original Einstein's work, for instance), we have objects called Γs, which are coefficients of a linear connection ∇ and thus determined by a parallel transport of tangent vectors.

The biggest advantage of treating O(3, 1) as an "explicit symmetry" of the theory is that we have obtained the possibility of defining a *principal connection*, which is the same kind of entity we have in an ordinary gauge theory[9].

3.1. Ehresmann Connection

If we consider a smooth fiber bundle $\pi : E \to M$, where fibers are differentiable manifolds, we can of course take tangent spaces at points $e \in E$. Having the tangent bundle TE, we may wonder if it is possible to separate the contributions coming from M to the ones from the fibers.

This cannot be done just by stating $TE = TM \oplus TF$, unless $E = M \times F$ is the trivial bundle. Namely, we cannot split directly vector fields on M from vector fields on the fibers F.

We can formalize this idea: use our projection π for constructing a tangent map $\pi_* = d\pi : TE \to TM$, and consider its kernel.

Definition 9 (Vertical bundle). *Let M be a differentiable manifold and $\pi : E \to M$ be a smooth fiber bundle.*

We call the sub-bundle $VE = \mathrm{Ker}(\pi_* : TE \to TM)$ the vertical bundle.

Following this definition, we have the natural extension to the complementary bundle of the vertical bundle, which is somehow the formalization of the idea we had of a bundle that takes care of tangent vector fields on M.

Definition 10 (Ehresmann connection). *Let M be a differentiable manifold and $\pi : E \to M$ be a smooth fiber bundle.*

Consider a complementary bundle HE such that $TE = HE \oplus VE$. We call this smooth sub-bundle HE the horizontal bundle or Ehresmann connection.

Thus, vector fields will be called *vertical* or *horizontal* depending on whether they belong to $\Gamma(VE)$ or $\Gamma(HE)$, respectively.

3.2. Ehresmann Connection and Horizontal Lift

We recall the case of the linear connection ∇; it was uniquely determined by a parallel transport procedure.

In the case of a principal connection, we have an analogous.

Definition 11 (Lift). *Let $\pi : E \to M$ be a fiber bundle, M be a differentiable manifold, $x \in M$ and $e \in E$ such that $\pi(e) = x$.*

Given a smooth curve $\gamma : \mathbb{R} \to M$ such that $\gamma(0) = x$, we define a lift of γ through e as the curve $\tilde{\gamma}$, satisfying

$$\tilde{\gamma}(0) = e \quad \text{and} \quad \pi(\tilde{\gamma}(t)) = \gamma(t) \quad \forall t. \tag{10}$$

If E is smooth, then a lift is horizontal if every tangent to $\tilde{\gamma}$ lies in a fiber of HE, namely

$$\dot{\tilde{\gamma}}(t) \in HE_{\tilde{\gamma}(t)} \ \ \forall t. \tag{11}$$

It can be shown that an Ehresmann connection uniquely determines a horizontal lift. Here, it is the analogy with parallel transport.

---

[9] Think of $(U(1), A_\mu)$ for electromagnetism.

3.3. Connection Form in a G-Principal Bundle

We now focus on the case where the smooth fiber bundle is a G-principal bundle with smooth action $\mathfrak{P}$.

Here, we need a group G, that we generally take to be a matrix Lie group. We then have the corresponding algebra $\mathfrak{g}$, a matrix vector space in the present case.

The action $\mathfrak{P}$ defines a map $\sigma : \mathfrak{g} \to \Gamma(VE)$ called the *fundamental map*[10], where at $p \in P$, for an element $\xi \in \mathfrak{g}$, it is given via the exponential map $Exp : \mathfrak{g} \to G$.

$$\sigma_p(\xi) = \frac{\mathrm{d}}{\mathrm{d}t}\mathfrak{P}_{e^{t\xi}}(p)\big|_{t=0}. \tag{12}$$

The map is vertical because

$$\pi_*\sigma_p(\xi) = \frac{\mathrm{d}}{\mathrm{d}t}\pi(\mathfrak{P}_{e^{t\xi}}(p))\big|_{t=0} = \frac{\mathrm{d}}{\mathrm{d}t}\pi(p) = 0. \tag{13}$$

Thus, the vector $\sigma_p(\xi)$ is vertical and it is called the *fundamental vector*.

Before proceeding, we need some Lie group theory.

Recall of Lie machinery: Let G be a Lie group (a differentiable manifold) with $\mathfrak{g}$ as its Lie algebra and $\forall g, h \in G$. We define:

- $L_g : G \to G$ and $R_g : G \to G$, such that $L_g h = gh$ and $R_g h = hg$ are the *left* and *right* actions, respectively;
- the adjoint map $\mathrm{Ad}_g : G \to G$ via such left and right actions is $\mathrm{Ad}_g := L_g \circ R_{g^{-1}}$, namely $\mathrm{Ad}_g h = ghg^{-1}$. It also acts on elements of the algebra $\xi \in \mathfrak{g}$ as $\mathrm{Ad}_g : \mathfrak{g} \to \mathfrak{g}$ via the exponential map[11]

$$\begin{aligned}\mathrm{Ad}_g\xi &= \frac{\mathrm{d}}{\mathrm{d}t}((L_g \circ R_{g^{-1}})(e^{t\xi}))\big|_{t=0} = \frac{\mathrm{d}}{\mathrm{d}t}\big|_{t=0} \\ &= g\xi g^{-1} \ \in \mathfrak{g},\end{aligned} \tag{14}$$

  where the last two equalities hold in the present case of matrix Lie groups. This is not to be confused with the adjoint action $\mathrm{ad} : \mathfrak{g} \times \mathfrak{g} \to \mathfrak{g}$, which is generated by the derivative of the adjoint map with $g = e^{t\chi}$ and $\chi \in \mathfrak{g}$, such that $\mathrm{ad}_\chi\xi = [\chi, \xi]$;
- the left invariant vector fields $v \in \Gamma(TG)$ as $L_{g*} \circ v = v$, namely $v(g) = L_{g*}v(e)$;
- the *Maurer–Cartan form* is the left invariant $\mathfrak{g}$-valued 1-form $\theta \in \Omega^1(G, \mathfrak{g})$ defined by its values at g.

$$\theta_g := L_{g^{-1}*} : T_gG \to T_eG \cong \mathfrak{g}. \tag{15}$$

  For any left invariant vector field v, it holds $\forall g \in G$ that $\theta_g(v(g)) = v(e)$. Therefore, left invariant vector fields are identified by their values over the identity thanks to the Maurer–Cartan form θ. So we can state ([8]) that this identification $v(e) \mapsto v$ defines an isomorphism between the space of left

---

10 It turns out that it is an isomorphism, since $\mathfrak{P}$ is regular.

11 We stress that the exponential map is not an isomorphism for all Lie groups; thus, the elements generated by the exponential map belong, in general, to a connected subgroup of the total group, which is usually homeomorphic to its simply connected double cover. More in general, the isomorphism is between a subset of the algebra containing 0 and a subset of the group containing the identity. Moreover, for a compact, connected, and simply connected Lie group, the algebra always generates the whole group via the exponential map.

invariant vector fields on G and the space of vectors in T_eG, thus, the Lie algebra $\mathfrak{g}$. For matrix Lie groups, it holds that $\theta_g = g^{-1}dg$.

By definition, the action of an element of the group on P is $\mathfrak{P}_g : P \to P$, and therefore, it defines a tangent map $\mathfrak{P}_{g*}$: $TP \to TP$, for which the following Lemma holds:

Lemma 1.

$$\mathfrak{P}_{g*} \circ \sigma(\xi) = \sigma(Ad_{g^{-1}}\xi). \tag{16}$$

Proof. At $p \in P$

$$\mathfrak{P}_{g*}\sigma_p(\xi) = \frac{\mathrm{d}}{\mathrm{d}t}((\mathfrak{P}_g \circ \mathfrak{P}_{e^{t\xi}})(p))\big|_{t=0} = \frac{\mathrm{d}}{\mathrm{d}t}((\mathfrak{P}_g \circ \mathfrak{P}_{e^{t\xi}} \circ \mathfrak{P}_{g^{-1}} \circ \mathfrak{P}_g)(p))\big|_{t=0}, \tag{17}$$

we then use the fact that $\mathfrak{P}_g \circ \mathfrak{P}_{e^{t\xi}} \circ \mathfrak{P}_{g^{-1}} = \mathfrak{P}_{g^{-1}e^{t\xi}g} = \mathfrak{P}_{\mathrm{Ad}_{g^{-1}}e^{t\xi}}$ and the identity for matrix groups $\mathrm{Ad}_g e^{t\xi} = e^{t\mathrm{Ad}_g\xi}$ to get the following:

$$\mathfrak{P}_{g*}\sigma_p(\xi) = \frac{\mathrm{d}}{\mathrm{d}t}(\mathfrak{P}_{e^{t(\mathrm{Ad}_{g^{-1}}\xi)}}(\mathfrak{P}_g(p)))\big|_{t=0} = \sigma_{\mathfrak{P}_g(p)}(\mathrm{Ad}_{g^{-1}}\xi). \tag{18}$$

□

It is time to define what we were aiming to define at the beginning of the section: the *connection form*.

Definition 12. *Let P be a smooth G-principal bundle and $HE \subset TP$ be an Ehresmann connection. We call the $\mathfrak{g}$-valued 1-form $\omega \in \Omega^1(P, \mathfrak{g})$, satisfying*

$$\omega(v) = \begin{cases} \xi & if \;\; v = \sigma(\xi),\ \xi \in C^\infty(P, \mathfrak{g}) \\ 0 & if \;\; v \;\; horizontal, \end{cases} \tag{19}$$

the connection 1-form.

Proposition 1.

$$\mathfrak{P}_g^*\omega = Ad_{g^{-1}} \circ \omega. \tag{20}$$

Proof. Suppose $v = \sigma(\xi)$, since the other case left is trivial.

We can carry out some calculations on the left-hand side, and following from the definition of pull-back and Lemma 1, we have

$$\left(\mathfrak{P}_g^*\omega\right)(\sigma(\xi)) = \omega\left(\mathfrak{P}_{g*} \circ \sigma(\xi)\right) = \omega\left(\sigma(\mathrm{Ad}_{g^{-1}}(\xi)\right) = \mathrm{Ad}_{g^{-1}}(\xi). \tag{21}$$

Then, we only need to manipulate the right-hand side as

$$\mathrm{Ad}_{g^{-1}}\left(\omega(\sigma(\xi))\right) = \mathrm{Ad}_{g^{-1}}(\xi). \tag{22}$$

Both times, we used just the given definition of connection 1-form (Equation (19)). □

Remark 1. *This last Proposition is called G-equivariance. It can be imposed instead of by assuming that HE is an Ehresmann connection, and then HE can be shown to be such an Ehresmann connection.*

Another fundamental concept is given in the following:

Definition 13 (Tensorial form). *Let $\rho : G \to \mathrm{Aut}(V)$ be a representation over a vector space V and $\alpha \in \Omega^k(P, V)$ be a vector valued differential form.*

We call α a tensorial form if it is the following:

- *horizontal, i.e., $\alpha(v_1, ..., v_k) = 0$ if at least one v_i is a vertical vector field, and*
- *equivariant, i.e., for all $g \in G$, $\mathfrak{P}_g^* \alpha = \rho(g^{-1}) \circ \alpha$.*

We define horizontal and equivariant forms as maps belonging to $\Omega_G^k(P, V)$.

Observation 5: The connection form ω is not, in general, horizontal; thus, it is not a tensorial form, $\omega \notin \Omega_G^1(P, \mathfrak{g})$. This will be clear when taking into account how the gauge field transforms under a change of trivialization in Section 4.

3.4. Curvature Forms

Given our connection 1-form ω, we can proceed in two ways: the first consists in taking a map called the *horizontal projection* and in defining the curvature as this projection applied on the exterior derivative of ω. In this way, we naturally see that curvature measures the displacement of the commutator of two vectors from being horizontal.

We will proceed in a different way though. We will define the curvature through a *structure equation.*

Definition 14. *Given $\omega \in \Omega^1(P, \mathfrak{g})$, a principal connection 1-form, the 2-form $\Omega \in \Omega_G^2(P, \mathfrak{g})$ satisfies the following:*

$$\Omega = d\omega + \frac{1}{2}[\omega \wedge \omega] \tag{23}$$

whic is called curvature 2-form.

In Equation (23), $[\omega \wedge \omega]$ denotes the bilinear operation on the Lie algebra $\mathfrak{g}$ called differential Lie bracket. It is defined as follows:

$$[\omega \wedge \eta](u, v) = \frac{1}{2}([\omega(u), \eta(v)] - [\omega(v), \eta(u)]), \tag{24}$$

where u and v are vector fields.

It follows straightforwardly that, if we take two general horizontal vector fields $u, v \in \Gamma(HE)$ and we use the ordinary formula[12] for the exterior derivative of a 1-form $d\omega(u, v) = u\omega(v) - v\omega(u) - \omega([u, v])$, since $\omega(u) = \omega(v) = 0$, we get

$$\Omega(u, v) = -\omega([u, v]). \tag{25}$$

We see that Ω measures how the commutator of two horizontal vector fields is far from being horizontal as well.

---

[12] Here, we regard $\omega(u)$ as a function $\omega(u) : P \to \mathfrak{g}$ belonging to the algebra of smooth functions to $\mathfrak{g}$, $C^\infty(P, \mathfrak{g})$.

4. Exterior Covariant Derivative

4.1. For an Ehresmann Connection HE

Observation 6: $\Omega^k_G(P,V)$ is not closed under the ordinary exterior derivative. In that sense, if $\alpha \in \Omega^k_G(P,V)$, then $d\alpha \notin \Omega^{k+1}_G(P,V)$. This is what a covariant differentiation will do instead.

The idea of a covariant exterior derivative for a connection HE is, given such an Ehresmann connection HE, the one of projecting vector fields onto this horizontal bundle and then feed our ordinary exterior derivative with such horizontal vector fields.

First of all, we define a map acting as a pull-back. Namely that, given a map $h : TP \to HE$ such that, for all vertical vector fields v, we get $h \circ v := hv = 0$ (called the *horizontal projection*), we define the dual map $h^* : T^*P \to HE^*$ such that, for $\alpha \in \Omega^1(P,V)$ and V a vector space, we have $h^* \circ \alpha := h^*\alpha = \alpha \circ h$.

Definition 15 (d^h). *Let P be a G-principal bundle, V be a vector space, and $\alpha \in \Omega^k(P,V)$ be an equivariant form. We define the exterior covariant derivative d^h as a map $d^h : \Omega^k(P,V) \to \Omega^{k+1}_G(P,V)$ such that*

$$d^h\alpha(v_0,...,v_k) := h^*d\alpha(v_0,...,v_k) = d\alpha(hv_0,...,hv_k), \tag{26}$$

where $v_0,...,v_k$ are vector fields.

It depends on the choice of our Ehresmann connection HE, which reflects onto the horizontal projection h; that is why we have the index ^{h}.

Observation 7: We can make our covariant derivative depend only on ω, if we restrict it to only forms in $\Omega^k_G(P,V)$ and if we consider the representation of the algebra induced by the derivative of ρ that we denote $d\rho : \mathfrak{g} \to \mathrm{End}(V)$. Then, we have $d\rho \circ \omega \in \Omega^k(P, \mathrm{End}(V))$.

4.2. For a Connection Form ω

Definition 16 (d_ω). *Let P be a G-principal bundle, V be a vector space, and $\alpha \in \Omega^k_G(P,V)$ be a tensorial form. We define the exterior covariant derivative d_ω as a map $d_\omega : \Omega^k_G(P,V) \to \Omega^{k+1}_G(P,V)$ such that*[13]

$$\begin{aligned} d_\omega\alpha &:= d\alpha + \omega \wedge_{d\rho} \alpha \\ &:= d\alpha + d\rho \circ \omega \wedge \alpha. \end{aligned} \tag{27}$$

Remark 2.

- *We observe that $d^2_\omega\alpha \neq 0$ for a general $\alpha \in \Omega^k_G(P,V)$, but it is easy to show that it holds*[14]

$$d^2_\omega\alpha = \Omega \wedge_{d\rho} \alpha, \tag{28}$$

Thus, for a flat connection such that $\Omega = 0$, we have $d^2_\omega\alpha = d^2\alpha = 0$.

---

[13] For a general k-form:

$$(\omega \wedge_{d\rho} \alpha)(v_1,...,v_{k+1}) = \frac{1}{(1+k)!}\sum_\sigma \mathrm{sign}(\sigma) d\rho\big(\omega(v_{\sigma(1)})\big)\big(\alpha(v_{\sigma(2)},...,v_{\sigma(k+1)})\big).$$

[14] See the first Bianchi identity in Equation (56) for the proof.

- *We have observed that $\omega \notin \Omega^1_G(P, \mathfrak{g})$. Therefore, $d_\omega \omega$ is not well defined. However, we can consider $d^h\omega \in \Omega^2_G(P, \mathfrak{g})$, and this is precisely our curvature $\Omega = d\omega + \frac{1}{2}[\omega \wedge \omega]$, where the anomalous $\frac{1}{2}$ factor comes from the "non-tensoriality" of ω. As a matter of fact, there is no representation that would make the $\frac{1}{2}$ term arise if we considered $d_\omega \omega$ instead.*
- *The fact that d_ω is not well defined for non-tensorial forms does not mean that ω defines a less general derivative than what d^h does. As a matter of fact, HE could be defined starting from ω, as we mentioned above, since $HE = Ker\,\omega$.*

5. Gauge Field and Field Strength

5.1. Make It Clear

Definition 17 (Gauge field). *Let $P \to M$ be a G-principal bundle, G be a Lie group with $\mathfrak{g}$ as the respective Lie algebra, $\{U_\beta\}$ be a cover of M, and $s_\beta : U_\beta \to P$ be a section.*

We define the gauge field as the pull-back of the connection form $\omega \in \Omega^1(P, \mathfrak{g})$ as

$$A_\beta = s^*_\beta \omega \ \in \Omega^1(U_\beta, \mathfrak{g}). \tag{29}$$

We notice that, under a change of trivialization, such a gauge field changes via the action of the adjoint map.

In fact, we have the following:

Lemma 2. *The restriction of ω to $\pi^{-1}(U_\beta)$ agrees with*

$$\omega_\beta = Ad_{g_\beta^{-1}} \circ \pi^* A_\beta + g^*_\beta \theta, \tag{30}$$

where $g_\beta : \pi^{-1}(U_\beta) \to G$ is the map induced by the inverse of the trivialization map φ_β defined in Equation (2), and with $Ad_{g_\beta^{-1}}$, we intend for the adjoint map at the group element given by $g_\beta(p)^{-1}$ at a point $p \in \pi^{-1}(U_\beta)$.

The proof comes from the observation that Equations (19) and (30) coincide in $\pi^{-1}(U_\beta)$ for both a horizontal (for which they are zero) and a vertical vector field.

Thanks to this, we easily have the following:

Proposition 2. *Let G be a matrix Lie group. Then it holds the following transformation for a gauge field:*

$$A_\beta = g_{\beta\gamma} A_\gamma g^{-1}_{\beta\gamma} - dg_{\beta\gamma} g^{-1}_{\beta\gamma}. \tag{31}$$

Proof. Using Equations (29) and (30) for all $x \in U_\beta \cap U_\gamma$,

$$\begin{aligned} A_\beta &= s^*_\beta \omega \\ &= s^*_\beta \omega_\beta = s^*_\beta \omega_\gamma \\ &= s^*_\beta (\mathrm{Ad}_{g_\gamma^{-1}} \circ \pi^* A_\gamma + g^*_\gamma \theta) \\ &= \mathrm{Ad}_{g_{\beta\gamma}^{-1}} \circ A_\gamma + g^*_{\gamma\beta}\theta \qquad &(\text{using } g_\gamma \circ s_\beta := g_{\beta\gamma} : U_\beta \cap U_\gamma \to G) \\ &= \mathrm{Ad}_{g_{\beta\gamma}} \circ (A_\gamma - g^*_{\beta\gamma}\theta) \qquad &(\mathrm{Ad}_{g_{\beta\gamma}} \circ g^*_{\beta\gamma}\theta = -g^*_{\gamma\beta}\theta), \end{aligned} \tag{32}$$

which reduces to the assert for matrix Lie groups. □

Observations 8:

i. We observe that a local gauge transformation of the gauge field corresponds to a change of trivialization chart.
ii. Non-tensoriality of ω was given by the fact that it is, in general, not horizontal. For the gauge field A, we can generalize to forms on M the concept of tensoriality/non-tensoriality by noticing that a form obtained by the pull-back of a tensorial form, denoted with $t \in \Omega^1_G(P,V)$, would transform differently compared to A, namely as

$$t_\beta := s_\beta^* t = g_{\beta\gamma} t_\gamma g_{\beta\gamma}^{-1}. \tag{33}$$

The Maurer–Cartan form θ reflects the non-horizontality of ω to the gauge field, from Equation (30).
iii. A difference of two gauge fields like $A - A'$ transforms as Equation (33). In fact, the transformation rule is one of a tensorial form, since the Maurer–Cartan forms simplify.
iv. We notice that (iii) is a particular case of a more general one. Indeed, it is possible to show with proof in Reference [2] (Chapter 5) that $\Omega^k_G(P,V) \cong \Omega^k(M, P\times_\rho V)$. This is essentially due to the fact that, thanks to the equivalence relation of the associated bundle and the gluing condition of sections on overlaps, the pull-backs by sections $s_\beta : U_\beta \to P$ give a one-to-one correspondence between these two spaces. Therefore, we can obtain forms with a tensorial transformation like Equation (33) just by taking the pull-back of tensorial forms on P; these will be forms on M with values into the associated bundle $P \times_\rho V$.
v. Observations (iii) and (iv) ensure that an object built with gauge fields $A_\beta \in \Omega^1(U_\beta, \mathfrak{g})$ (which transform on overlaps by Equation (31)) will be in $\Omega^2(M, P\times_{\mathrm{Ad}}\mathfrak{g})$; see Observation 9.

Claim. *This gauge field defines an exterior covariant derivative for bundle-valued forms on M. We denote such a map with*

$$d_A : \Omega^k(M, P\times_\rho V) \to \Omega^{k+1}(M, P\times_\rho V). \tag{34}$$

Anyway, we will further develop this argument in Section 6.1.

We can proceed analogously and can define the pull-back of the curvature:

Definition 18 (Field strength). *Let $P \to M$ be a G-principal bundle, G be a Lie group with $\mathfrak{g}$ as the respective Lie algebra, $\{U_\beta\}$ be a cover of M, and $s_\beta : U_\beta \to P$ be a section.*

We define the field strength as the pull-back of the curvature form $\Omega \in \Omega^2_G(P,\mathfrak{g})$ as

$$F_\beta = s_\beta^*\Omega \in \Omega^2_G(U_\beta, \mathfrak{g}), \tag{35}$$

which, by definition of Ω, is

$$F_\beta = dA_\beta + \frac{1}{2}[A_\beta \wedge A_\beta]. \tag{36}$$

Similarly to what we have done for the gauge field, we can show[15] that the field strength transforms as

$$F_\beta = \mathrm{Ad}_{g_{\beta\gamma}} \circ F_\gamma = g_{\beta\gamma} F_\gamma g_{\beta\gamma}^{-1}, \tag{37}$$

15 Using the Cartan structure equation for θ, $d\theta = -\frac{1}{2}[\theta,\theta]$.

where the last equality holds for matrix Lie groups with g and g^{-1} in G. This is indeed the transformation of a tensorial form, as in Equation (33).

Observation 9: Thanks to our previous observation, i.e., there is a canonical isomorphism between $\Omega^k_G(P, V)$ and $\Omega^k(M, P \times_\rho V)$, we can relate Ω and F_β with a form[16] $F_A \in \Omega^2(M, \mathrm{ad}P)$. Namely there is a canonical isomorphism sending $\Omega \in \Omega^2_G(P, \mathfrak{g})$ to $F_A \in \Omega^2(M, \mathrm{ad}P)$. Indeed, given the transformation law for the field strength in Equation (37), we see that $\{F_\beta\}$ is horizontal and equivariant and, thus, forms a global section belonging to $\Omega^2(M, \mathrm{ad}P)$, which is usually denoted as F_A.

The notation F_A stresses that it is obtained from gauge fields in $\Omega^1(U_\beta, \mathfrak{g})$.

In the case of a trivial bundle, it is also possible to define a global gauge field $A \in \Omega^1(M, \mathfrak{g})$.

5.2. 2nd Bianchi Identity

Consider $d_A : \Omega^k(M, P \times_\rho V) \to \Omega^{k+1}(M, P \times_\rho V)$ as the exterior covariant derivative and $F_A \in \Omega^2(M, \mathrm{ad}\, P)$ as the field strength.

Then, we have the following:

Proposition 3.

$$d_A F_A = 0. \tag{38}$$

This is the second Bianchi identity.

Proof. Given

$$F_A = dA + \frac{1}{2}[A \wedge A], \tag{39}$$

then

$$\begin{aligned} d_A F_A &= dF_A + [A \wedge F_A] \\ &= d^2 A + \frac{1}{2} d[A \wedge A] + [A \wedge dA] + \frac{1}{2}[A \wedge [A \wedge A]] \\ &= \frac{1}{2}[A \wedge [A \wedge A]] && (d^2 A = 0 \text{ and } \frac{1}{2} d[A \wedge A] = -[A \wedge dA]) \\ &= 0. && (\text{because of Jacobi identity}) \end{aligned} \tag{40}$$

□

6. Affine Formulation

In the usual formulation of General Relativity, one defines a covariant derivative ∇, which is a map among tensors. Then, one can define curvature and torsion and eventually get the field equations for ECSK theory or General Relativity by setting torsion to zero.

One may wonder if this latter formulation is equivalent to the one we have been implementing through principal bundles and principal connection.

The answer is positive and is given in the next two sections.

---

[16] Where we have introduced the notation $\Omega^k(M, P \times_{\mathrm{Ad}} \mathfrak{g}) := \Omega^k(M, \mathrm{ad}P)$.

6.1. Affine Covariant Derivative

We have built our setup by taking ρ to be the fundamental representation of $O(3,1)$, $P = F_O(M)$, and $\mathcal{V} = F_O(M) \times_\rho V$ to be the Minkowski bundle, as in (ii) of Observations 4. Therefore, as mentioned above, the isomorphism between $\Omega^k_G(P,V)$ and $\Omega^k(M, P\times_\rho V)$ allows to define an exterior covariant derivative of forms in $\Omega^1(M,\mathcal{V})$:

$$d_A : \Omega^k(M,\mathcal{V}) \to \Omega^{k+1}(M,\mathcal{V}). \tag{41}$$

In this way, we have a covariant differentiation for $\mathcal{V}$-valued differential forms on M and, thus, also for tetrads, since $e \in \Omega^1(M,\mathcal{V})$.

Since we note that $d\rho$ induces a one form $d\rho(A) \in \Omega^1(M, \mathrm{End}(V))$, we can further define another kind of derivative that "takes care" of internal indices only; in particular, this will not be necessarily a map between differential forms.

This derivative in components reads, for $\phi \in \Gamma(\mathcal{V})$,

$$(D_A\phi)^a_\mu = (\partial_\mu \phi^a + A^{ac}_\mu \eta_{cb}\phi^b) \tag{42}$$

and, for $\alpha \in \Omega^k(M,\mathcal{V})$,

$$(D_A\alpha)^a_{\mu\nu_1\ldots\nu_k} = (\partial_\mu \alpha^a_{\nu_1\ldots\nu_k} + A^{ac}_\mu \eta_{cb}\alpha^b_{\nu_1\ldots\nu_k}), \tag{43}$$

which shows that it does not map α to a differential form.

Now, we immediately apply the inverse of a tetrad to $D_A\phi$ and identify it with ∇.

In fact, we take a vector field $X \in \Gamma(TM)$, feed the tetrad e with it, then apply[17] D_A to get $D_A(\iota_X e)$, and finally pull it back with the inverse of the tetrad $\bar{e}$.

In components, this reads as follows:

$$\left(D_A(\iota_X e)\right)^a_\mu = D_\mu(e^a_\nu X^\nu) = \partial_\mu(e^a_\nu X^\nu) + \omega^{ab}_\mu \eta_{bc} e^c_\nu X^\nu, \tag{44}$$

where, for reasons of metric compatibility[18] with η, we have the only antisymmetric part of the gauge field, for which we used the notation $A^a_{\mu c} = \omega^{ab}_\mu \eta_{bc}$.

Be aware: Do not get confuse. We shall refer to $\omega \in \Omega^1(M, \Lambda^2\mathcal{V})$ as the *spin connection*. To stress that we want D_A to depend on the spin connection only, we shall denote it with D_ω.

Pulling back via $\bar{e}$, we obtain

$$\bar{e}^\sigma_a\left(D_\mu(e^a_\nu X^\nu)\right) = \bar{e}^\sigma_a\left(\partial_\mu(e^a_\nu X^\nu) + \omega^{ab}_\mu \eta_{bc} e^c_\nu X^\nu\right). \tag{45}$$

We define the Christoffel symbols $\Gamma^\sigma_{\mu\nu}$ as

$$\begin{aligned}\Gamma^\sigma_{\mu\nu} &= \bar{e}^\sigma_a(D_\mu e^a_\nu) \\ &= \bar{e}^\sigma_a(\partial_\mu e^a_\nu + \omega^{ab}_\mu \eta_{bc} e^c_\nu)\end{aligned} \tag{46}$$

---

17 Here, we use the so-called *interior product*, i.e., a map $\iota_\xi : \Omega^k(M) \to \Omega^{k-1}(M)$, such that $(\iota_\xi\alpha)(X_1,\ldots,X_{k-1}) = \alpha(\xi, X_1,\ldots,X_{k-1})$, for vector fields $\xi, X_1,\ldots X_{k-1}$. Furthermore it respects $\iota_\xi(\alpha\wedge\beta) = (\iota_\xi\alpha)\wedge\beta + (-1)^k\alpha\wedge(\iota_\xi\beta)$, where $\alpha \in \Omega^k(M)$. Therefore, it forms an antiderivation. The relation with the Lie derivative is given by the formula $\mathcal{L}_\xi\alpha = d(\iota_\xi\alpha) + \iota_\xi d\alpha$, called the Cartan identity. The interior product of a commutator satisfies $\iota_{[X,Y]} = [\mathcal{L}_X, \iota_Y]$, with X and Y vector fields.

18 In fact, imposing the condition of $D_A\eta = 0$ implies the antisymmetry of the gauge field.

and, thus, we get

$$\nabla_\mu X^\sigma := \bar{e}^\sigma_a \left(D_\mu (e^a_\nu X^\nu)\right) = \partial_\mu X^\sigma + \Gamma^\sigma_{\mu\nu} X^\nu, \tag{47}$$

which is the covariant derivative well known in General Relativity.

We can also see what the curvature form is in terms of the commutator of two derivatives, given by the only antisymmetric part of the connection.

Then,

$$F_{A^{antis}} := F_\omega \tag{48}$$

and it is given by

$$\left(D_{[\mu} D_{\nu]} \phi\right)^a = \left(\partial_{[\mu} \omega_{\nu]}{}^{ab} + \omega^{ad}_{[\mu} \eta_{de} \omega^{eb}_{\nu]}\right) \eta_{bc} \phi^c = F^{ab}_{\mu\nu} \eta_{bc} \phi^c, \tag{49}$$

where $A_{[\mu} B_{\nu]} = A_\mu B_\nu - A_\nu B_\mu$ is our convention for the antisymmetrization. The fact that F_ω is a 2-form shows that $F^{ab}_{\mu\nu} = -F^{ab}_{\nu\mu}$; furthermore, metric compatibility ensures also $F^{ab}_{\mu\nu} = -F^{ba}_{\mu\nu}$, therefore $F_\omega \in \Omega^2_G(M, \Lambda^2 \mathcal{V})$.

Observation 10: We see[19] that, here, the bundle metric η acts as a map $\eta : \Omega^2(M, \Lambda^2 \mathcal{V}) \to \Omega^2(M, \mathrm{End}(V))$ isomorphically; thus, it permits to identify elements of the second exterior power $\Lambda^2 \mathcal{V}$ with linear maps given by the fundamental representation of the algebra $\mathfrak{g} = \mathfrak{so}(3,1)$. We can introduce the notation for the wedge product in the fundamental representation as $\wedge_f$; namely for, say, an $\alpha \in \Omega^1(M, \mathcal{V})$, we have $(\omega \wedge_f \alpha)^a = \omega^{ab} \eta_{bc} \wedge \alpha^c$.

6.2. Riemann Curvature Tensor

We can now consider the commutator of two affine covariant derivatives and use Equation (47), getting

$$(\nabla_{[\mu} \nabla_{\nu]} X)^\sigma = \bar{e}^\sigma_a \left(D_{[\mu} D_{\nu]} (\iota_X e)\right)^a = \bar{e}^\sigma_a F^{ab}_{\mu\nu} \eta_{bc} e^c_\omega X^\omega. \tag{50}$$

We identify the Riemann tensor

$$R_{\mu\nu\omega}{}^\sigma = \bar{e}^\sigma_a F^{ab}_{\mu\nu} \eta_{bc} e^c_\omega, \tag{51}$$

the Ricci curvature tensor

$$R_{\mu\omega} = R_{\mu\sigma\omega}{}^\sigma = \bar{e}^\sigma_a F^{ab}_{\mu\sigma} \eta_{bc} e^c_\omega \tag{52}$$

and thus the Ricci scalar

$$R = g^{\mu\omega} R_{\mu\omega} = \bar{e}^\mu_d \bar{e}^\omega_e \eta^{de} \bar{e}^\sigma_a F^{ab}_{\mu\sigma} \eta_{bc} e^c_\omega = -\bar{e}^\mu_a \bar{e}^\omega_b F^{ab}_{\mu\omega}. \tag{53}$$

It follows the antisymmetry of the Riemann tensor in the indices $\mu\nu$ and $\omega\sigma$, but it is important to note that we cannot ensure any symmetry in the Ricci curvature instead due to the presence of torsion.

7. Torsion

Here, we start focusing on the importance of torsion, which arises quite naturally as curvature does.

19 We denote the metric acting on the bundle and the metric acting on the fibers in the same way; however, we will usually deal with elements of the fibers.

7.1. Torsion Form

Definition 19 (Solder form/soldering of a G-principal bundle). *Let $\pi : P \to M$ be a smooth G-principal bundle over a differentiable manifold M, $\rho : G \to \mathrm{Aut}(V)$ be a representation, and G be a Lie group.*

We define the solder form, or soldering, as the vector-valued 1-form $\theta \in \Omega^1_G(P, V)$ such that $\tilde{\theta} : TM \to P \times_\rho V$ is a bundle isomorphism, where $\tilde{\theta} \in \Omega^1(M, P \times_\rho V)$ is the associated bundle map induced by the isomorphism of $\Omega^1_G(P, V) \cong \Omega^1(M, P \times_\rho V)$.

Observations 11:

- The choice of the solder form is not unique, in general.
- We can observe that, taking $P = F_O(M)$, ρ as the fundamental representation of $O(3,1)$, and V as the vector space with reference metric η, $\tilde{\theta}$ corresponds to our definition of tetrads. The different choices of soldering give rise to different tetrads.
- In the case that $P = F_O(M)$ and that the associated bundle is simply chosen to be TM, the solder form is called *canonical* or *tautological*. Since the associated bundle TM sets the bundle isomorphism $\tilde{\theta}$ to be the identity map $\mathrm{id} : TM \to TM$.
- In Observations 4, we mentioned that the Minkowski bundle cannot be canonically identified with the tangent bundle itself; indeed, we fixed a reference metric η, which cannot be pulled back by the identity map to give the metric on TM in general, and thus, the solder form is not canonical.

The soldering of the principal frame bundle allows us to define the torsion form[20].

Definition 20 (Torsion form). *Let $P = F_O(M)$, $\rho : O(3,1) \to \mathrm{Aut}(V)$ be the fundamental representation, V be a vector space with reference metric η, and $\theta \in \Omega^1_G(P, V)$ be a solder form.*
We define the torsion form $\Theta \in \Omega^2_G(P, V)$ as follows:

$$\Theta = d_\omega \theta = d\theta + \omega \wedge_f \theta. \tag{54}$$

7.2. Torsion in a Local Basis

We would like to express the torsion form in terms of tetrads and the gauge field.

In Reference[9], a formula is given and it is obtained by applying the previous definition of the torsion form under the canonical isomorphism $\Omega^k_G(P, V) \cong \Omega^k(M, \mathcal{V})$; therefore yielding

$$\tilde{\Theta}^a = (d_A e)^a = de^a + A^{ab}\eta_{bc} \wedge e^c. \tag{55}$$

7.3. 1st Bianchi Identity

Proposition 4. *Following our previous definitions, we have*

$$d_\omega \Theta = \Omega \wedge_f \theta, \tag{56}$$

which is called the first Bianchi identity.

Proof. For this proof, we prefer using Equation (26).

We consider three vector fields $u, v, w \in \Gamma(TP)$. By definition, it follows

[20] Torsion can be defined for every principal bundle, but physics arises when considering the frame bundle.

$$\begin{aligned} d^h\Theta(u,v,w) &= d\Theta(hu,hv,hw) \\ &= (d\omega \wedge_f \theta - \omega \wedge_f d\theta)(hu,hv,hw) \quad \text{(because of Equation (54))} \\ &= d\omega \wedge_f \theta(hu,hv,hw) \quad \text{(because of Equation (19))} \\ &= \Omega \wedge_f \theta(u,v,w). \end{aligned} \tag{57}$$

The last equality holds because of tensoriality of θ and the second remark in Remark 2. □

This proposition is a natural consequence of the property of the covariant differential expressed in Equation (28).

7.4. Torsion Tensor

Definition 21 (Torsion tensor). *Given two vector fields $X, Y \in \Gamma(TM)$ and a 1-form $\tau \in \Omega^1(M)$, we define the torsion tensor field Q as the tensor field of type-$\binom{1}{2}$ such that*

$$Q(X,Y;\tau) := \tau(Q(X,Y)) = \tau\big(\bar{e}(d_A e(X,Y))\big). \tag{58}$$

It is evidently antisymmetric in X, Y, by definition.

Proposition 5. *We have the following formula:*

$$Q(X,Y) = \nabla_X Y - \nabla_Y X - [X,Y] \tag{59}$$

and, in components, it reads

$$Q_{\mu\nu}{}^{\sigma} = \Gamma^{\sigma}_{\mu\nu} - \Gamma^{\sigma}_{\nu\mu} - C^{\sigma}_{\mu\nu}, \tag{60}$$

where $C^{\sigma}_{\mu\nu} = 0$ in a holonomic basis for X and Y and ∇ is the covariant derivative[21].

Proof. Recalling the definition of torsion

$$Q = \bar{e} \cdot (d_\omega e) = \bar{e}_a (d_\omega e)^a, \tag{61}$$

it follows

$$\begin{aligned} \bar{e}_a (d_\omega e)^a &= \bar{e}^{\sigma}_a \big(\partial_{[\mu} e^a{}_{\nu]} + \omega^a_{[\mu b} e^b_{\nu]}\big) dx^\mu \wedge dx^\nu \otimes \partial_\sigma \\ &= (\Gamma^{\sigma}_{\mu\nu} - \Gamma^{\sigma}_{\nu\mu}) dx^\mu \wedge dx^\nu \otimes \partial_\sigma \quad (\Gamma^{\sigma}_{\mu\nu} = \bar{e}^{\sigma}_a (D_\mu e^a_\nu).) \\ &= Q_{\mu\nu}{}^{\sigma} dx^\mu \wedge dx^\nu \otimes \partial_\sigma, \end{aligned} \tag{62}$$

then $Q_{\mu\nu}{}^{\sigma} = \Gamma^{\sigma}_{\mu\nu} - \Gamma^{\sigma}_{\nu\mu}$. □

We have now set up all the background for building our theory and for discussing field equations of ECSK theory.

---

[21] See Reference [10] for references about this.

8. Field Equations and Conservation Laws

We present here field equations for ECSK theory[22]. Thus, we will neither assume the possibility of a propagating torsion (and we will always keep non-identically vanishing Riemann curvature [14]) nor display a lagrangian for a totally independent torsion field; rather, we will only set the Palatini–Cartan lagrangian for gravity, as done in Reference [15], and a matter lagrangian as the source. This theory is known as Einstein–Cartan–Sciama–Kibble gravity (ECSK).

In the present case, torsion reduces to an algebraic constraint. This is a consequence of making torsion join the action of the theory as only contained in the Ricci scalar because of a non-torsion-free connection and not with an independent coupling coefficient. In works like References [16–19], torsion is present as an independent part (independent coupling coefficient) of the action and it does propagate.

This is why the ECSK is considered as the most immediate generalization of General Relativity with the presence of torsion.

Therefore, we wish to eventually obtain an action of two independent objects, tetrads and connection, where this latter action should give rise to equations for curvature when varying tetrads and for torsion when varying the connection.

We will focus more on the geometrical side of these equations and we will not dwell on deepening matter interaction (couplings, symmetry breaking, etc.), as done for instance in References [20–24].

8.1. ECSK Equations

ECSK theory with cosmological constant belongs to the Lovelock–Cartan family, which describes the most general action in four dimensions such that this action is a polynomial on the tetrads and the spin connection (including derivatives), is invariant under diffeomorphisms and local Lorentz transformations, and is constructed without the Hodge dual[23].

Recalling that we will refer to A as ω and stressing that it must be only the antisymmetric part, the notation for d_A becomes d_ω.

We will be dealing with a variational problem given by an action of the kind

$$S = S_{PC} + S_{matter}, \tag{63}$$

where the Palatini–Cartan action is

$$S_{PC}[e,\omega] = \int_M \mathrm{Tr}\,\Big[\frac{1}{2} e \wedge e \wedge F_\omega + \frac{\Lambda}{4!} e^4\Big]. \tag{64}$$

The wedge product is defined over both space–time and internal indices as a map[24] $\wedge : \Omega^k(M, \Lambda^p\mathcal{V}) \times \Omega^l(M, \Lambda^q\mathcal{V}) \to \Omega^{k+l}(M, \Lambda^{p+q}\mathcal{V})$ and the trace is a map $\mathrm{Tr} : \Lambda^4\mathcal{V} \to \mathbb{R}$, normalized such that (for v_i elements of a basis in $\mathcal{V}$) $\mathrm{Tr}[v_i \wedge v_j \wedge v_k \wedge v_l] = \varepsilon_{ijkl}$. The choice of the normalization of the trace works as a choice of orientation for M (since the determinant of a matrix in $O(3,1)$ may be ± 1). Therefore, we reduce the total improper Lorentz group $\mathrm{O}(3,1)$ to the only orientation preserving part, which is still not connected, $\mathrm{SO}(3,1)$. This gives an invariant volume form on M. In this way, we consider sections of $\Lambda^k T^*M \otimes \Lambda^p\mathcal{V}$.

---

[22] Some classical works about ECSK theory and General Relativity with torsion, like References [11–13].

[23] See Reference [25] for details.

[24] Such that, for $\alpha \in \Omega^k(M, \Lambda^p\mathcal{V})$ and $\beta \in \Omega^l(M, \Lambda^q\mathcal{V})$, we have $\alpha \wedge \beta = (-1)^{(k+p)(l+q)} \beta \wedge \alpha$.

Be aware: Later on, we will make explicit some indices and keep implicit some others; for this purpose, we will specify what kind of wedge product we are dealing with, even though it will be evident because it will be among the implicit indices.

We recall the definition of F_ω and deduce the identity for its variation

$$\delta_\omega F_\omega = d_\omega \delta\omega, \tag{65}$$

where we stress that, despite ω being non-tensorial, $\delta\omega$ is instead, and therefore, it transforms under the adjoint action (as F_ω does) like in Equation (33). That holds because $\delta\omega$ may be regarded as a difference of two spin connections.

The action for the matter is of the kind

$$S_{matter}[e,\omega,\varphi] = \kappa \int_M \mathrm{Tr}[L(e,\omega,\varphi)], \tag{66}$$

where L is an invariant lagrangian density form with the proper derivative order in our variables, φ is a matter field, and κ is a constant.

Such matter lagrangian is supposed to be source for both curvature and torsion equations, namely it will be set for fulfilling some conditions fitting the theory.

Therefore, varying the actions in Equations (64) and (66) and considering Equation (65), we have[25]

$$\begin{aligned} \int_M \mathrm{Tr}[\delta e \wedge (e \wedge F_\omega + \tfrac{\Lambda}{3!}e^3)] &= \int_M \mathrm{Tr}[\kappa \tfrac{\delta L}{\delta e} \wedge \delta e] \\ \int_M \mathrm{Tr}[\tfrac{1}{2} d_\omega (e \wedge e) \wedge \delta\omega] &= \int_M \mathrm{Tr}[\kappa \tfrac{\delta L}{\delta \omega} \wedge \delta\omega], \end{aligned} \tag{67}$$

which is equivalent to

$$\begin{aligned} \varepsilon_{abcd} e^b \wedge F_\omega^{cd} + \tfrac{\Lambda}{3!}\varepsilon_{abcd} e^b \wedge e^c \wedge e^d &= \kappa \tfrac{\delta \mathrm{Tr}[L]}{\delta e^a} := \kappa T_a \\ \tfrac{1}{2}\varepsilon_{abcd} d_\omega (e^c \wedge e^d) &= \kappa \tfrac{\delta \mathrm{Tr}[L]}{\delta \omega^{ab}} := \kappa \Sigma_{ab} \end{aligned} \tag{68}$$

where the wedge product here is only between differential forms.

Setting $\Lambda = 0$ and in performing the derivative, Equation (68) can be rewritten as

$$\begin{aligned} \varepsilon_{abcd} e^b \wedge F_\omega^{cd} &= \kappa T_a \\ \varepsilon_{abcd}\, \tilde{Q}^c \wedge e^d &= \kappa \Sigma_{ab}, \end{aligned} \tag{69}$$

where we have set $\tilde{Q} = d_\omega e$.

These are equations for the ECSK theory in their implicit form[26], where T and Σ are related to, respectively, the *energy momentum* and the *spin* tensor, once pulled back.

By making all the indices explicit, as given in Reference [20], and properly setting κ according to natural units[27], Equation (69) takes the following form

$$\begin{aligned} G_{\mu\nu} &= 8\pi T_{\mu\nu} \\ Q_{\mu\nu}{}^{\sigma} &= -16\pi \Sigma_{\mu\nu}{}^{\sigma}. \end{aligned} \tag{70}$$

---

25 Omitting equations of motion $\frac{\delta L}{\delta \varphi} = 0$ for the matter field, which have to be satisfied for conservation laws anyway.

26 Without making space–time indices explicit.

27 All fundamental constants $= 1$.

Observations 12:

i. $T_{\mu\nu}$ is not symmetric, as expected from the non-symmetry of the Ricci curvature $R_{\mu\nu}$.
ii. We stress that, even though e is an isomorphism, the map $e \wedge \cdot : \Omega^k(M, \Lambda^p \mathcal{V}) \to \Omega^{k+1}(M, \Lambda^{p+1}\mathcal{V})$ is not an isomorphism, in general. In fact, taking $\frac{\delta L}{\delta e} = 0$ (with $\Lambda = 0$) in Equation (67) does not imply $F_\omega = 0$, which would imply a flat connection.
iii. Setting $\frac{\delta L}{\delta \omega} = 0$ in Equation (67), one recovers the condition of vanishing torsion (hence, a Levi–Civita connection) and, therefore, the Einstein equations.
iv. It is interesting to note that, requiring a totally antisymmetric spin tensor, sets the total antisymmetry of the torsion tensor. Namely, in the case of a totally antisymmetric Σ, we need to couple the only totally antisymmetric part of torsion into the geometrical lagrangian. This is further discussed in Reference [20].

8.2. Conservation Laws

We have two symmetries, i.e., local Lorentz transformations and diffeomorphisms. They are continuous symmetries, and as such, we expect two conservation laws. Since we are dealing with local symmetries, we shall not find two conserved currents but rather two relations for the variations of the matter lagrangian w.r.t. e and ω.

These relations directly imply the Bianchi identities of Equations (38) and (56), but we could also do the converse, namely assuming Equations (38) and (56) and then deriving such conservation laws. This means that conservation laws are a consequence of symmetry on the one hand, implemented via the following symmetries (respectively diffeomorphisms and local SO(3,1))

$$\begin{aligned} \delta_\xi e^a &= \mathcal{L}_\xi e^a = \iota_\xi de^a + d\iota_\xi e^a \\ \delta_\xi \omega^{ab} &= \mathcal{L}_\xi \omega^{ab} = \iota_\xi d\omega^{ab} + d\iota_\xi \omega^{ab}, \end{aligned} \tag{71}$$

where ξ is the generator vector field,

$$\begin{aligned} \delta_\Lambda e^a &= \Lambda^{ab}\eta_{bc}e^c \\ \delta_\Lambda \omega^{ab} &= -d_\omega \Lambda^{ab} \quad \Lambda \in \mathfrak{so}(3,1), \end{aligned} \tag{72}$$

or a direct consequence if we impose field equations and, thus, gravitational dynamics and Bianchi identities on the other hand.

We will follow the shortest derivation, namely to implement the Bianchi identities of Equations (38) and (56) on field Equation (69).

Thanks to Bianchi identities, left hand side of field Equation (69) can be rewritten in the following way:

$$\begin{aligned} d_\omega(\varepsilon_{abcd}e^b \wedge F_\omega^{cd}) &= \iota_a\tilde{Q}^b \wedge (\varepsilon_{bcde}e^c \wedge F_\omega^{de}) + \iota_a F_\omega^{bc} \wedge (\varepsilon_{bcde}\tilde{Q}^d \wedge e^e) \\ d_\omega(\varepsilon_{abcd}\tilde{Q}^c \wedge e^d) &= -\tfrac{1}{2}(\varepsilon_{acde}e^c \wedge F_\omega^{de} \wedge e_b - \varepsilon_{bcde}e^c \wedge F_\omega^{de} \wedge e_a), \end{aligned} \tag{73}$$

where $\iota_a = \iota_{\bar{e}_a}$ and $e_b = \eta_{bc}e^c$.

However, because of the same field in Equation (69), they reduce to

$$\begin{aligned} d_\omega T_a &= \iota_a\tilde{Q}^b \wedge T_b + \iota_a F_\omega^{bc} \wedge \Sigma_{bc} \\ d_\omega \Sigma_{ab} &= \tfrac{1}{2}T_{[a} \wedge e_{b]}, \end{aligned} \tag{74}$$

In References [26,27], a more detailed discussion can be found. These are conservation laws for ECSK theory.

In components, as given in Reference [20], they read

$$\begin{aligned} \nabla_\mu T^{\mu\nu} + T_{\sigma\rho} Q^{\sigma\rho\nu} - \Sigma_{\mu\sigma\rho} R^{\mu\sigma\rho\nu} &= 0 \\ \nabla_\mu \Sigma_{\sigma\omega}{}^{\mu} + \tfrac{1}{2} T_{[\sigma\omega]} &= 0. \end{aligned} \tag{75}$$

9. Conclusions

We have set up all the mathematical background for building ECSK theory, eventually achieving field equations and conservation laws.

In ECSK theory, torsion is only an algebraic constraint and it does not propagate. This is a natural consequence of inserting torsion into the theory without an independent coupling coefficient but simply generalizing the Einstein–Hilbert action (or Palatini action in our formulation) $\int R\sqrt{-g}d^4x$ to a non-torsion-free connection ∇ (or spin connection in our case). In this case, the Ricci scalar contains both curvature and torsion.

It is possible to immediately recover General Relativity by imposing the zero torsion condition, which, in the considered theory, translates to letting the matter field φ generate a null contribution to the spin tensor $\Sigma_{\mu\nu}{}^{\sigma}$. The most natural matter fields which would fit with the theory are spinors; indeed, spinors are the way in which we can have a non-vanishing spin tensor which is also dynamical because of equations of motion for the spinor field.

This review does not want to substitute the well-known literature but to just give a self-contained and mathematically rigorous introduction to ECSK theory, providing also some references for deepening knowledge in the present subjects. Also, we intentionally did not dive too deeply into physical applications to cosmology (like done in References [28–33]), that might be a valid argument for writing another review article.

Funding: The APC was funded by Professor Matthias Gaberdiel (Institute for Theoretical Physics, ETH Zürich).

Conflicts of Interest: The authors declare no conflict of interest. The funders had no role in the design of the study; in the collection, analyses, or interpretation of data; in the writing of the manuscript; or in the decision to publish the results.

References

1. Baez, J.; Muniain, J.P. *Gauge Fields, Knots and Gravity*; World Scientific Publishing Company: Singapore, 1994.
2. Kobayashi, S.; Nomizu, K. *Foundations of Differential Geometry*; Wiley and Sons: New York, NY, USA, 1969; Volume 2.
3. Michor, P.W.; Kolar, I.; Slovak, J. *Natural Operations in Differential Geometry*; Springer: Berlin/Heidelberg, Germany, 1993.
4. Morita, S. *The Geometry of Differential Forms*; American Mathematical Society: Providence, RI, USA, 2001; Volume 201.
5. Brzeziński, T. On synthetic interpretation of quantum principal bundles. *arXiv* **2009**, arXiv:0912.0213.
6. Sharpe, R.W. *Differential Geometry: Cartan's Generalization of Klein's Erlangen Program*; Springer: New York, NY, USA, 1997.
7. Gielen, S.; Wise, D.K. Lifting General Relativity to Observer Space. *J. Math. Phys.* **2013**, 54, 052501. [CrossRef]
8. Fecko, M. *Differential Geometry and Lie Groups for Physicists*; Cambridge University Press: Cambridge, UK, 2011.
9. Lott, J. The Geometry of Supergravity Torsion Constraints. *arXiv* **2001**, arXiv:math/0108125.
10. Straumann, N. *General Relativity*; Springer: Dordrecht, The Netherlands, 2012.
11. Hehl, F.W.; Obukhov, Y.N. Elie Cartan's torsion in geometry and in field theory, an essay. *Ann. Fond. Broglie* **2007**, 32, 157.

12. Hehl, F.W.; Von der Heyde, P.; Kerlick, G.D.; Nester, J.M. General relativity with spin and torsion: Foundations and prospects. *Rev. Mod. Phys.* **1976**, *48*, 393–416. [CrossRef]
13. Chakrabarty, S.; Lahiri, A. Different types of torsion and their effect on the dynamics of fields. *Eur. Phys. J. Plus* **2018**, *133*, 242. [CrossRef]
14. Nester, J.M.; Yo, H.J. Symmetric teleparallel general relativity. *Chin. J. Phys.* **1999**, *37*, 113.
15. Cattaneo, A.S.; Schiavina, M. The reduced phase space of Palatini-Cartan-Holst theory. *Ann. Henri Poincare* **2019**, *20*, 445. [CrossRef]
16. Fabbri, L. Singularity-free spinors in gravity with propagating torsion. *Mod. Phys. Lett. A* **2017**, *32*, 1750221. [CrossRef]
17. Fabbri, L.; Tecchiolli, M. Restrictions on torsion-spinor field theory. *Mod. Phys. Lett.* **2019**, 1950311. [CrossRef]
18. Fabbri, L. A discussion on the most general torsion-gravity with electrodynamics for Dirac spinor matter fields. *Int. J. Geom. Meth. Mod. Phys.* **2015**, *12*, 1550099. [CrossRef]
19. Fabbri, L.; da Rocha, R. Torsion Axial Vector and Yvon-Takabayashi Angle: Zitterbewegung, Chirality and all that. *Eur. Phys. J. C* **2018**, *78*, 207. [CrossRef]
20. Fabbri, L. Foundations Quadrilogy. *arXiv* **2017**, arXiv:1703.02287.
21. Gies, H.; Lippoldt, S. Fermions in gravity with local spin-base invariance. *Phys. Rev. D* **2014**, *89*, 064040. [CrossRef]
22. Cabral, F.; Lobo, F.S.N.; Rubiera-Garcia, D. Einstein-Cartan-Dirac gravity with $U(1)$ symmetry breaking. *arXiv* **2019**, arXiv:1902.02222.
23. Inglis, S.; Jarvis, P. The self-coupled Einstein–Cartan–Dirac equations in terms of Dirac bilinears. *J. Phys. A* **2019**, *52*, 045301. [CrossRef]
24. Diether, C.F.; Christian, J. Existence of Matter as a Proof of the Existence of Gravitational Torsion. *Prespace. J.* **2019**, *10*, 610.
25. Mardones, A.; Zanelli, J. Lovelock-Cartan theory of gravity. *Class. Quantum Grav.* **1991**, *8*, 1545. [CrossRef]
26. Bonder, Y.; Corral, C. Unimodular Einstein–Cartan gravity: Dynamics and conservation laws. *Phys. Rev. D* **2018**, *97*, 084001. [CrossRef]
27. Jiang, Y. Angular momentum conservation law in Einstein-Cartan space–time. *arXiv* **2000**, arXiv:gr-qc/0010047.
28. Poplawski, N.J. Cosmology with Torsion: An alternative to cosmic inflation. *Phys. Lett. B* **2010**, *694*, 181. [CrossRef]
29. Poplawski, N.J. Universe in a Black Hole with Spin and Torsion. *arXiv* **2014**, arXiv:1410.3881.
30. Medina, S.B.; Nowakowski, M.; Batic, D. Einstein-Cartan Cosmologies. *Ann. Phys.* **2019**, *400*, 64. [CrossRef]
31. Mehdizadeh, M.R.; Ziaie, A.H. Einstein-Cartan Wormhole Solutions. *Phys. Rev. D* **2017**, *95*, 064049. [CrossRef]
32. Pesmatsiou, K.; Tsagas, C.G.; Barrow, J.D. Kinematics of the Einstein-Cartan universes. *Phys. Rev. D* **2017**, *95*, 104007. [CrossRef]
33. Kranas, D.; Tsagas, C.G.; Barrow, J.D.; Iosifidis, D. Friedmann-like universes with torsion. *Eur. Phys. J. C.* **2019**, *79*, 341. [CrossRef]

MDPI

Article

1. Introduction

Einstein's theory of gravitation, or more in general any of its extensions with higher-order derivative terms, have acquired over the years the status of superb physical theories: established upon one single principle, that is that the space-time is a $(1+3)$-dimensional manifold, whose curvature is determined by the energy content, they have produced many of the most surprising predictions in all of physics, starting with the bending of light rays in 1916 and ending with the detection of gravitational waves one century later. In fact, if Dark Matter is indeed a form of matter, there is not a single prediction left to confirm.[1]

Quantum Field Theory has done no less: its predictions are among the most precise in history, and while still in need of rigorous mathematical formalization very few doubt that this task will be accomplished in the foreseeable future.

There are, nevertheless, still two problems that appear to cloud the beauty of these two pillars of modern physics, both related to the issue of singularities. QFT's range of applicability, while very vast, is limited to the assumption that particles be structureless point-like objects. Einstein gravity, on the other hand, seems to be doomed by what is known as the Hawking–Penrose theorem, stating that gravitational formation of singularities of matter distributions is an unavoidable consequence for all known fields.

In detail, the HP theorem tells that if some conditions on the energy density tensor hold, as they do for all physical fields, then the overall gravitational pull would never stop and the whole material distribution would inevitably collapse into a singular point. Since the HP theorem, in its essence, is a result of Einstein gravity, a first attempt at avoiding singularities might come for extensions of Einstein gravity, where higher-order derivative field equations would entirely change the shape of the problem. It is indeed the case that at least in some of these extensions there are solutions of the gravitational field equations that are without singularities [1]. Because the theory presented in [1] is the only extension of gravity that is renormalizable [2], one may be led to think that all singularity issues are avoidable also in QFT and therefore that nothing more need be done. Just the same, even if this were to be true, one would still have to rely upon the theory presented in [1]. Granted that such an extension of Einstein gravity is the only one compatible with renormalizability, we still do not know if such a theory is the description of gravity at high energy densities.

In lack of any prediction for a high-energy density gravitational effect, one might wish to stay on the more comfortable environment provided by Einstein gravity. However, it in, the problem of singularity formation would be still far from being solved. So a possible, alternative way out of the problem might be to look for different manners in which Einstein gravity could be enlarged. As Einstein gravity is indeed a remarkably stringent theory, there are not many possibilities. In fact, in the perspective of staying on $(1+3)$-dimensional manifolds, and without tampering with the differential order of the field equations, the only possibility that remains is not to neglect the torsion.

The torsion of space-time is a very natural ingredient of differential geometry, one that is always present in the most general setting. For a comprehensive review on torsion we invite the interested reader to have a look at the introductory chapter on the present Special Issue [3] and at some of the references therein, either for a general review [4] or for the seminal papers [5,6]. As the torsional completion of Einstein gravity has the same differential order of Einstein gravity itself then the HP theorem applies as usual and the presence of torsion can only change the conditions on the energy density. This was the initial hope, as it was worked out by Kerlick in [7]. The initial hope, however, was soon

Citation: . *Universe* **2023**, *9*, 73. https://doi.org/10.3390/universe9020073

Academic Editor: Matteo Luca Ruggiero

Received: 12 January 2023
Revised: 27 January 2023
Accepted: 28 January 2023
Published: 30 January 2023

to be lost, since Kerlick himself pointed out that torsion, instead of solving this problem, was actually worsening it [7,8]. The catch was that the model used was just the torsional completion of Einstein gravitation known as Einstein–Sciama–Kibble theory, in which the torsion–spin interaction is repulsive [9]. While such a repulsive character may at first look appealing, in view of avoiding singularities, in reality repulsive forces give positive energy contributions that increase the gravitational pull and so the tendency to form these singularities. This is the physical reason why Kerlick and Inomata find that singularities are enhanced in [7,8]. In [9] Popławski points out that, despite this, singularities still do not appear at least in some cosmological scenarios.

It is difficult to understand why this should be the case, and instead of a theory in which singularities are avoided despite being enhanced, it would be better to have a theory where singularities are avoided because they are dissipated. To this purpose, one must then violate all of the energy conditions. This can only be done by lowering the energy contributions via the inclusion of negative potentials, and these can only be given by some attractive force. Thus, as paradoxical as it may be, one has to look for attractive torsionally-induced spin–spin interaction in the Dirac equation. This is precisely what many have done, for example in [10–12] and references therein.

The idea is that the ESK theory, of all torsional completions of Einstein gravity, is only the simplest, the one in which torsion is added only implicitly through the curvature tensor in the Lagrangian. That is, from Einstein gravity $L=R(g)$ the ESK theory is given by $L=R(g,Q)$ with Q being the torsion tensor. In [10] the authors take instead $L=R(g,Q)+Q^2$ where Q^2 is formally any of the three scalar contractions of the squared torsion tensor. Thus, with the freedom granted by three new universal constants, the authors of [10] are able to obtain an effective constant for torsion whose sign is undetermined and therefore with the possibility of being inverted with respect to the ESK theory in its simplest form, and as a result they indeed get to avoid the singularity of the early universe.

Others, however, took into account yet another generalization. Because the Q^2 term can be read as a combination of three mass terms, one for each of the three irreducible components of torsion, we could then make sure that its sign be that of a positive mass, then add the three dynamical terms of the three parts of torsion. In reality, however, of the three irreducible components of torsion, only the completely antisymmetric part seems to be acceptable for consistency arguments connected to the metric-compatibility of the manifold and related properties [13–16]. Because a torsion that is completely antisymmetric is the Hodge dual of an axial-vector, the resulting model is that of an axial-vector torsion, namely a model in which torsion is a parity-odd Proca field. In it there may be effects that are due to torsion propagation [17]. Some of these have been studied for the avoidance of singularity [18].

In the present work, we will recall all results of [18] but employing the polar form of spinors [19] so to get cleaner and more powerful results for the problem of singularities in gravitational systems. In addition, we will extend all of the above results to the case of singularities that appear also in absence of gravity and for general systems.

2. The Polar Form of Spinors

To do what we intend to do we first introduce the polar formulation of Dirac spinors, for which we shall always refer to the fundamental work [19] and references therein.

However, we recall here a few basic conventions. The set of Clifford matrices $\boldsymbol{\gamma}_a$ verifying $\{\boldsymbol{\gamma}_a,\boldsymbol{\gamma}_b\}=2\eta_{ab}\mathbb{I}$ are used to set $\boldsymbol{\sigma}_{ab}=\frac{1}{4}[\boldsymbol{\gamma}_a,\boldsymbol{\gamma}_b]$ as the generators of the complex Lorentz transformation. Then $2i\boldsymbol{\sigma}_{ab}=\varepsilon_{abcd}\boldsymbol{\pi}\boldsymbol{\sigma}^{cd}$ is used to define the matrix $\boldsymbol{\pi}$ so that $\mathbb{I},\boldsymbol{\gamma}_a,\boldsymbol{\sigma}_{ab},\boldsymbol{\gamma}_a\boldsymbol{\pi},\boldsymbol{\pi}$ are a basis for the space of 4×4 complex matrices. As a consequence, we have that the identities $\boldsymbol{\gamma}_i\boldsymbol{\gamma}_j\boldsymbol{\gamma}_k=\boldsymbol{\gamma}_i\eta_{jk}-\boldsymbol{\gamma}_j\eta_{ik}+\boldsymbol{\gamma}_k\eta_{ij}+i\varepsilon_{ijkq}\boldsymbol{\pi}\boldsymbol{\gamma}^q$ hold in general.

The product of a complex Lorentz transformation and a unitary phase is the spin transformation $\boldsymbol{S}$ and spinors ψ are defined as what transforms according to these spin transformations. Adjoint spinors are related by $\overline{\psi}=\psi^\dagger\boldsymbol{\gamma}^0$ and with these one can construct all spinor bi-linears and prove their Fierz identities. Before doing that however we will give the following theorem. If $|\overline{\psi}\psi|^2+|i\overline{\psi}\pi\psi|^2\neq 0$ one can always write the spinor field as

$$\psi=\phi e^{-\frac{i}{2}\beta\pi}\boldsymbol{L}^{-1}\begin{pmatrix}1\\0\\1\\0\end{pmatrix} \tag{1}$$

up to $\psi\rightarrow\pi\psi$ and for some $\boldsymbol{L}$ having the structure of a spin transformation whose parameters are recognized to be the spinorial Goldstone fields and where ϕ and β are a real scalar and a real pseudo-scalar field called module and chiral angle. The above is the polar form of the spinor field and with it we can write the spinor bi-linears as

$$i\overline{\psi}\pi\psi=2\phi^2\sin\beta \tag{2}$$
$$\overline{\psi}\psi=2\phi^2\cos\beta \tag{3}$$

as well as

$$\overline{\psi}\gamma^a\pi\psi=2\phi^2 s^a \tag{4}$$
$$\overline{\psi}\gamma^a\psi=2\phi^2 u^a \tag{5}$$

known as spin axial-vector and velocity vector verifying

$$u_{[\mu}s_{\nu]}\sigma^{\mu\nu}\pi\psi+\psi=0 \tag{6}$$
$$u_\mu\gamma^\mu\psi=-s_\mu\gamma^\mu\pi\psi=e^{i\beta\pi}\psi \tag{7}$$

and

$$u_a u^a=-s_a s^a=1 \tag{8}$$
$$u_a s^a=0 \tag{9}$$

known as Fierz identities. Consequently, when the spinor field is re-written in polar form its eight real components are re-arranged so that the two degrees of freedom, the module and chiral angle, remain isolated from the six components that can always be transferred into the frame, and that is what is known to be the Goldstone fields.

To better see the role of these Goldstone fields, we can write them explicitly according to the formula

$$\boldsymbol{L}^{-1}\partial_\mu\boldsymbol{L}=iq\partial_\mu\xi\mathbb{I}+\frac{1}{2}\partial_\mu\xi^{ab}\sigma_{ab} \tag{10}$$

in which ξ and ξ^{ab} are respectively the gauge and space-time Goldstone fields. Together with the gauge potential and spin connection A_μ and $C_{ij\mu}$ we can define

$$q(\partial_\mu\xi-A_\mu)\equiv P_\mu \tag{11}$$
$$\partial_\mu\xi_{ij}-C_{ij\mu}\equiv R_{ij\mu} \tag{12}$$

which can be proven to be real tensors respectively called gauge and space-time tensorial connection. With them

$$\boldsymbol{\nabla}_\mu\psi=\left(\nabla_\mu\ln\phi\mathbb{I}-\tfrac{i}{2}\nabla_\mu\beta\pi-iP_\mu\mathbb{I}-\tfrac{1}{2}R_{ij\mu}\sigma^{ij}\right)\psi \tag{13}$$

is the polar form of the covariant derivative of the spinor field. As a consequence

$$\nabla_\mu s_i=R_{ji\mu}s^j \tag{14}$$
$$\nabla_\mu u_i=R_{ji\mu}u^j \tag{15}$$

as general identities. As is clear, after that the Goldstone fields are transferred into the phase and the frame, they combine with gauge potential and spin connection hence becoming the longitudinal parts of the P_μ and $R_{ij\mu}$ tensors.

To write the Dirac spinor field equations for the polar variables we introduce the pair of dual potentials

$$\Sigma_{ij\mu} = R_{ij\mu} - 2P_\mu u^a s^b \varepsilon_{ijab} \tag{16}$$
$$M^{ab}{}_{\mu} = \tfrac{1}{2} R_{ij\mu} \varepsilon^{ijab} + 2P_\mu u^{[a} s^{b]} \tag{17}$$

and their contractions $\Sigma_{\alpha\nu\pi} g^{\nu\pi} = \Sigma_\alpha$ and $M_{\alpha\nu\pi} g^{\nu\pi} = M_\alpha$ in terms of which we have

$$\nabla_\mu \ln \phi^2 + \Sigma_\mu + 2m s_\mu \sin\beta = 0 \tag{18}$$
$$\nabla_\mu \beta - 2XW_\mu + M_\mu + 2m s_\mu \cos\beta = 0 \tag{19}$$

with m being the mass of the spinor, X the spin–torsion coupling constant and W_μ the torsion axial-vector [3,13–16].

These matter field equations have to be complemented with the geometric field equations given by

$$\nabla_\rho (\partial W)^{\rho\mu} + M^2 W^\mu = 2X\phi^2 s^\mu \tag{20}$$

where M is the mass of torsion together with

$$\begin{aligned} R^{\rho\sigma} - \tfrac{1}{2} R g^{\rho\sigma} - \Lambda g^{\rho\sigma} = \tfrac{1}{2} [\tfrac{1}{4} F^2 g^{\rho\sigma} - F^{\rho\alpha} F^{\sigma}{}_{\alpha} + \\ + \tfrac{1}{4} (\partial W)^2 g^{\rho\sigma} - (\partial W)^{\sigma\alpha} (\partial W)^{\rho}{}_{\alpha} + \\ + M^2 (W^\rho W^\sigma - \tfrac{1}{2} W^2 g^{\rho\sigma}) + \\ + \phi^2 [P^\rho u^\sigma + P^\sigma u^\rho + \\ + (\nabla^\rho \beta/2 - XW^\rho) s^\sigma + (\nabla^\sigma \beta/2 - XW^\sigma) s^\rho - \\ - \tfrac{1}{4} R_{\alpha\nu}{}^{\sigma} s_\kappa \varepsilon^{\rho\alpha\nu\kappa} - \tfrac{1}{4} R_{\alpha\nu}{}^{\rho} s_\kappa \varepsilon^{\sigma\alpha\nu\kappa}]] \end{aligned} \tag{21}$$

for the gravitational field equations and

$$\nabla_\sigma F^{\sigma\mu} = 2q\phi^2 u^\mu \tag{22}$$

for the electrodynamic field equations.

As a concluding remark we also give the expression

$$\begin{aligned} L = -\tfrac{1}{4} (\partial W)^2 + \tfrac{1}{2} M^2 W^2 - R - 2\Lambda - \tfrac{1}{4} F^2 \\ + \tfrac{i}{2} (\overline{\psi} \gamma^\mu \boldsymbol{\nabla}_\mu \psi - \boldsymbol{\nabla}_\mu \overline{\psi} \gamma^\mu \psi) - \\ - XW_\sigma \overline{\psi} \gamma^\sigma \boldsymbol{\pi} \psi - m \overline{\psi} \psi \end{aligned} \tag{23}$$

which is the Lagrangian of the full theory.

The field equations above are just the polar form of the field equations one would have in the torsional completion of gravity with electrodynamics [3].

We have maintained the Lagrangian in its standard form to easily compare it to the Lagrangians of known theories, as we will do next.

3. Singularities Avoidance in Spinorial Average

We begin a more in-depth treatment of the theory, useful for future developments, by introducing the concept of effective approximation. As we have just seen, torsion can be massive, and since we have not observed it so far, it may also be argued that its mass should be quite large indeed. So it is reasonable to assume that there might be regimes in which the mass term dominates all dynamical terms. When the dynamical terms can be suppressed in favour of the mass term we are in the so-called effective approximation. The torsion field equations reduce to

$$M^2 W^\mu \approx 2X\phi^2 s^\mu \tag{24}$$

and so by substituting torsion everywhere in terms of the spin we should expect torsion–spin interactions to convert into spin–spin interactions of non-linear character.

When this is done in the Lagrangian we get

$$\begin{aligned} L = & -R - 2\Lambda - \tfrac{1}{4}F^2 \\ & +\tfrac{i}{2}(\overline{\psi}\gamma^\mu \boldsymbol{\nabla}_\mu \psi - \boldsymbol{\nabla}_\mu \overline{\psi}\gamma^\mu \psi) - \\ & -\tfrac{1}{2}\tfrac{X^2}{M^2}\overline{\psi}\gamma^\sigma \pi \psi \overline{\psi}\gamma_\sigma \pi \psi - m\overline{\psi}\psi \end{aligned} \tag{25}$$

or by employing Fierz re-arrangements

$$\begin{aligned} L = & -R - 2\Lambda - \tfrac{1}{4}F^2 \\ & +\tfrac{i}{2}(\overline{\psi}\gamma^\mu \boldsymbol{\nabla}_\mu \psi - \boldsymbol{\nabla}_\mu \overline{\psi}\gamma^\mu \psi) + \\ & +\tfrac{1}{2}\tfrac{X^2}{M^2}\overline{\psi}\gamma^\sigma \psi \overline{\psi}\gamma_\sigma \psi - m\overline{\psi}\psi \end{aligned} \tag{26}$$

or yet again

$$\begin{aligned} L = & -R - 2\Lambda - \tfrac{1}{4}F^2 \\ & +\tfrac{i}{2}(\overline{\psi}\gamma^\mu \boldsymbol{\nabla}_\mu \psi - \boldsymbol{\nabla}_\mu \overline{\psi}\gamma^\mu \psi) + \\ & +\tfrac{1}{2}\tfrac{X^2}{M^2}|i\overline{\psi}\pi\psi|^2 + \tfrac{1}{2}\tfrac{X^2}{M^2}|\overline{\psi}\psi|^2 - m\overline{\psi}\psi \end{aligned} \tag{27}$$

revealing that torsionally-induced spin–spin interactions are attractive. In fact if we were to further split into the left-handed and right-handed projections then we would see that such an attraction takes place between these two chiral parts. This is expected, since the above Lagrangian is just the Lagrangian of the Nambu–Jona-Lasinio model [20].

The effective approximation within the gravitational field Equation (21) gives instead the expression

$$\begin{aligned} & R^{\rho\sigma} - \tfrac{1}{2}Rg^{\rho\sigma} - \Lambda g^{\rho\sigma} = \tfrac{1}{2}[\tfrac{1}{4}F^2 g^{\rho\sigma} - F^{\rho\alpha}F^\sigma{}_\alpha + \\ & +\phi^2(P^\rho u^\sigma + P^\sigma u^\rho + \nabla^\rho\beta/2s^\sigma + \nabla^\sigma\beta/2s^\rho - \\ & -\tfrac{1}{4}R_{\alpha\nu}{}^\sigma s_\kappa \varepsilon^{\rho\alpha\nu\kappa} - \tfrac{1}{4}R_{\alpha\nu}{}^\rho s_\kappa \varepsilon^{\sigma\alpha\nu\kappa} + 2\tfrac{X^2}{M^2}\phi^2 g^{\rho\sigma})] \end{aligned} \tag{28}$$

while in the Dirac Equation (19) we get

$$\nabla_\mu\beta - 2(2\tfrac{X^2}{M^2}\phi^2 - m\cos\beta)s_\mu + M_\mu = 0 \tag{29}$$

with all other equations remaining unchanged. Contracting (28) and employing the scalar product of (29) along the spin we get the particularly simple

$$R + 4\Lambda = -2\tfrac{X^2}{M^2}\phi^4 - m\phi^2\cos\beta \tag{30}$$

which can be substituted back into the initial field equations to give the gravitational field equations

$$\begin{aligned} & R^{\rho\sigma} = -\Lambda g^{\rho\sigma} + \tfrac{1}{2}[\tfrac{1}{4}F^2 g^{\rho\sigma} - F^{\rho\alpha}F^\sigma{}_\alpha + \\ & +\phi^2(P^\rho u^\sigma + P^\sigma u^\rho + \nabla^\rho\beta/2s^\sigma + \nabla^\sigma\beta/2s^\rho - \\ & -\tfrac{1}{4}R_{\alpha\nu}{}^\sigma s_\kappa \varepsilon^{\rho\alpha\nu\kappa} - \tfrac{1}{4}R_{\alpha\nu}{}^\rho s_\kappa \varepsilon^{\sigma\alpha\nu\kappa} - m\cos\beta g^{\rho\sigma})] \end{aligned} \tag{31}$$

so that all spin–spin interactions disappear as source for the dynamics of the Ricci tensor.

This however does not mean that torsion has no impact, as we will now see.

The gravitational field equations written for the Ricci tensor are the most adapted to investigate the dominant energy condition since this is given as

$$R^{\rho\sigma}u_{\rho}u_{\sigma}\geqslant 0 \tag{32}$$

with u^{α} a normalized time-like vector. Imposing this condition on (31) we obtain

$$\begin{aligned}&\phi^2(-2\tfrac{X^2}{M^2}\phi^2-s^{\mu}\nabla_{\mu}\beta/2+\tfrac{1}{4}R_{ijb}s_a\varepsilon^{ijba}-\\&-\tfrac{1}{4}R_{\alpha\nu\sigma}u^{\sigma}u_{\rho}s_{\kappa}\varepsilon^{\alpha\nu\rho\kappa}+\tfrac{1}{2}m\cos\beta)-\Lambda\geqslant 0\end{aligned} \tag{33}$$

where (29) was again used and in which electrodynamics has been neglected for simplicity.

This expression is still quite general for the purpose that we have in mind.

In the treatment done by Kerlick and further generalizations [7–12], this is the moment where some additional assumptions are made by these authors. They are either specific symmetries or special properties for the material distributions. Because most of these assumptions appear ad hoc and some of them even unphysical, we will assume none of them in the following. In what is next we employ the only assumption that is physically adequate and very general, namely for gravitating systems such as the Big Bang and Black Holes the random distribution of particles allows the spin average to vanish all terms in which spin appears linearly. This means that (33) becomes

$$-2\tfrac{X^2}{M^2}\phi^4+\tfrac{1}{2}m\phi^2\cos\beta-\Lambda\geqslant 0 \tag{34}$$

in spin-average. When no matter field is present we have that (34) reduces to $-\Lambda\geqslant 0$ which is obviously invalid if the cosmological constant is positive (as is in our convention). This is what we expect for a universe in continuous accelerated expansion. Neglecting the cosmological constant

$$-4\tfrac{X^2}{M^2}\phi^2+m\cos\beta\geqslant 0 \tag{35}$$

as easy to see. In absence of torsion this is always verified whenever $m\cos\beta\geqslant 0$ as is always the case for a standard matter distribution (for which the chiral angle is assumed to vanish albeit not in a justified way [19]). This is what recovers the common arguments about the gravitational formation of singularities and their inevitability. In presence of torsion, however, we have that for densities that are larger and larger, regardless of the values of the chiral angle, we can always approximate

$$-\tfrac{X^2}{M^2}\geqslant 0 \tag{36}$$

again as is clear. This is of course always violated in a very dramatic way (see for instance [18]). So failing the hypothesis there is no implication from the HP theorem.

The conclusion is that for systems given by statistically distributed spinor fields, such as the Big Bang or Black Holes, the formation of gravitational singularities is not a necessity.

It is instructive to look back at the previous attempts in a comparative way and draw the differences. That is, it is now the time to ask, given that the torsionally-induced spin–spin interaction renders the gravitational formation of singularities not a necessary occurrence, how is it that in the past previous attempts did not come up with the same result? The answer is that there are many different ways to let torsion take its place in the Lagrangian: one, the simplest, is via the minimal coupling, as done by the Kerlick school [7,9]; another is via the generalization that takes into account also explicit squared torsion terms in the Lagrangian, as in [10]; the final is to include also all squared derivative torsion terms within the most general dynamically-consistent Lagrangian [18]. In the first one, which is just the ESK gravitation, the torsionally-induced spin–spin interaction has a constant equal to 3/16 when measured in natural units, since its value is rigidly linked to that of the Newton constant; in the generalizations of the ESK gravity the constant has an indefinite value, as it results from combining the three constants that appear in front of each of the three squared torsion terms that may be added; in the most general case the constant is $-X^2/M^2$ when the effective approximation is implemented because in a theory in which torsion propagates we must require the energy density and the mass to be positive. Therefore the gravitational formation of singularities that cannot be avoided in Einstein gravity, and that is worsened in ESK gravity, may become avoidable in further generalizations, and it is certainly avoidable in the most general case of propagating torsion (in effective approximation).

The mechanism for gravitational singularity avoidance can only be effectively enforced in a theory of propagating torsion, that is in the most general physical situation.

The interpretation we can give is that in such a case the torsionally-induced spin–spin interactions turn attractive, which have a negative potential, and thus the overall energy contribution is decreased. If the negative potential becomes dominant, the total energy turns negative, then the curvature reverts sign, and gravity becomes repulsive.

It is the fact that torsion is an attraction between chiral parts what causes the relaxation of the attraction of the gravitational field of the overall material distribution.

4. Singularities Avoidance for Single Particles

In the previous section we have investigated the problem of the singularity formation in gravitational systems, constituted by statistically distributed spinor fields. The vanishing of the spin-average in linear terms was our only hypothesis.

What if we have no random distributions? Worse, what if we have a single particle as in QFT? In this case gravitation would always be negligible and we could not use the above argument to avoid singularities, so what is there to be done in this case?

Of course, the full analysis has to be re-done without the help of any of the previous hypotheses, and when only torsion appears in the dynamics of spinor fields. As a consequence, in the following we will consider only torsion and Dirac spinor field equations and see what information we can extract for the singularity problem. Torsion will still be taken in its effective approximation and for the Dirac spinor field equations we will do some formal manipulations to put them in a form that is more suitable for our purposes.

One of the advantages of the polar formulation of the Dirac theory is that, with it, it is very easy to see that the second-order derivative field equations are given by

$$\begin{aligned}&\nabla^{\mu}(\phi^2\nabla_{\mu}\beta)-(8X^2/M^2\phi^2 m\sin\beta-\\&-2XW^{\mu}\Sigma_{\mu}-\nabla_{\mu}M^{\mu}+M_{\mu}\Sigma^{\mu})\phi^2=0\end{aligned}\tag{37}$$

as a continuity equation for the chiral angle and

$$\begin{aligned}&|\nabla\beta/2|^2-m^2-\phi^{-1}\nabla^2\phi+\tfrac{1}{4}(-2\nabla_{\mu}\Sigma^{\mu}+\\&+\Sigma^{\mu}\Sigma_{\mu}-M_{\mu}M^{\mu}+4XM_{\mu}W^{\mu}-4X^2W^{\mu}W_{\mu})=0\end{aligned}\tag{38}$$

as a Hamilton–Jacobi equation for the module. Singularities form at high density, and so what really interests us is the Hamilton–Jacobi equation for the module (38), for which we are now going to assume torsion in its effective approximation, resulting into the simpler expression

$$\begin{aligned}&\nabla^2\phi-4\tfrac{X^4}{M^4}\phi^5-2\tfrac{X^2}{M^2}(M_{\mu}s^{\mu})\phi^3+(\tfrac{1}{2}\nabla_{\mu}\Sigma^{\mu}-\\&-\tfrac{1}{4}\Sigma^{\mu}\Sigma_{\mu}+\tfrac{1}{4}M_{\mu}M^{\mu}-|\nabla\beta/2|^2+m^2)\phi=0\end{aligned}\tag{39}$$

in which as expected non-linearities have arisen. In it, the quintic term is always negative, so that the highest-order self-interaction is always attractive. Then the cubic term is negative so long as $M_{\mu}s^{\mu}>0$ and in this case also the lower-order self-interaction is attractive. The linear term is positive when $2\nabla_{\mu}\Sigma^{\mu}-\Sigma^{\mu}\Sigma_{\mu}+M_{\mu}M^{\mu}-|\nabla\beta|^2+4m^2>0$ and in this case it behaves as a regular mass term. In case of singularity formation, we should expect the density to increase, leading to larger values of the module. There is no need to evaluate the signs of these two terms because both are negligible with respect to the highest-order term and therefore we end up having the definite form

$$\nabla^2\phi - 4\frac{X^4}{M^4}\phi^5 = 0 \tag{40}$$

which does have a non-singular behaviour. In fact it is a straightforward operation to see that in stationary cases

$$\vec{\nabla}\cdot\vec{\nabla}\phi + 4\frac{X^4}{M^4}\phi^5 = 0 \tag{41}$$

and for spherical symmetry

$$\frac{1}{r^2}\partial_r(r^2\partial_r\phi) + 4\frac{X^4}{M^4}\phi^5 = 0 \tag{42}$$

which is exactly one of the instances of the Lane–Emden equation that can be solved. The explicit solution is

$$\phi = \sqrt{\frac{3M^4}{4X^4 + 3M^4r^2}} \tag{43}$$

which is non-singular. Notice that when the torsion constant tends to zero then the $1/r$ singular behaviour is recovered as a general feature of the solution. Therefore we can say that it is precisely the presence of torsion what forbids singularities.

The formation of singularities recovered with a vanishing torsion constant can also be seen as due to a too large torsion mass. It is therefore tempting to speculate that it is the torsion mass what gives the scale of the cut-off beyond which divergences are negligible in QFT. In more detail, renormalization schemes are based on the idea for which ultraviolet divergences appear because we do not actually know what happens at very high energy, but we postulate that in these regimes physics is different and free of singularities. As it stands one may think that this new physics is just what we would have always had, had we never neglected torsion from the beginning.

How would QFT change if we were to employ the non-singular solutions we have in presence of torsion? Would we still need any of the known renormalization protocols?

We conclude this section by noticing that if in Equation (39) both conditions on the cubic and linear terms hold then (39) has the structure of a solitonic equation.

5. One Specific Circumstance: Walking Droplets

Let us focus more on (39) and ask the following question: what happens if in it the linear term is $2\nabla_\mu\Sigma^\mu - \Sigma^\mu\Sigma_\mu + M_\mu M^\mu - |\nabla\beta|^2 + 4m^2 < 0$ instead? For such a situation the linear term becomes that of a mass with imaginary value. At high density solutions are still regular. However far from the origin there no longer is an exponentially decreasing behaviour with radial distance but rather an oscillating behaviour. Solutions would no longer be solitons but something more similar to a material distribution with a solid core surrounded by wavelets. It is tempting to see in these the bouncing droplets used for hydrodynamic analogs of quantum systems [21,22].

Let us try to give an explicit example. Consider $P_t = m$ and a tensorial connection with component

$$R_{r\varphi\varphi} = -r|\sin\theta|^2 \tag{44}$$

$$R_{r\theta\theta} = -r \tag{45}$$

$$R_{\varphi\theta t} = -2\varepsilon r^2 \sin\theta \tag{46}$$

where ε is a constant due to the Riemann curvature being equal to zreo. These are compatible with $s_r = 1$ and $u_t = 1$ from the constraints (14) and (15). The dual potentials (16) and (17) are given by the expressions

$$\Sigma_r = 2/r \tag{47}$$

$$M_r = -2(m+\varepsilon) \tag{48}$$

from which we can now see that for $\beta = 0$ they give

$$s^\mu M_\mu = 2(m+\varepsilon) \tag{49}$$

$$2\nabla_\mu \Sigma^\mu - \Sigma^\mu \Sigma_\mu + M^\mu M_\mu + 4m^2 = -4(2m+\varepsilon)\varepsilon \tag{50}$$

to be discussed. If ε is negative with $m > |\varepsilon| > 0$ we obtain the self-interaction to be attractive and the mass term to be a real mass term, so that the conditions to have solitonic solutions are ensured. If instead ε is positive we always obtain a self-interaction of attractive character but now with mass of imaginary type, so that the solution has the peripheral oscillations proper of wavelets. Equation (38) in this case is in fact given according to

$$\nabla^2 \phi - 4\tfrac{X^4}{M^4}\phi^5 - 4\tfrac{X^2}{M^2}(m+\varepsilon)\phi^3 - (2m+\varepsilon)\varepsilon\phi = 0 \tag{51}$$

so that at high densities it reduces to (40) with solution (43) but at lower densities it becomes

$$\nabla^2 \phi - (2m+\varepsilon)\varepsilon\phi = 0 \tag{52}$$

which in the stationary spherical case has

$$\phi = \frac{\sin\left(r\sqrt{\varepsilon|\varepsilon+2m|}\right)}{r\sqrt{\varepsilon|\varepsilon+2m|}} \quad \text{and} \quad \phi = \frac{\cos\left(r\sqrt{\varepsilon|\varepsilon+2m|}\right)}{r\sqrt{\varepsilon|\varepsilon+2m|}} \tag{53}$$

as solutions. This is the solution of a second-order derivative equation for the module with no chiral angle and as such it does not strictly describe a complete solution for the spinor field. It does however provide valuable insight into the dynamics of matter distributions that behave as bouncing droplets. The droplet itself would be the high-density part and the surrounding bath would be the low-density part of the module of the material field.

Could this really mean that the behaviour of quantum particles, so accurately mimicked by a bouncing droplet, is ultimately the result of the presence of torsionally induced tensions within matter distributions?

Quantum mechanics has always had difficulties in interpreting what a particle actually is. From Louis de Broglie on, the idea of particles behaving as waves was ubiquitously accepted, although still today we have no idea how exactly such a particle-wave duality really describes what happens in nature. It was de Broglie in a number of articles during the 1920s who put forward the possible interpretation known as pilot-wave model, with physical waves propagating in space and then guiding all particles around. Later known also as the double-solution model due to the fact that in it waves and particles are solution of two different equations [23,24], this interpretation has also been re-discovered by Bohm [25], and it is nowadays one of the few models still compatible with all restrictions imposed by Bell-like inequalities, from the non-locality to the contextuality constraints.

The fact that in the dBB interpretation of quantum mechanics contextuality, and more specifically non-locality, can be compatible with relativistic invariance has been recently shown be presenting a fully covariant version (with spin) of the dBB model. The key issue was to convert the Dirac theory into a consistent hydrodynamic formulation, which can be done by employing the polar form of spinor fields [26].

In this hydrodynamic formulation, quantum mechanics is naturally written in a manner that makes it easy to visualize and interpret the motion of particles as the result of an underlying wave dynamics, precisely as mimicked by bouncing droplets.

In the dBB model, waves and particles are seen as solution of two fields equations, a linear equation which possesses wave solutions and a non-linear equation which has soliton solutions. However, one is left with the uncomfortable feeling that it is inelegant to have such a duplicity, and on top of it one may still wonder why one equation should be linear and the other should be non-linear. A possible escape from this conundrum might come from the simple observation that extended waves and localized solitons do not need to be solutions of two different field equations, but just two different solutions of the same field equation with the wave behaviour emerging for low densities and the solitonic behaviour emerging for high densities. In this way, the most elegant solution to the problem raised by de Broglie one century ago would merely be to consider spinorial matter field equations in presence of torsion. As also commented in the previous section, the torsion mass, providing a natural scale for the matter distribution, would give the scale beyond which the quantum field would start to convert its wave behaviour into a localized bulk of matter. Then, the size of the particle would give the cut-off scale presently used in all renormalization schemes of QFT.

Is this interpretation sensible? One good aspect is that all it seems to need is the presence of torsion, which is a natural ingredient of the most general Dirac spinorial field theory. On the other hand, more work must necessarily be done to see whether complete exact solutions to the non-linear field equation can actually be found.

We are confident that very soon we might include some preliminary assessment on this problem in a following paper.

6. Conclusions

In this paper, we have considered the polar formulation of the Dirac spinor field theory fully coupled to geometric general backgrounds to face the problem of singularities.

As a first step, we have discussed the effective approximation of torsion and the fact that within this approximation the torsion can be effectively substituted in terms of the spin axial-vector in all field equations. The result is torsionally-induced spin–spin interactions of attractive character, exactly as we have in the Nambu–Jona-Lasinio model. We have discussed how, for the Hawking–Penrose theorem, this model provides the environment to violate the dominant energy condition, resulting in the avoidance of singularities for gravitational systems constituted by a large number of particles. We have compared our results with previous attempts and given a physical justification.

For single particles, where all gravitational fields can be neglected, such singularity avoidance has been discussed by using the field equations for the module of the matter distribution. We have seen that second-order differential equations of Hamilton-Jacobi type for the module, in the effective approximation of torsion, and for large densities of matter, become one of the Lane–Emden equations for which an exact solution exists, and it is found to be non-singular. General considerations in QFT related to some ultraviolet divergences have been discussed in terms of a cut-off scale that is derived from the torsion mass.

Finally, general comments about solitons and bouncing droplets have been given in terms of covariant conditions. The relevance of these results for quantum mechanics, like the de Broglie double-solution model or the Bohmian dynamics, has also been addressed.

We are certain of the fact that the foundations of physics as a whole could benefit from the effects that would arise if torsion were considered as a fundamental field.

Funding: This research received no external funding.

Data Availability Statement: This research has no data available.

Conflicts of Interest: The author declares no conflict of interest.

Notes

[1] Dark Energy is also a problem that has to be faced, but in this case the treatment of the zero-point energy and the phase-shift due to spontaneous symmetry breaking are more pertinent to the domain of high-energy particle physics, and as such Dark Energy is more of an issue that belongs to the interface of cosmology and the fundamental constituents of matter.

References

1. Fabbri, L. A Note on Singularity Avoidance in Fourth-Order Gravity. *Universe* **2022**, *8*, 51. [CrossRef]
2. Stelle, K.S. Renormalization of higher-derivative quantum gravity. *Phys. Rev. D* **1977**, *16*, 953. [CrossRef]
3. Fabbri, L. Fundamental Theory of Torsion Gravity. *Universe* **2021**, *7*, 305. [CrossRef]
4. Hehl, F.W.; Von Der Heyde, P.; Kerlick, G.D.; Nester, J.M. General relativity with spin and torsion: Foundations and prospects. *Rev. Mod. Phys.* **1976**, *48*, 393. [CrossRef]
5. Sciama, D.W. The Physical Structure of General Relativity. *Rev. Mod. Phys.* **1964**, *36*, 463. [CrossRef]
6. Kibble, T.W.B. Lorentz Invariance and the Gravitational Field. *J. Math. Phys.* **1961**, *2*, 212. [CrossRef]
7. Kerlick, G.D. Cosmology and particle pair production via gravitational spin-spin interaction in the Einstein-Cartan-Sciama-Kibble theory of gravity. *Phys. Rev. D* **1975**, *12*, 3004. [CrossRef]
8. Inomata, A. Effect of the Selfinduced Torsion of the Dirac Sources on Gravitational Singularities. *Phys. Rev. D* **1978**, *18*, 3552. [CrossRef]
9. Poplawski, N.J. Nonsingular, big-bounce cosmology from spinor-torsion coupling. *Phys. Rev. D* **2012**, *85*, 107502. [CrossRef]
10. Magueijo, J.; Zlosnik, T.G.; Kibble, T.W.B. Cosmology with a spin. *Phys. Rev. D* **2013**, *87*, 063504. [CrossRef]
11. Khriplovich, I.B.; Rudenko, A.S. Gravitational four-fermion interaction and dynamics of the early Universe. *J. High Energy Phys.* **2013**, *1311*, 174. [CrossRef]
12. Alexander, S.; Bambi, C.; Marciano, A.; Modesto, L. Fermi-bounce Cosmology and scale invariant power-spectrum. *Phys. Rev. D* **2014**, *90*, 123510. [CrossRef]
13. Fabbri, L. On a completely antisymmetric Cartan torsion tensor. *Annales Fond. Broglie.* **2007**, *32*, 215.
14. Hayashi, K. Restrictions on gauge theory of gravitation. *Phys. Lett. B* **1976**, *65*, 437. [CrossRef]
15. Audretsch, J.; Lämmerzahl, C. Constructive axiomatic approach to spacetime torsion. *Class. Quant. Grav.* **1988**, *5*, 1285. [CrossRef]
16. Lämmerzahl, C.; Macias, A. On the dimensionality of space-time. *J. Math. Phys.* **1993**, *34*, 4540. [CrossRef]
17. Carroll, S.M.; Field, G.B. Consequences of propagating torsion in connection dynamic theories of gravity. *Phys. Rev. D* **1994**, *50*, 3867. [CrossRef] [PubMed]
18. Fabbri, L. Singularity-free spinors in gravity with propagating torsion. *Mod. Phys. Lett. A* **2017**, *32*, 1750221. [CrossRef]
19. Fabbri, L. Spinors in Polar Form. *Eur. Phys. J. Plus* **2021**, *136*, 354. [CrossRef]
20. Nambu, Y.; Jona-Lasinio, G. Dynamical Model of Elementary Particles Based on an Analogy with Superconductivity. *Phys. Rev.* **1961**, *122*, 345. [CrossRef]
21. Moláček, J.; Bush, J. Drops bouncing on a vibrating bath. *J. Fluid Mech.* **2013**, *727*, 582.
22. Couder, Y.; Fort, E. Single-Particle Diffraction and Interference at a Macroscopic Scale. *Phys. Rev. Lett.* **2006**, *97*, 154101. [CrossRef] [PubMed]
23. de Broglie, L. Les principes de la nouvelle mécanique ondulatoire. *J. Phys. Radium* **1926**, *7*, 321. [CrossRef]
24. de Broglie, L. La mécanique ondulatoire. In *Mémorial des Sciences Physiques;* Gauthier-Villars: Paris, France, 1928.
25. Bohm, D. A suggested interpretation of the quantum theory in terms of hidden variables. *Phys. Rev.* **1952**, *85*, 166. [CrossRef]
26. Fabbri, L. de Broglie-Bohm Formulation of Dirac Fields. *Found. Phys.* **2022**, *52*, 116. [CrossRef]

Article

Quantum Hydrodynamics of Spinning Particles in Electromagnetic and Torsion Fields

Mariya Iv. Trukhanova [1,2] and Yuri N. Obukhov [2,*]

1 Faculty of Physics, M. V. Lomonosov Moscow State University, Leninskie Gory, 119991 Moscow, Russia; trukhanova@physics.msu.ru
2 Theoretical Physics Laboratory, Nuclear Safety Institute, Russian Academy of Sciences, B. Tulskaya 52, 115191 Moscow, Russia
* Correspondence: obukhov@ibrae.ac.ru

Abstract: We develop a many-particle quantum-hydrodynamical model of fermion matter interacting with the external classical electromagnetic and gravitational/inertial and torsion fields. The consistent hydrodynamical formulation is constructed for the many-particle quantum system of Dirac fermions on the basis of the nonrelativistic Pauli-like equation obtained via the Foldy–Wouthuysen transformation. With the help of the Madelung decomposition approach, the explicit relations between the microscopic and macroscopic fluid variables are derived. The closed system of equations of quantum hydrodynamics encompasses the continuity equation, and the dynamical equations of the momentum balance and the spin density evolution. The possible experimental manifestations of the torsion in the dynamics of spin waves is discussed.

Keywords: quantum hydrodynamics; spin; gauge gravity; torsion

Citation: Trukhanova, M.I.; Obukhov, Y.N. Quantum Hydrodynamics of Spinning Particles in Electromagnetic and Torsion Fields. *Universe* **2021**, *7*, 498. https://doi.org/10.3390/universe7120498

Academic Editor: Marcello Abbrescia

Received: 22 November 2021
Accepted: 13 December 2021
Published: 15 December 2021

1. Introduction

Spin (an intrinsic angular momentum) is an important physical property of matter, associated with rotation, which is considered as the source of the gravitational field [1–10] in the framework of the gauge gravity approach. The spin angular momentum, together with the energy-momentum current, determines the geometrical structure of the spacetime manifold and predicts nontrivial post-Riemannian deviations from Einstein's general relativity (GR) theory. The development of the gravitational theory with torsion has a long history, going back to 1922 when Élie Cartan [1] introduced the corresponding geometrical formalism. Since the 1960s, the interest in the theory of gravitation with spin and torsion based on the Riemann–Cartan geometry had considerably grown, and the consistent formalism was developed [2–7]. A further generalization of the gauge gravitational theory takes into account the additional microstructural physical properties of matter (encompassing the intrinsic shear and the dilation currents along with the spin) as the sources of gravity, that results in the extension of the spacetime structure to the metric-affine geometry [8]. An exhaustive overview of historic developments can be found in [9,10].

The study of dynamics of the spinning matter on the Riemann–Cartan spacetime represents a nontrivial problem which is of interest both theoretically and experimentally. Quoting Einstein [11], "... the question whether this continuum has a Euclidean, Riemannian, or any other structure is a question of physics proper which must be answered by experience, and not a question of a convention to be chosen on grounds of mere expediency." It is now well established that the spacetime torsion can only be detected with the help of the spin [12–14]. The early theoretical analysis of the possible experimental manifestations of the torsion field at low energies can be found in [15].

By noticing that the spin and the energy-momentum tensors are the two Noether currents for the Poincaré group, one can develop a natural formulation of the theory of gravity with torsion as a local gauge theory for the Poincaré spacetime symmetry [16–18]. This underlies the previous study of the quantum dynamics of a Dirac particle in the Poincaré gauge gravitational field [19,20], where both the minimal and nonminimal coupling of the Dirac fermion with the electromagnetic and the gauge gravitational fields was comprehensively analysed for an arbitrary spacetime geometry with the curvature and torsion. It was demonstrated that the Pauli-like equation for the spinning particles contains new torsion-dependent terms which could give rise to the physical effects competing with the electromagnetic ones.

The study of the spin-torsion coupling obviously fits into the general context of the theoretical and experimental research of spin-dependent long-range forces [21–40]. Certain extensions of the Standard Model in the high energy particle physics predict the existence of new particles, in particular, of the light pseudoscalar bosons (such as goldstones, axions [41–43], arions [44,45], etc.) that may give rise to the spin–spin interactions of various kinds. An exchange via such light bosons between two fermions is qualitatively described by a magnetic dipole-dipole type potential. Different methods were proposed to detect these spin–spin interactions, including ferromagnetic detectors with a highly sensitive two-channel UHF receiver [45], paramagnetic salt with a dc SQUID used in a rotating copper mass [22,23], examining the hyperfine resonances for $^9Be^+$ ions stored in the Penning ion trap [24], and even treating the Earth as a polarized spin source [25]. In the recent experiment [26], the transversely polarized slow neutrons were used in an attempt to observe a possible spin rotation of neutrons that traversed a meter of liquid ^4He under the action of the torsion field. Ultracold neutrons provide a convenient tool, with the quantum gravitational states of ultracold neutrons being sensitive to the post-Riemannian contributions [38]. On the theoretical side, the covariant multipolar technique was used for the analysis of the equations of motion of test bodies with spin for a very general class of gravitational theories with the minimal and nonminimal coupling [27,28]. An interesting practical realization of theoretical findings has been recently proposed as a new Gravity Probe Spin space mission using mm-scale ferromagnetic gyroscopes in orbit around the Earth [29]. Typically, the predicted spin-torsion effects are expected to be quite small and difficult to observe experimentally, however, one can set the experimental bounds on the spin-torsion coupling constants and on the torsion field as well [30–40].

Here we for the first time develop the quantum hydrodynamics for the many-particle system of massive Dirac fermion spin-1/2 particles interacting with external electromagnetic, metric gravitational/inertial and torsion fields on the basis of the earlier analysis [19].

This article is organized as follows. In Section 2, we formulate a Pauli-type one-particle equation for a Dirac fermion moving on the background of gravitational and electromagnetic fields. In Section 3, we introduce the many-particle Pauli-like equation and construct the many-particle quantum hydrodynamics for the non-relativistic particles in the external classical fields. We derive the system of hydrodynamical equations and analyze the structure of force fields in these equations. In Section 4, we apply Madelung's decomposition for the spinor wave function to get the basic physical quantities in macroscopic form. Finally, in Section 5 we discuss possible experimental manifestations of the results obtained in this article, and Section 6 contains our conclusions.

Our basic conventions and notations are the same as in Reference [8]. In particular, the world indices are labeled by Latin letters $i, j, k, \dots = 0, 1, 2, 3$ (for example, the local spacetime coordinates x^i and the holonomic coframe dx^i), whereas we reserve Greek letters for tetrad indices, $\alpha, \beta, \dots = 0, 1, 2, 3$ (e.g., the anholonomic coframe $\vartheta^\alpha = e_i^\alpha dx^i$). In order to distinguish separate tetrad indices we put hats over them. Finally, spatial indices are denoted by Latin letters from the beginning of the alphabet, $a, b, c, \dots = 1, 2, 3$. The metric of the Minkowski spacetime reads $g_{\alpha\beta} = \mathrm{diag}(c^2, -1, -1, -1)$, and the totally antisymmetric Levi-Civita tensor $\eta_{\alpha\beta\mu\nu}$ has the only nontrivial component $\eta_{\hat{0}\hat{1}\hat{2}\hat{3}} = c$, so that $\eta_{\hat{0}abc} = c\varepsilon_{abc}$ with the three-dimensional Levi-Civita tensor ε_{abc}. The spatial components of the tensor objects are raised and lowered with the help of the Euclidean 3-dimensional metric δ_{ab}.

2. Pauli Equation for the System with Spin-Torsion Coupling

2.1. Poincaré Gauge Gravity Theory: The Basics

Recalling that the Standard Model in the fundamental particle physics is formulated as a gauge theory for the internal unitary symmetry groups, one may say that, apart from the gravitational interaction, the gauge-theoretic approach underlies the modern physics. There exist, however, a natural extension of Einstein's GR that is based on the Poincaré symmetry group $G = T_4 \rtimes SO(1,3)$, the semi-direct product of the four-parameter translation group T_4 and the six-parameter Lorentz group $SO(1,3)$, with the energy-momentum current and the spin angular momentum current as the sources of the gravitational field [9,10,16–18].

The gauge fields act as mediators of physical interactions for the fermion matter source. Specializing to the electromagnetic and gravitational interactions, in the framework of the standard Yang–Mills–Sciama–Kibble approach [10], one then describes electromagnetism by the $U(1)$ gauge field potential A_i, and in similar way, one describes gravity by the Poincaré gauge potentials e_i^α and $\Gamma_i{}^{\alpha\beta}$. Geometrically, the 4 potentials e_i^α of the translation subgroup T_4 are naturally interpreted as the coframe (or the tetrad) field of a physical observer on the spacetime manifold M_4, whereas the 6 potentials $\Gamma_i{}^{\alpha\beta} = -\Gamma_i{}^{\beta\alpha}$ for the Lorentz subgroup $SO(1,3)$ are identified with the local connection that introduces the parallel transport on the spacetime M_4.

The multiplet of gauge potentials,

$$\Big\{ A_i, \quad e_i^\alpha, \quad \Gamma_i{}^{\alpha\beta} \Big\}, \tag{1}$$

determines the corresponding multiplet of the "Yang–Mills" gauge field strengths:

$$F_{ij} = \partial_i A_j - \partial_j A_i, \tag{2}$$

$$T_{ij}{}^\alpha = \partial_i e_j^\alpha - \partial_j e_i^\alpha + \Gamma_{i\beta}{}^\alpha e_j^\beta - \Gamma_{j\beta}{}^\alpha e_i^\beta, \tag{3}$$

$$R_{ij}{}^{\alpha\beta} = \partial_i \Gamma_j{}^{\alpha\beta} - \partial_j \Gamma_i{}^{\alpha\beta} + \Gamma_{i\gamma}{}^\beta \Gamma_j{}^{\alpha\gamma} - \Gamma_{j\gamma}{}^\beta \Gamma_i{}^{\alpha\gamma}. \tag{4}$$

Thereby, we derive the Maxwell tensor F_{ij} as the $U(1)$ gauge field strength for the electromagnetic field, and the spacetime torsion tensor $T_{ij}{}^\alpha$ and the curvature tensor $R_{ij}{}^{\alpha\beta} = -R_{ij}{}^{\beta\alpha}$ as the two Poincaré (T_4 "translational" and $SO(1,3)$ "rotational", respectively) gauge field strengths for the gravitational field.

The nontrivial "mixed" form of the torsion (3) is explained by the semi-direct structure of the Poincaré symmetry group. The resulting Riemann–Cartan geometry on the spacetime M_4 is characterized by the nonvanishing torsion and curvature, whereas in the special case $T_{ij}{}^\alpha = 0$ we recover the Riemannian geometry, and for $R_{ij}{}^{\alpha\beta} = 0$ one finds the Weitzenböck space of distant parallelism.

Here we do not discuss the construction of the complete dynamical scheme of the Poincaré gauge theory that requires the introduction of the corresponding gravitational field Lagrangian, and consider the electromagnetic and the gravitational fields as a non-dynamical background. It is important to recall, though, that the variation of the Lagrange density of matter with respect to the gauge field potentials (1) gives rise to the corresponding dynamical currents: the electric current, the canonical energy-momentum tensor, and the spin angular momentum tensor, respectively. Further details can be found in [9,10,16–18], and we conclude this section with the following technical points which are needed for the subsequent discussion.

One can decompose the local Lorentz connection into the sum:

$$\Gamma_i{}^{\alpha\beta} = \widetilde{\Gamma}_i{}^{\alpha\beta} - K_i{}^{\alpha\beta} \tag{5}$$

of the Riemannian connection (denoted by the tilde), which is torsionless $\partial_i e_j^\alpha - \partial_j e_i^\alpha + \widetilde{\Gamma}_{i\beta}{}^\alpha e_j^\beta - \widetilde{\Gamma}_{j\beta}{}^\alpha e_i^\beta = 0$ and metric-compatible, plus the post-Riemannian contortion tensor,

$$K_{i\alpha\beta} = \frac{1}{2}\left(T_{\alpha\beta i} - T_{i\alpha\beta} + T_{i\beta\alpha}\right). \tag{6}$$

On the other hand, the torsion tensor $T_{\mu\nu}{}^\alpha = e_\mu^i e_\nu^j T_{ij}{}^\alpha$ can be decomposed into the three irreducible parts:

$$T_{\mu\nu}{}^\alpha = \frac{1}{3}(\delta_\mu^\alpha T_\nu - \delta_\nu^\alpha T_\mu) + \frac{1}{3}\eta_{\mu\nu\lambda}{}^\alpha \check{T}^\lambda + \mathcal{T}_{\mu\nu}{}^\alpha, \tag{7}$$

where $\mathcal{T}_{\mu\nu}{}^\alpha$ is the trace-free and axial trace-free tensor, the torsion trace vector $T_\mu = T_{\alpha\mu}{}^\alpha$, and the axial trace vector:

$$\check{T}^\alpha = -\frac{1}{2}\eta^{\alpha\mu\nu\lambda}T_{\mu\nu\lambda}, \tag{8}$$

with the totally antisymmetric Levi-Civita tensor $\eta^{\alpha\mu\nu\lambda}$.

2.2. Hamiltonian for the Dirac Fermion

The Pauli-like equation for a fermion particle, moving under the action of the torsion field had been derived in [15] for the flat Minkowski spacetime, and in [19] for an arbitrary curved space background. The relativistic dynamics of the Dirac particle with spin 1/2, electric charge q, and mass m minimally coupled to the gravitational and electromagnetic fields is described by the invariant action:

$$S = \int d^4x\sqrt{-g}L, \tag{9}$$

where the Lagrangian of the spinor wave function ψ and $\overline{\psi} = \psi^\dagger\gamma^0$ has the form:

$$L = \frac{i\hbar}{2}(\overline{\psi}\gamma^\alpha D_\alpha\psi - D_\alpha\overline{\psi}\gamma^\alpha\psi) - mc\,\overline{\psi}\psi. \tag{10}$$

The spinor covariant derivative describes the minimal coupling of the charged Dirac particle with the external electromagnetic and gravitational gauge fields (1):

$$D_\alpha = e_\alpha^i\left(\partial_i - \frac{iq}{\hbar}A_i + \frac{i}{4}\Gamma_i{}^{\beta\gamma}\sigma_{\beta\gamma}\right). \tag{11}$$

Here, c and $\hbar$ are the speed of light and Planck's constant, respectively, the 4-potential of the electromagnetic field $A_i = (-\phi, \boldsymbol{A})$ encompasses the scalar ϕ and vector $\boldsymbol{A}$ potentials, and $\sigma_{\alpha\beta} = \frac{i}{2}(\gamma_\alpha\gamma_\beta - \gamma_\beta\gamma_\alpha)$ are the Lorentz algebra generators, where the flat Dirac matrices γ^α are defined in local Lorentz frames.

We denote the local spatial and time coordinates by $x^i = (t, x^a)$, $a, b, c = 1,2,3$. An orthonormal coframe (tetrad) is needed to attach spinor spaces at every point of the space–time manifold. Then the dynamics of the Dirac particle can be investigated in an arbitrary Poincaré gauge field $(e_i^\alpha, \Gamma_i{}^{\alpha\beta})$, where the components of tetrads in the Schwinger gauge [19] read:

$$e_i^{\widehat{0}} = V\,\delta_i^0, \qquad e_i^{\widehat{a}} = W^{\widehat{a}}{}_b\left(\delta_i^b - cK^b\,\delta_i^0\right), \qquad a, b = 1,2,3. \tag{12}$$

As was shown in reference [19], the Hermitian Hamiltonian of the fermion particle has the form:

$$\begin{aligned}\mathcal{H} \;=\;& \beta mc^2 V + q\Phi + \frac{c}{2}\left(\pi_b\,\mathcal{F}^b{}_a\alpha^a + \alpha^a\mathcal{F}^b{}_a\pi_b\right)\\ &+ \frac{c}{2}(\boldsymbol{K}\cdot\boldsymbol{\pi} + \boldsymbol{\pi}\cdot\boldsymbol{K}) + \frac{\hbar c}{4}(\boldsymbol{\Xi}\cdot\boldsymbol{\Sigma} - \Upsilon\gamma_5),\end{aligned} \tag{13}$$

where the kinetic 3-momentum operator $\pi_a = -i\hbar\partial_a - qA_a = p_a - qA_a$ accounts of the interaction with the electromagnetic field, and we denoted:

$$\mathcal{F}^b{}_a = VW^b{}_{\widehat{a}}, \qquad \Upsilon = V\epsilon^{\widehat{a}\widehat{b}\widehat{c}}\Gamma_{\widehat{a}\widehat{b}\widehat{c}}, \qquad \Xi^a = \frac{V}{c}\epsilon^{\widehat{a}\widehat{b}\widehat{c}}(\Gamma_{\widehat{0}\widehat{b}\widehat{c}} + \Gamma_{\widehat{b}\widehat{c}\widehat{0}} + \Gamma_{\widehat{c}\widehat{0}\widehat{b}}). \tag{14}$$

As usual, $\alpha^a = \beta\gamma^a$ $(a, b, c, \cdots = 1, 2, 3)$ and the spin matrices $\Sigma^1 = i\gamma^{\hat{2}}\gamma^{\hat{3}}, \Sigma^2 = i\gamma^{\hat{3}}\gamma^{\hat{1}}, \Sigma^3 = i\gamma^{\hat{1}}\gamma^{\hat{2}}$ and $\gamma_5 = i\alpha^{\hat{1}}\alpha^{\hat{2}}\alpha^{\hat{3}}$. Boldface notation is used for 3-vectors $\boldsymbol{K} = \{K^a\}$, $\boldsymbol{\pi} = \{\pi_a\}$, $\boldsymbol{\alpha} = \{\alpha^a\}$, $\boldsymbol{\Sigma} = \{\Sigma^a\}$.

Taking into account the decomposition of the connection (5) into the Riemannian and post-Riemannian parts, we find that the Pauli-like equation with the Hermitian Hamiltonian (13) encompasses the spin-torsion coupling:

$$\Upsilon = \widetilde{\Upsilon} + Vc\check{T}^{\hat{0}}, \qquad \Xi^{\widehat{a}} = \widetilde{\Xi}^{\widehat{a}} - V\check{T}^{\widehat{a}}. \tag{15}$$

The tilde denotes the Riemannian quantities. The post-Riemannian contributions come from the components $\check{T}^\alpha = (\check{T}^{\hat{0}}, \check{T}^{\widehat{a}})$ of the axial torsion vector (8). Accordingly, the spin-torsion coupling terms read explicitly

$$-\frac{\hbar cV}{4}\left(\boldsymbol{\Sigma}\cdot\check{\boldsymbol{T}} + c\gamma_5\check{T}^{\hat{0}}\right). \tag{16}$$

The above general formalism can be applied to the study of fermion's dynamics in arbitrary external electromagnetic and gravitational (including the post-Riemannian one) fields.

Let us now specialize to the analysis of the possible physical effects of the spacetime torsion and the inertial forces on the non-relativistic particle in the rotating reference frame (such as the Earth), [46]:

$$V = 1, \quad W^{\widehat{a}}{}_b = \delta^a_b, \quad K^a = -\frac{(\boldsymbol{\omega}\times\boldsymbol{r})^a}{c}, \qquad \Gamma_{\hat{0}}{}^{\hat{a}\hat{b}} = -\frac{\varepsilon^{abc}\omega_c}{c}, \quad \Gamma_{\hat{0}}{}^{\hat{a}\hat{0}} = 0. \tag{17}$$

Substituting this configuration into the Hamiltonian (13) we derive:

$$\mathcal{H} = \beta mc^2 + c\boldsymbol{\alpha}\cdot\boldsymbol{\pi} - \boldsymbol{\omega}\cdot(\boldsymbol{r}\times\boldsymbol{\pi}) - \frac{\hbar}{2}\boldsymbol{\omega}\cdot\boldsymbol{\Sigma} - \frac{\hbar c}{4}\left(\check{T}^{\hat{0}}c\gamma_5 + \check{\boldsymbol{T}}\cdot\boldsymbol{\Sigma}\right). \tag{18}$$

In order to reveal the physical contents of the Schrödinger equation, we need to go to the Foldy–Wouthuysen (FW) representation. Applying the methods developed in [19], we find the FW Hamiltonian:

$$\begin{aligned} H \;=\; & \beta\epsilon + q\phi - \boldsymbol{\omega}\cdot(\boldsymbol{r}\times\boldsymbol{\pi}) - \frac{\hbar}{2}\boldsymbol{\omega}\cdot\boldsymbol{\Sigma} - \frac{q\hbar c^2}{4}\left\{\frac{1}{\epsilon}, \boldsymbol{B}\cdot\boldsymbol{\Pi}\right\} \\ & + \frac{\hbar c^3}{8}\left\{\frac{\boldsymbol{\pi}\cdot\boldsymbol{\Pi}}{\epsilon}, \check{T}^{\hat{0}}\right\} - \frac{\hbar c}{8}\left\{\frac{mc^2}{\epsilon}, \check{\boldsymbol{T}}\cdot\boldsymbol{\Sigma}\right\} \\ & - \frac{\hbar c^3}{8}\left[\frac{\boldsymbol{\Sigma}\cdot\boldsymbol{\pi}}{\epsilon(\epsilon+mc^2)}\boldsymbol{\pi}\cdot\check{\boldsymbol{T}} + \check{\boldsymbol{T}}\cdot\boldsymbol{\pi}\frac{\boldsymbol{\Sigma}\cdot\boldsymbol{\pi}}{\epsilon(\epsilon+mc^2)}\right] \\ & - \frac{q\hbar c}{8}\left\{\frac{1}{\epsilon(\epsilon+mc^2)}, \boldsymbol{\Sigma}\cdot(\boldsymbol{\mathfrak{E}}\times\boldsymbol{\pi} - \boldsymbol{\pi}\times\boldsymbol{\mathfrak{E}})\right\}. \end{aligned} \tag{19}$$

Here, $\boldsymbol{\Pi} = \beta\boldsymbol{\Sigma}$, $\{\ ,\ \}$ denotes anticommutators, $\epsilon = \sqrt{m^2c^4 + c^2\boldsymbol{\pi}^2}$, and $\boldsymbol{\mathfrak{E}} = \boldsymbol{E} + \boldsymbol{B}\times(\boldsymbol{\omega}\times\boldsymbol{r})$ is the physical electric field as seen in the noninertial rotating reference frame.

Under ordinary conditions we assume $|e\hbar B| \ll m^2c^2$, that is the magnetic field is much smaller than the critical field $|B| \ll B_c = m^2c^2/e\hbar$, and particle's velocity is much smaller than the speed of light, $|\pi|/m \ll c$. Then $\epsilon = mc^2 + \boldsymbol{\pi}^2/2m$, and in the semiclassical limit of (19) we finally obtain the Pauli-type equation $i\hbar\frac{\partial\psi}{\partial t} = H^{\rm nr}\psi$ with the non-relativistic Hamiltonian:

$$H^{\rm nr} = \frac{\pi^2}{2m} + q\phi - \omega\cdot(r\times\pi) - \frac{\hbar}{2}\omega\cdot\sigma - \frac{q\hbar}{2m}B\cdot\sigma + \frac{\hbar c}{8m}\left\{\pi\cdot\sigma, \check{T}^{\hat{0}}\right\} - \frac{\hbar c}{4}\check{T}\cdot\sigma. \tag{20}$$

This result is consistent with an alternative analysis based on the method of exact FW transformations [20], see the relevant discussion in [47,48].

In the physically important situations, the torsion pseudovector is spacelike, and $|\check{T}| \gg c\check{T}^{\hat{0}}$. Taking this into account, we now switch to the physically interesting case $\check{T}^{\hat{0}} = 0$. This is the true for the fermions (10) and (11) minimally coupled to gravity.

3. Quantum Hydrodynamics for Spin-Torsion Coupling

In this section, we derive the many-particle quantum hydrodynamics (MPQHD) equations from the many-particle Pauli-like equation for the system of charged particles with spin-1/2. The method of MPQHD allows to present the dynamics of a system of interacting quantum particles in terms of the functions defined in the three-dimensional physical space. This is important for the study of wave process, which take place in a three-dimensional physical space [49,50]. In flat spacetime, the MPQHD formalism for many-particles fermion systems was previously developed in [51–53], whereas the case of the noninertial reference frames was considered in [54]. The methods of MPQHD can be used for the analysis of a wide variety of systems of many interacting particles. In particular, the finite temperature hydrodynamic model has been derived recently in [55] for the spin-1 ultracold bosons. In Reference [56] the method was applied to the study of the polarization dynamics in a system of quantum particles with nontrivial electric dipole moments.

After applying the Foldy–Wouthuysen transformation for Dirac particle in combination with the method of many-particle quantum hydrodynamics, we arrive at the many-particle Pauli-like equation:

$$i\hbar\frac{\partial\psi_s}{\partial t} = \hat{H}\psi_s, \tag{21}$$

where the many-particle wave function of the system of N spinning particles:

$$\psi_s(R,t) = \psi_s(r_1, r_2, \dots, r_N, t) \tag{22}$$

is a spinor function in the $3N$-dimensional configuration space (s is the spin index), and the many-particle Hamiltonian reads:

$$\hat{H} = \sum_{p=1}^{N}\left(\frac{\hat{\pi}_p^2}{2m_p} + \frac{\hbar}{2}\sigma\cdot\Omega_p + \phi_p\right). \tag{23}$$

Here, we introduced:

$$\Omega_p = -\omega - \frac{q_p}{m_p}B_p - \frac{c}{2}\check{T}_p, \tag{24}$$

$$\phi_p = q_p\,\phi(r_p) - \frac{m_p}{2}[\omega\times r_p]^2, \tag{25}$$

$$\hat{\pi}_p = -i\hbar\nabla_p - q_pA_p - m_p\,\omega\times r_p\,, \tag{26}$$

and m_p and q_p denote the mass and the charge of p-th particle, respectively. In particular, q_p stands for the charge of electrons $q_e = -e$, or for the charge of ions $q_i = e$. The electromagnetic vector and scalar potentials $A_p = A(r_p)$ and $\phi = \phi(r_p)$ are taken at the positions r_p of the p-th particle, and the same applies to the external magnetic $B_p = B(r_p)$ and the torsion $\check{T}_p = \check{T}(r_p)$ fields. The last terms in (25) and (26) manifest the inertial contributions in the rotating reference frame with the angular velocity ω.

As compared to the standard case of a system in an external electromagnetic field, the many-particle Hamiltonian (23) includes the torsion effects, encoded in the second term $\sim \boldsymbol{\sigma} \cdot \check{\boldsymbol{T}}_p$, that has the same form as the Zeeman energy in the magnetic field. In addition, this Hamiltonian includes Mashhoon's spin-rotation contribution $\sim \boldsymbol{\sigma} \cdot \boldsymbol{\omega}$, see [57–59].

The state of the system is characterized by the concentration of particles in the neighborhood of a point $\boldsymbol{r}$ in the physical space as:

$$n(\boldsymbol{r},t) = \int dR \sum_{p=1}^{N} \delta(\boldsymbol{r} - \boldsymbol{r}_p)\psi_s^*(R,t)\psi_s(R,t) = \langle \psi^\dagger \hat{n} \psi \rangle. \tag{27}$$

Here the integration measure reads $dR = \prod_p d^3r_p$. The function $n(\boldsymbol{r},t)$ is thus determined as the quantum average of the concentration operator $\hat{n} = \sum_p \delta(\boldsymbol{r} - \boldsymbol{r}_p)$ in the coordinate representation. The spin density vector of fermions is determined in a similar way:

$$\boldsymbol{S}(\boldsymbol{r},t) = \int dR \sum_{p=1}^{N} \delta(\boldsymbol{r} - \boldsymbol{r}_p)\psi_s^*(R,t)(\hat{\boldsymbol{s}}_p)_{ss'}\psi_{s'}(R,t) = \langle \psi^\dagger \boldsymbol{\mathfrak{s}} \psi \rangle, \tag{28}$$

as the quantum average of the spin operator $\boldsymbol{\mathfrak{s}} = \sum_p \delta(\boldsymbol{r} - \boldsymbol{r}_p)\hat{\boldsymbol{s}}_p$, with $\hat{\boldsymbol{s}}_p = \frac{\hbar}{2}\boldsymbol{\sigma}_p$.

The continuity equation for the concentration of the particles $n(\boldsymbol{r},t)$ can be derived by taking the time derivative of the definition (27) and making use of the many-particle Pauli-like Equation (21):

$$\partial_t n(\boldsymbol{r},t) + \boldsymbol{\nabla} \cdot \boldsymbol{J}(\boldsymbol{r},t) = 0, \tag{29}$$

where the current density is defined as the microscopic average $\boldsymbol{J}(\boldsymbol{r},t) = \frac{1}{2}\langle \psi^\dagger \boldsymbol{\mathfrak{J}} \psi + \text{c.c.}\rangle$ of the operator

$$\boldsymbol{\mathfrak{J}} = \sum_p \delta(\boldsymbol{r} - \boldsymbol{r}_p)\frac{\hat{\boldsymbol{\pi}}_p}{m_p}. \tag{30}$$

Here the generalized momentum operator is defined by (26).

3.1. Spin Density Evolution

In a similar way, the dynamical equation for the spin density can be obtained by differentiating the definition (28) with respect to time and making use of the many-particle Pauli-like Equation (21):

$$\partial_t S^a(\boldsymbol{r},t) + \partial_b \Lambda^{ba}(\boldsymbol{r},t) = \varepsilon^{abc}\Omega^b S^c(\boldsymbol{r},t). \tag{31}$$

Here the spin precession angular velocity is defined as:

$$\boldsymbol{\Omega} = -\boldsymbol{\omega} - \frac{q}{m}\boldsymbol{B} - \frac{c}{2}\check{\boldsymbol{T}}, \tag{32}$$

cf. (24), whereas the spin current density tensor is introduced as a microscopic average $\Lambda^{ba}(\boldsymbol{r},t) = \frac{1}{2}\langle \psi^\dagger \mathfrak{L}^{ba} \psi + \text{c.c.}\rangle$ of the operator

$$\mathfrak{L}^{ba} = \sum_p \delta(\boldsymbol{r} - \boldsymbol{r}_p)\frac{\hat{\pi}_p^a \hat{s}_p^b}{m_p}. \tag{33}$$

3.2. Equation of Motion

Along the same lines, the derivation of the equation of motion of a hydrodynamic system in an external electromagnetic and torsion fields is based on differentiating the expression for the current density $\boldsymbol{J}(\boldsymbol{r},t)$ with respect to time and using the Pauli-like Equation (21) with the Hamiltonian (23). The result reads:

$$m\partial_t J^a(\boldsymbol{r},t) + \partial_b \Pi^{ab}(\boldsymbol{r},t) = qnE^a(\boldsymbol{r},t) + q\varepsilon^{abc}J^b(\boldsymbol{r},t)B^c(\boldsymbol{r},t) - S^b(\boldsymbol{r},t)\partial^a\Omega^b + F^a_{\text{iner}}, \tag{34}$$

where the momentum flux tensor appears in fluid dynamics as a quantum average $\Pi^{ab}(\boldsymbol{r},t) = \frac{1}{2}\langle\psi^{\dagger}\mathfrak{P}^{ab}\psi + \text{c.c.}\rangle$ of the operator

$$\mathfrak{P}^{ab} = \sum_{p} \delta(\boldsymbol{r}-\boldsymbol{r}_p)\,\frac{\hat{\pi}_p^{(a}\,\hat{\pi}_p^{b)}}{m_p}. \tag{35}$$

The equation of motion (34) describes the influence of the external electromagnetic and torsion fields on the fermion matter in terms of the Lorentz and the Stern-Gerlach forces.

As a next step, one can move from the microscopic representation of the particle current density and the spin current density to their corresponding macroscopic variables by making use of an explicit representation of the spinor wave function. Such an explicit representation of the wave function is known as the Madelung decomposition.

4. Madelung Decomposition

The microscopic many-particle wave function or the Madelung decomposition [60] of the N-particle wave function can be represented in terms of the amplitude $a(R,t)$, the phase $\xi(R,t)$ and the local spinor $\mathcal{Z}(R,\boldsymbol{r},t)$, defined in the local rest frame and normalized so that $\mathcal{Z}^{\dagger}\mathcal{Z} = 1$:

$$\psi(R,t) = a(R,t)\,\varphi(R,\boldsymbol{r},t), \qquad \varphi(R,\boldsymbol{r}.t) = e^{\frac{i}{\hbar}\xi(R,t)}\,\mathcal{Z}(R,\boldsymbol{r},t). \tag{36}$$

Applying the decomposition (36) to the p-th particle, we can introduce a microscopic velocity and a microscopic spin as $\boldsymbol{v}_p := \frac{1}{m_p}\varphi^{\dagger}\,\hat{\boldsymbol{\pi}}_p\,\varphi$ and $\boldsymbol{s}_p := \varphi^{\dagger}\,\hat{\boldsymbol{s}}_p\,\varphi = \frac{\hbar}{2}\,\varphi^{\dagger}\,\boldsymbol{\sigma}_p\,\varphi$, respectively. Explicitly, we then find:

$$\boldsymbol{v}_p(R,\boldsymbol{r},t) = \frac{1}{m_p}\Big(\boldsymbol{\nabla}_p\,\xi - q\boldsymbol{A}_p - i\hbar\mathcal{Z}^{\dagger}\boldsymbol{\nabla}_p\,\mathcal{Z} - m_p\,\boldsymbol{\omega}\times\boldsymbol{r}_p\Big), \tag{37}$$

$$\boldsymbol{s}_p(R,\boldsymbol{r},t) = \frac{\hbar}{2}\,\mathcal{Z}^{\dagger}\boldsymbol{\sigma}_p\mathcal{Z}. \tag{38}$$

The velocity field of the p-th particle can be decomposed $\boldsymbol{v}_p(R,\boldsymbol{r},t) = \boldsymbol{v}(\boldsymbol{r},t) + \boldsymbol{\eta}_p(R,\boldsymbol{r},t)$ into a sum of the macroscopic average $\boldsymbol{v}(\boldsymbol{r},t)$ and the thermal fluctuations part $\boldsymbol{\eta}_p(R,\boldsymbol{r},t)$ of the velocity. In a similar way, the spin of the p-th particle can be represented as the sum $\boldsymbol{s}_p(R,\boldsymbol{r},t) = \boldsymbol{s}(\boldsymbol{r},t) + \boldsymbol{\tau}_p(R,\boldsymbol{r},t)$ of the macroscopic average $\boldsymbol{s}(\boldsymbol{r},t)$ and the thermal fluctuations part $\boldsymbol{\tau}_p(R,\boldsymbol{r},t)$ of the spin. By definition, the averages of the fluctuations vanish, $\langle a^2\boldsymbol{\eta}_p\rangle = 0$ and $\langle a^2\boldsymbol{\tau}_p\rangle = 0$. We assume that the particle system is closed and not placed in a thermostat. Recalling that the temperature is the average kinetic energy of the chaotic motion of the particles of the system, we consider deviations of the velocity and spin of quantum particles from the local average values, which correspond to the ordered motion of the particles.

Combining Equations (27), (30) and (28) with (36)–(38), we can derive the macroscopic concentration, the macroscopic current density and the macroscopic spin density from the corresponding microscopic variables:

$$n(\boldsymbol{r},t) = \int dR\sum_{p=1}^{N}\delta(\boldsymbol{r}-\boldsymbol{r}_p)\,a^2(R,t), \tag{39}$$

$$\boldsymbol{J}(\boldsymbol{r},t) = \int dR\sum_{p=1}^{N}\delta(\boldsymbol{r}-\boldsymbol{r}_p)\,a^2(R,t)\,\boldsymbol{v}_p(R,\boldsymbol{r},t) = n(\boldsymbol{r},t)\,\boldsymbol{v}(\boldsymbol{r},t), \tag{40}$$

$$\boldsymbol{S}(\boldsymbol{r},t) = \int dR\sum_{p=1}^{N}\delta(\boldsymbol{r}-\boldsymbol{r}_p)\,a^2(R,t)\,\boldsymbol{s}_p(R,\boldsymbol{r},t) = n(\boldsymbol{r},t)\,\boldsymbol{s}(\boldsymbol{r},t). \tag{41}$$

After the Madelung decomposition procedure for the basic physical variables in the microscopic representation, the spin current density (33) and the momentum flux (35) can be recast in terms of the fluid variables into:

$$\Lambda^{ba}(\boldsymbol{r},t) = \int dR \sum_{p=1}^{N} \delta(\boldsymbol{r}-\boldsymbol{r}_p)\left(a^2 s_p^a v_p^b - \frac{a^2}{m_p}\varepsilon^{acd} s_p^c \partial_p^b s_p^d\right), \tag{42}$$

$$\begin{aligned}\Pi^{ab}(\boldsymbol{r},t) &= \int dR \sum_{p=1}^{N} \delta(\boldsymbol{r}-\boldsymbol{r}_p)\left(\frac{\hbar^2}{2m_p}\left(\partial_p^a a \partial_p^b a - a\partial_p^a \partial_p^b a\right)\right. \\ &\quad \left. + m_p a^2 v_p^a v_p^b + \frac{a^2}{m_p}\,\partial_p^a s_p^c \partial_p^b s_p^c\right),\end{aligned} \tag{43}$$

respectively. We are now ready to write down the complete set of dynamical equations for the quantum system of spinning particles explicitly in terms of the fluid variables. This set encompasses the continuity equation

$$\partial_t n + \boldsymbol{\nabla}\cdot(n\,\boldsymbol{v}) = 0, \tag{44}$$

and the momentum balance equation

$$\begin{aligned}\left(\partial_t + v^b\partial_b\right) v^a &= \frac{q}{m}E^a + \frac{q}{m}(\boldsymbol{v}\times\boldsymbol{B})^a - \frac{1}{n}\partial_b p^{ab} + \frac{\hbar^2}{2m^2}\partial^a\left(\frac{\Delta\sqrt{n}}{\sqrt{n}}\right) \\ &\quad + \frac{1}{2m^2}\partial^a\left(\partial^b \boldsymbol{s}\cdot\partial^b \boldsymbol{s}\right) - \frac{1}{m}\boldsymbol{s}\cdot\partial^a\hat{\boldsymbol{\Omega}} + f^a_{\rm iner} - \frac{1}{m}Q^a_{\rm therm},\end{aligned} \tag{45}$$

whereas the spin evolution Equation (31) reads:

$$\left(\partial_t + v^b\partial_b\right)\boldsymbol{s} = \hat{\boldsymbol{\Omega}}\times\boldsymbol{s} - \boldsymbol{\Theta}_{\rm therm}, \tag{46}$$

where the spin precession angular velocity is modified, cf. (32),

$$\hat{\boldsymbol{\Omega}} = -\,\boldsymbol{\omega} - \frac{q}{m}\boldsymbol{B} - \frac{c}{2}\check{\boldsymbol{T}} - \frac{1}{mn}\,\partial_b(n\partial^b\boldsymbol{s}). \tag{47}$$

This modification arises from the interaction of spin with the surrounding spin-texture of the fluid, and one can formally interpret this in terms of an effective magnetic field defined as a sum of an external magnetic field and the emergent field:

$$\hat{\boldsymbol{B}} = \boldsymbol{B} + \frac{1}{qn}\,\partial_b(n\partial^b\boldsymbol{s}). \tag{48}$$

In fact, one can also view the first term on the right-hand side of (47), which is due to Mashhoon's spin-rotation coupling term and describes the Barnett effect, and the third term on the right-hand side of (47), representing the spin-torsion coupling, as the two additional contributions to the "effective" magnetic field

$$\boldsymbol{B}_\omega = \frac{m}{q}\,\boldsymbol{\omega}, \qquad \boldsymbol{B}_T = \frac{mc}{2q}\,\check{\boldsymbol{T}}. \tag{49}$$

The dynamical Equation (46) describes the precession of spin under the action of the torque produced by the external magnetic field and the emergent fields, leading to the Zeeman type effect. The additional torque in (46) arises from thermal-spin interactions:

$$\boldsymbol{\Theta}_{\rm therm} = \partial_b \int dR \sum_{p=1}^{N} \delta(\boldsymbol{r}-\boldsymbol{r}_p)\left(a^2\eta_p^b \boldsymbol{s}_p - \frac{a^2}{m_p}\,\boldsymbol{s}_p\times\partial_p^b\boldsymbol{\tau}_p\right), \tag{50}$$

that is also responsible for the last force term in the momentum balance Equation (45)

$$\begin{aligned} Q^a_{\text{therm}} &= \frac{1}{n}\partial_b \int dR \sum_{p=1}^{N} \delta(\boldsymbol{r}-\boldsymbol{r}_p)\,\frac{a^2}{m_p}\Big(\partial^b_p \boldsymbol{\tau}_p \cdot \partial^a_p \boldsymbol{s}_p + \partial^b_p \boldsymbol{s}_p \cdot \partial^a_p \boldsymbol{\tau}_p - \partial^a_p \boldsymbol{\tau}_p \cdot \partial^b_p \boldsymbol{\tau}_p\Big) \\ &- \partial^a \left\{ \frac{1}{n}\partial_b \left[n\,\partial^b \Big(\frac{1}{n} \int dR \sum_{p=1}^{N} \delta(\boldsymbol{r}-\boldsymbol{r}_p)\,\frac{a^2}{2m_p}\,\boldsymbol{\tau}_p \cdot \boldsymbol{\tau}_p \Big) \right] \right\}. \end{aligned} \tag{51}$$

Analysing the structure of the equation of motion (45), we identify the first two terms on the right hand side with the Lorenz force determined by the external electric and magnetic fields $\boldsymbol{E}$ and $\boldsymbol{B}$, while the third term is the divergence of the kinetic pressure tensor

$$p^{ab}(\boldsymbol{r},t) = \int dR \sum_{p=1}^{N} \delta(\boldsymbol{r}-\boldsymbol{r}_p)\, a^2\, m_p\, \eta^a_p\, \eta^b_p. \tag{52}$$

The fifth term on the right hand side of the Equation (45) represents the effect of spin–spin interactions inside the fluid, the interaction of the spin with the spin background texture. The sixth term describes the Stern–Gerlach force that characterizes the influence of the non-uniform effective magnetic and the torsion field. In the non-inertial frame, an additional contribution encompasses the Coriolis force, the centrifugal force and Euler force field:

$$f_{\text{iner}} = -\,2\,\boldsymbol{\omega}\times\boldsymbol{v} - \boldsymbol{\omega}\times(\boldsymbol{\omega}\times\boldsymbol{\mathfrak{R}}) - \frac{\partial\boldsymbol{\omega}}{\partial t}\times\boldsymbol{\mathfrak{R}}, \tag{53}$$

where the vector of center of mass is defined as:

$$\boldsymbol{\mathfrak{R}}(\boldsymbol{r},t) = \frac{1}{n}\int dR \sum_{p=1}^{N} \delta(\boldsymbol{r}-\boldsymbol{r}_p)\psi^*_s(R,t)\boldsymbol{r}_p\psi_s(R,t). \tag{54}$$

Our derivations are consistent with the earlier analysis [54].

5. Experimental Manifestations of Spin-Torsion Coupling

Experimental search for the nontrivial torsion effects is naturally embedded into the broader framework of the studies of the spin-dependent long-range forces [21–40]. By making use of the corresponding experimental techniques, it is possible to find strong limits on the values of the gauge gravity spin-torsion coupling constants and on the torsion field itself. A good example of an efficient approach in this respect gives an observation of the nuclear spin precession in gaseous spin polarized ^{3}He or ^{129}Xe samples with the help of a highly sensitive low-field magnetometer [61–63] detecting a sidereal variation of the relative spin precession frequency in a new type of ^{3}He/^{129}Xe clock comparison test. In a similar experiment [64], the ratio of nuclear spin-precession frequencies of ^{199}Hg and ^{201}Hg atoms was measured in the magnetic field and the Earth's gravitational field. Based on the corresponding experimental data from [64], the analysis of dynamics of the minimally coupled Dirac fermion [19] in external electromagnetic and gravitational fields revealed the strong bounds on the possible background space-time torsion:

$$\frac{c}{2}\,|\check{\boldsymbol{T}}|\cdot|\cos\theta| < 6.45\times10^{-6}s^{-1}, \tag{55}$$

where θ is the angle between the magnetic $\boldsymbol{B}$ and torsion $\check{\boldsymbol{T}}$ fields. On the other hand, by making use of the experimental data from [61], one finds the restriction:

$$\frac{c}{2}\,|\check{\boldsymbol{T}}|\cdot|\cos\theta| < 3.59\times10^{-7}s^{-1}. \tag{56}$$

These results are consistent with the alternative empirical estimates for the torsion limits [30–40]. As another powerful tool one can mention the use of the quantum interferometry to probe the spacetime structure, including the search for possible post-Riemannian deviations, focusing on the detection of the phase shift and polarization rotation effects for the neutron and atom beams [65–69].

As an application of the quantum hydrodynamics formalism, let us investigate a simple model of a continuous medium of particles with spin and consider the dynamics of the spin waves in such a particle system. Neglecting the spin-thermal coupling in the spin dynamical Equation (46) and assuming the small perturbations of the spin $\boldsymbol{s} = \boldsymbol{s}_0 + \delta\boldsymbol{s}$ around an undisturbed value $|\boldsymbol{s}_0| = \hbar/2$, we find, in the first order,

$$\partial_t \delta\boldsymbol{s} = \hat{\boldsymbol{\Omega}}^{(0)} \times \delta\boldsymbol{s} + \hat{\boldsymbol{\Omega}}^{(1)} \times \boldsymbol{s}_0. \quad (57)$$

Here, the equilibrium values of the external background fields are encoded in $\hat{\boldsymbol{\Omega}}^{(0)} = -\boldsymbol{\omega}_0 - \frac{q}{m}\boldsymbol{B}_0 - \frac{c}{2}\check{\boldsymbol{T}}$, where $\boldsymbol{B}_0$, $\boldsymbol{\omega}_0$ and $\check{\boldsymbol{T}}$ are the external uniform magnetic field, the Earth's angular velocity and the background torsion, respectively, and the small disturbance reads:

$$\hat{\boldsymbol{\Omega}}^{(1)} = -\frac{\Delta\delta\boldsymbol{s}}{m}, \quad (58)$$

Assuming that the perturbations of the spin vary as $\delta\boldsymbol{s} \sim \exp(-i\omega_s t + i\boldsymbol{k}\cdot\boldsymbol{r})$, we then derive the dispersion law relating the wave frequency ω_s and the wave vector $\boldsymbol{k}$ for the spin waves excited in the external magnetic and torsion fields:

$$\omega_s^2 = \Omega_c^2 + \frac{c^2}{4}\check{T}^2 + \omega_0^2 + 2\Omega_c\omega_0\cos\theta_1 + c\,\Omega_c\check{T}\cos\theta_2 + c\,\omega_0\check{T}\cos(\theta_2 - \theta_1). \quad (59)$$

Here, $\Omega_c = \frac{qB_0}{m} + \frac{\hbar k^2}{2m}$, whereas θ_1 and θ_2 are, respectively, the angle between the external magnetic field and Earth's angular velocity, and the angle between $\boldsymbol{B}_0$ and the background torsion $\check{\boldsymbol{T}}$. Equation (59) is a generalization for the dispersion relation of spin waves found in Reference [70] and it takes into account the contribution of the spin part of the quantum Bohm potential as an additional spin torque due to the self-action inside the system of particles, which leads to the propagation of spin waves. The square of the frequency ω_s^2 encompasses a contribution proportional to the square of the modulus of the wave vector $\sim k^2$. As we can see from the dispersion relation (59) the torsion effect is maximal when the pseudovector field $\check{\boldsymbol{T}}$ is aligned along the external magnetic field.

6. Discussion and Conclusions

In this paper, we for the first time developed the quantum hydrodynamics for the many-particle system of massive Dirac fermion spin-1/2 particles interacting with external electromagnetic, metric gravitational/inertial and torsion fields. This essentially extends the single-particle quantum hydrodynamical approach which was developed for the flat spacetime, see [71–76] and the references therein. Taking as the basis of the earlier general formalism [19], the consistent hydrodynamical formulation was constructed for the many-particle quantum system of fermions, and the explicit relations between the microscopic and macroscopic fluid variables were derived with help of the Madelung decomposition approach. In the present study, we have focused on the physically important situation with $\check{T}^{\hat{0}} = 0$, a more exotic case with $\check{T}^{\hat{0}} \neq 0$ (which may be realized in cosmology, for example, see the earlier work [15], or in the more general models) will be considered elsewhere.

The resulting system of hydrodynamical equations consists of the continuity Equation (44), the momentum balance Equation (45) and the spin dynamics Equation (46). The momentum balance equation includes the contributions in the form of the quantum Bohm potential and a new spin part of the Bohm quantum potential which are proportional to the square of Planck's constant. In addition, the dynamical equations take into account the thermal effects resulting from the fluctuations of the spin and velocity near their average values. As an application of the formalism, we evaluated the possible effects of the spacetime torsion and the spin part of quantum Bohm potential on the dispersion characteristics of the spin waves (59) excited in the many-particle fermion system. The developed hydrodynamical model can be used in the future studies of various types of transport phenomena in spinning matter with an account of external electromagnetic and gravitational fields.

Author Contributions: All authors discussed the results and contributed to the final manuscript. All authors have read and agreed to the published version of the manuscript.

Funding: The work of MIT was supported by the Russian Science Foundation under the grant 19-72-00017, and the research of YNO was supported in part by the Russian Foundation for Basic Research (Grant No. 18-02-40056-mega).

Conflicts of Interest: The authors declare no conflict of interest.

References

1. Cartan, É. *On Manifolds with an Affine Connection and the Theory of General Relativity*; Magnon, A., Ashtekar, A., Eds.; Bibliopolis: Napoli, Italy, 1986.
2. Sciama, D.W. On the analogy between charge and spin in general relativity. In *Recent Developments in General Relativity, Festschrift for Infeld*; Pergamon Press: Oxford, UK; PWN: Warsaw, Poland, 1962; pp. 415–439.
3. Kibble, T.W.B. Lorentz invariance and the gravitational field. *J. Math. Phys.* **1961**, *2*, 212–221. [CrossRef]
4. Trautman, A. Einstein–Cartan theory. In *Encyclopedia of Mathematical Physics*; Françoise, J.-P., Naber, G.L., Tsou, S.T., Eds.; Elsevier: Oxford, UK, 2006; pp. 189–195. Available online: https://arxiv.org/abs/gr-qc/0606062 (accessed on 1 November 2021).
5. Hehl, F.W.; von der Heyde, P.; Kerlick, G.D. General relativity with spin and torsion and its deviations from Einstein's theory. *Phys. Rev. D* **1974**, *10*, 1066–1069. [CrossRef]
6. Hehl, F.W.; Lemke, J.; Mielke, E.W. Two lectures on fermions and gravity. In *Geometry and Theoretical Physics*; Debrus, J., Hirshfeld, A.C., Eds.; Springer: Heidelberg, Germany, 1991; pp. 56–140.
7. Hehl, F.W.; von der Heyde, P.; Kerlick, G.D.; Nester, J. General relativity with spin and torsion: Foundations and prospects. *Rev. Mod. Phys.* **1976**, *48*, 393–416. [CrossRef]
8. Hehl, F.W.; McCrea, J.D.; Mielke, E.W.; Ne'eman, Y. Metric-affine gauge theory of gravity: Field equations, Noether identities, world spinors, and breaking of dilation invariance. *Phys. Rept.* **1995**, *258*, 1–171 . [CrossRef]
9. Blagojević, M. *Gravitation and Gauge Symmetries*; IOP: Bristol, UK, 2002.
10. Blagojević, M., Hehl, F.W. (Eds.) *Gauge Theories of Gravitation. A Reader with Commentaries*; Imperial College Press: London, UK, 2013. [CrossRef]
11. Einstein, A. Geometrie und Erfahrung. *Sitzungsber. Preuss. Akad. Wiss. Phys.-Math. Klasse* **1921**, *1*, 123–130.
12. Yasskin, P.B.; Stoeger, W.R. Propagating equations for test bodies with spin and rotation in theories of gravity with torsion. *Phys. Rev. D* **1980**, *21*, 2081–2094. [CrossRef]
13. Hehl, F.W.; Obukhov, Y.N.; Puetzfeld, D. On Poincaré gauge theory of gravity, its equations of motion, and Gravity Probe B. *Phys. Lett. A* **2013**, *377*, 1775–1781. [CrossRef]
14. Obukhov, Y.N.; Puetzfeld, D. Multipolar test body equations of motion in generalized gravity theories. In *Fundamental Theories of Physics*; Springer: Cham, Switzerland, 2015; Volume 179, pp. 67–119. [CrossRef]
15. Shapiro, I.L. Physical aspects of the space-time torsion. *Phys. Rept.* **2002**, *357*, 113–213. [CrossRef]
16. Obukhov, Y.N. Poincaré gauge gravity: Selected topics. *Int. J. Geom. Meth. Mod. Phys.* **2006**, *3*, 95–137. [CrossRef]
17. Obukhov, Y.N. Poincaré gauge gravity: An overview. *Int. J. Geom. Meth. Mod. Phys.* **2018**, *15* (Suppl. 1), 1840005. [CrossRef]
18. Ponomarev, V.N.; Barvinsky, A.O.; Obukhov, Y.N. *Gauge Approach and Quantization Methods in Gravity Theory*; Nauka: Moscow, Russia, 2017; 360p.
19. Obukhov, Y.N.; Silenko, A.J.; Teryaev, O.V. Spin-torsion coupling and gravitational moments of Dirac fermions: Theory and experimental bounds. *Phys. Rev. D* **2014**, *90*, 124068. [CrossRef]
20. Gonçalves, B.; Obukhov, Y.N.; Shapiro, I.S. Exact Foldy–Wouthuysen transformation for a Dirac spinor in torsion and other *CPT* and Lorentz violating backgrounds. *Phys. Rev. D* **2009**, *80*, 125034. [CrossRef]
21. Hou, L.S.; Ni, W.-T.; Li, Y.C.M. Test of cosmic spatial isotropy for polarized electrons using a rotatable torsion balance. *Phys. Rev. Lett.* **2003**, *90*, 201101. [CrossRef]
22. Ni, W.-T.; Pan, S.S.; Yeh, H.C.; Hou, L.S.; Wan, J. Search for an axionlike spin coupling using a paramagnetic salt with a dc SQUID. *Phys. Rev. Lett.* **1999**, *82*, 2439–2442. [CrossRef]
23. Chui, T.C.P.; Ni, W.-T. Experimental search for an anomalous spin–spin interaction between electrons. *Phys. Rev. Lett.* **1993**, *71*, 3247–3250. [CrossRef] [PubMed]
24. Wineland, D.J.; Bollinger, J.J.; Heinzen, D.J.; Itano, W.M.; Raizen, M.G. Search for anomalous spin-dependent forces using stored-ion spectroscopy. *Phys. Rev. Lett.* **1991**, *67*, 1735–1738. [CrossRef] [PubMed]
25. Hunter, L.; Gordon, J.; Peck, S.; Ang, D.; Lin, J.-F. Using the Earth as a polarized electron source to search for long-range spin–spin interactions. *Science* **2013**, *339*, 928–932. [CrossRef]
26. Lehnert, R.; Snow, W.M.; Yan, H. A first experimental limit on in-matter torsion from neutron spin rotation in liquid ^{4}He. *Phys. Lett. B* **2014**, *730*, 353–356. [CrossRef]

27. Puetzfeld, D.; Obukhov, Y.N. Prospects of detecting spacetime torsion. *Int. J. Mod. Phys. D* **2014**, *23*, 1442004. [CrossRef]
28. Obukhov, Y.N.; Puetzfeld, D. Conservation laws in gravitational theories with general nonminimal coupling. *Phys. Rev. D* **2013**, *87*, 081502(R). [CrossRef]
29. Fadeev, P.; Wang, T.; Band, Y.B.; Budker, D.; Graham, P.W.; Sushkov, A.O.; Kimball, D.F.J. Gravity Probe Spin: Prospects for measuring general-relativistic precession of intrinsic spin using a ferromagnetic gyroscope. *Phys. Rev. D* **2021**, *103*, 044056. [CrossRef]
30. Lämmerzahl, C. Constraints on space-time torsion from Hughes-Drever experiments. *Phys. Lett. A* **1997**, *228*, 223–231. [CrossRef]
31. Mohanty, S.; Sarkar, U. Constraints on background torsion field from *K*-physics. *Phys. Lett. B* **1998**, *433*, 424–428. [CrossRef]
32. Belyaev, A.S.; Shapiro, I.L.; do Vale, M.A.B. Torsion phenomenology at the CERN LHC. *Phys. Rev. D* **2007**, *75*, 034014. [CrossRef]
33. Kostelecký, V.A.; Russell, N.; Tasson, J.D. Constraints on torsion from bounds on Lorentz violation. *Phys. Rev. Lett.* **2008**, *100*, 111102. [CrossRef] [PubMed]
34. Kostelecký, V.A.; Li, Z. Searches for beyond-Riemann gravity. *Phys. Rev. D* **2021**, *104*, 044054. [CrossRef]
35. Heckel, B.R.; Adelberger, E.G.; Cramer, C.E.; Cook, T.S.; Schlamminger, S.; Schmidt, U. Preferred-frame and *CP*-violation tests with polarized electrons. *Phys. Rev. D* **2008**, *78*, 092006. [CrossRef]
36. Kimball, D.F.J.; Boyd, A.; Budker, D. Constraints on anomalous spin–spin interactions from spin-exchange collisions. *Phys. Rev. A* **2010**, *82*, 062714. [CrossRef]
37. Ivanov, A.N.; Wellenzohn, M. Spin precession of slow neutrons in Einstein–Cartan gravity with torsion, chameleon, and magnetic field. *Phys. Rev. D* **2016**, *93*, 045031. [CrossRef]
38. Ivanov, A.N.; Wellenzohn, M.; Abele, H. Quantum gravitational states of ultracold neutrons as a tool for probing of beyond-Riemann gravity. *Phys. Lett. B* **2021**, *822*, 136640. [CrossRef]
39. Fabbri, L. Fundamental theory of torsion gravity. *Universe* **2021**, *7*, 305. [CrossRef]
40. Ni, W.-T. Searches for the role of spin and polarization in gravity. *Rep. Prog. Phys.* **2010**, *73*, 056901. [CrossRef]
41. Moody, J.; Wilczek, F. New macroscopic forces? *Phys. Rev. D* **1984**, *30*, 130–138. [CrossRef]
42. Graham, P.W.; Irastorza, I.G.; Lamoreaux, S.K.; Lindner, A.; van Bibber, K.A. Experimental searches for the axion and axion-like particles. *Annu. Rev. Nucl. Part. Sci.* **2015**, *65*, 485–514. [CrossRef]
43. Choi, K.; Im, S.H.; Shin, C.S. Recent progress in the physics of axions and axion-like particles. *Annu. Rev. Nucl. Part. Sci.* **2021**, *71*, 225–252. [CrossRef]
44. Anselm, A.A.; Uraltsev, N.G. Long range "arion" field in the radio frequency band. *Phys. Lett. B* **1982**, *116*, 161–164. [CrossRef]
45. Vorobyov, P.V.; Kakhidze, A.I.; Kolokolov, I.V. Axion wind: A search for cosmological axion condensate. *Phys. Atom. Nuclei* **1995**, *58*, 959–963.
46. Hehl, F.W.; Ni, W.-T. Inertial effects of a Dirac particle. *Phys. Rev. D* **1990**, *42*, 2045–2048. [CrossRef]
47. Obukhov, Y.N. Spin, gravity, and inertia. *Phys. Rev. Lett.* **2001**, *86*, 192–195. [CrossRef] [PubMed]
48. Obukhov, Y.N. Obukhov replies. *Phys. Rev. Lett.* **2002**, *89*, 068903. [CrossRef]
49. Trukhanova, M.I. Quantum hydrodynamics approach to the research of quantum effects and vorticity evolution in spin quantum plasmas. *Prog. Theor. Exp. Phys.* **2013**, *2013*, 111I01. [CrossRef]
50. Trukhanova, M.I.; Andreev, P. Hydrodynamic description of Weyl fermions in condensed state of matter. *Eur. Phys. J. B* **2021**, *94*, 170. [CrossRef]
51. Kuzmenkov, L.S.; Maksimov, S.G.; Fedoseev, V.V. Microscopic quantum hydrodynamics of systems of fermions: I. *Theor. Math. Phys.* **2001**, *126*, 110–120. [CrossRef]
52. Kuzmenkov, L.S.; Maksimov, S.G.; Fedoseev, V.V. Microscopic quantum hydrodynamics of systems of fermions: II. *Theor. Math. Phys.* **2001**, *126*, 212–223. [CrossRef]
53. Andreev, P.A.; Kuzmenkov, L.S. On equations for the evolution of collective phenomena in fermion systems. *Russ. Phys. J.* **2007**, *50*, 1251–1258. [CrossRef]
54. Trukhanova, M.I. Quantum hydrodynamics in the rotating reference frame. *Phys. Plasmas* **2016**, *23*, 112114. [CrossRef]
55. Andreev, P.A.; Mosaki, I.N.; Trukhanova, M.I. Quantum hydrodynamics of the spinor Bose–Einstein condensate at non-zero temperatures. *Phys. Fluids* **2021**, *33*, 067108. [CrossRef]
56. Andreev, P.A.; Kuzmenkov, L.S.; Trukhanova, M.I. Quantum hydrodynamics approach to the formation of waves in polarized two-dimensional systems of charged and neutral particles. *Phys. Rev. B* **2011**, *84*, 245401. [CrossRef]
57. Mashhoon, B. General covariance and quantum theory. *Found. Phys.* **1986**, *16*, 619–635. [CrossRef]
58. Mashhoon, B. *Quantum Theory in aCcelerated Frames of Reference*; Lecture Notes in Physics; Springer: Berlin, Germany, 2006; Volume 702, pp. 112–132. [CrossRef]
59. Mashhoon, B. Toward nonlocal electrodynamics of accelerated systems. *Universe* **2020**, *6*, 229. [CrossRef]
60. Madelung, E. Quantentheorie in hydrodynamischer Form. *Z. Physik* **1927**, *40*, 322–326. [CrossRef]
61. Gemmel, C.; Heil, W.; Karpuk, S.; Lenz, K.; Ludwig, C.; Sobolev, Y.; Tullney, K.; Burghoff, M.; Kilian, W.; Knappe-Grüneberg, S.; et al. Ultra-sensitive magnetometry based on free precession of nuclear spins. *Eur. Phys. J. D* **2010**, *57*, 303–320. [CrossRef]
62. Allmendinger, F.; Heil, W.; Karpuk, S.; Kilian, W.; Scharth, A.; Schmidt, U.; Schnabel, A.; Sobolev, Y.; Tullney, K. New limit on Lorentz-invariance- and *CPT*-violating neutron spin interactions using a free-spin-precession ^{3}He-^{129}Xe comagnetometer. *Phys. Rev. Lett.* **2014**, *112*, 110801. [CrossRef] [PubMed]

63. Burghoff, M.; Gemmel, C.; Heil, W.; Karpuk, S.; Kilian, W.; Knappe-Grüneberg, S.; Lenz, K.; Müller, W.; Tullney, K.; Schmidt, U.; et al. Probing Lorentz invariance and other fundamental symmetries in ^{3}He-^{129}Xe clock-comparison experiments. *J. Phys. Conf. Ser.* **2011**, *295*, 012017. [CrossRef]
64. Venema, B.J.; Majumder, P.K.; Lamoreaux, S.K.; Heckel, B.R.; Fortson, E.N. Search for a coupling of the Earth's gravitational field to nuclear spins in atomic mercury. *Phys. Rev. Lett.* **1992**, *68*, 135–138. [CrossRef] [PubMed]
65. de Sabbata, V.; Pronin, P.I.; Sivaram, C. Neutron interferometry in gravitational field with torsion. *Int. J. Theor. Phys.* **1991**, *30*, 1671–1678. [CrossRef]
66. Hammond, R.T. Upper limit on the torsion coupling constant. *Phys. Rev. D* **1995**, *52*, 6918–6921. [CrossRef] [PubMed]
67. Kiefer, C.; Weber, C. On the interaction of mesoscopic quantum systems with gravity. *Ann. Phys. (Leipzig)* **2005**, *14*, 253–278. [CrossRef]
68. Lämmerzahl, C. Interferometry as a universal tool in physics. In *Planck Scale Effects in Astrophysics and Cosmology*; Kowalski-Glikman, J., Amelino-Camelia, G., Eds.; Lecture Notes in Physics; Springer: Berlin, Germany, 2005; Volume 669, pp. 161–198. [CrossRef]
69. Sponar, S.; Sedmik, R.I.P.; Pitschmann, M.; Abele, H.; Hasegawa, Y. Tests of fundamental quantum mechanics and dark interactions with low-energy neutrons. *Nat. Rev. Phys.* **2021**, *3*, 309–327. [CrossRef]
70. Misra, A.; Brodin, G.; Marklund, M.; Shukla, P. Circularly polarized modes in magnetized spin plasmas. *J. Plasma Phys.* **2010**, *76*, 857–864. [CrossRef]
71. Takabayasi, T. Hydrodynamical formalism of quantum mechanics and Aharonov-Bohm effect. *Prog. Theor. Phys.* **1983**, *69*, 1323–1344. [CrossRef]
72. Takabayasi, T. Vortex, spin and triad for quantum mechanics of spinning particle I: General theory. *Prog. Theor. Phys.* **1983**, *70*, 1–17. [CrossRef]
73. Salesi, G.; Recami, E. Hydrodynamical reformulation and quantum limit of the Barut-Zanghi theory. *Found. Phys. Lett.* **1997**, *10*, 533–546. [CrossRef]
74. Fabbri, L. Spinors in polar form. *Eur. Phys. J. Plus* **2021**, *136*, 354. [CrossRef]
75. Hestenes, D. Zitterbewegung structure in electrons and photons. *arXiv* **2020**, arXiv:1910.11085. Available online: https://arxiv.org/abs/1910.11085 (accessed on 1 November 2021).
76. Holland, P.R. *The Quantum Theory of Motion: An Account of the de Broglie-Bohm Causal Interpretation of Quantum Mechanics*; Cambridge University Press: Cambridge, UK, 1995. [CrossRef]

 universe

Article

Search for Manifestations of Spin–Torsion Coupling

Mariya Iv. Trukhanova [1,2], Pavel Andreev [1] and Yuri N. Obukhov [2,*]

[1] Faculty of Physics, M. V. Lomonosov Moscow State University, Leninskie Gory, 119991 Moscow, Russia
[2] Theoretical Physics Laboratory, Nuclear Safety Institute, Russian Academy of Sciences, B. Tulskaya 52, 115191 Moscow, Russia
* Correspondence: obukhov@ibrae.ac.ru

Abstract: We investigate the axial vector spin–torsion coupling effects in the framework of the Poincaré gauge theory of gravity with the general Yang–Mills type Lagrangian. The dynamical equations for the "electric" and "magnetic" components of the torsion field variable are obtained in the general form and it is shown that the helicity density and the spin density of the electromagnetic field appear as the physical sources. The modified Maxwell's equations for the electromagnetic field are derived, and the electromagnetic wave propagation under the action of the uniform homogeneous torsion field is considered. We demonstrate the Faraday effect of rotation of the polarization for such a wave and establish the strong bound on the possible cosmic axial torsion field from the astrophysical data.

Keywords: spin; gauge gravity; torsion; electromagnetic wave; Faraday effect

Citation: Trukhanova, M.I.; Andreev, P.; Obukhov, Y.N. Search for Manifestations of Spin–Torsion Coupling. *Universe* **2023**, 9, 38. https://doi.org/10.3390/universe9010038

Academic Editor: Luca Fabbri

Received: 30 November 2022
Revised: 28 December 2022
Accepted: 30 December 2022
Published: 6 January 2023

1. Introduction

The search for new spin-dependent interactions between fundamental particles apart from the magnetic dipole interaction is an important area of high-energy physics research beyond the Standard Model [1]. É. Cartan was the first who proposed, at the beginning of the 20th century, the post-Riemannian geometrical structures generated by the microstructural properties of the physical matter, particularly to analyze the coupling of the torsion of spacetime to the intrinsic spin [2].

The interest in the theory of gravitation with spin and torsion based on the Riemann–Cartan geometry had considerably grown in the second half of the 20th century after the consistent gauge theory formalism was developed [3–7]. It is now well established that the spacetime torsion can only be detected with the help of the spin [8–10]. The early theoretical analysis of the possible experimental manifestations of the torsion field can be found in [11–13]. The so-called Einstein–Cartan theory [14–16] with the linear Hilbert–Einstein gravitational Lagrangian represents a degenerate version of the Poincaré gauge gravity. In this model, the torsion couples to spin algebraically and, therefore, it vanishes outside the matter sources, but essentially modifies the physical structure inside astrophysical compact objects, see [17–19].

The torsion becomes a dynamical propagating field in the Poincaré gauge gravity theory with a Yang–Mills type Lagrangian [20], and the most general gravitational model with the Lagrangian (which is quadratic in the curvature and torsion) was considered in Reference [21] with an emphasis on the consistency of the gauge theory of gravity with experimental observations at the macroscopic level. Accordingly, its probing should be essentially confined to the microscopic level and focus on the study of the dynamics of fundamental particles, atoms, and molecules. It is worthwhile to note that it has not yet been possible to create a source of spin density that could generate torsion to be detected in the laboratory. However, one can establish the constraints on the spin–torsion coupling, in particular from the experimental search for the Lorentz and CPT violation. The bounds on new spin-dependent interactions were found [22,23] with the help of a torsion pendulum technique, which was also used in the search for CP-violating interactions between the pendulum's electrons and unpolarized matter in laboratory surroundings or the Sun. Among other physical effects, the contribution of the interacting vector and pseudovector of the torsion to the hyperfine splitting of the ground state of the hydrogen atom was evaluated in [24,25], whereas a possible manifestation of the spin–torsion coupling in the scattering of polarized photons in a medium of sodium vapor was analyzed in [26]. The experimental upper bounds on the spin–torsion interactions were reported in [27–29].

In the context of the growing interest in fundamental physics at the sub-eV scale, Moody and Wilczek [30] analyzed new fields, generating dipole couplings between fermions that can be detectable in laboratory experiments, paying special attention to axions. An axion as a hypothetical particle was postulated in the Peccei–Quinn theory to resolve the strong CP problem in quantum chromodynamics, and it can produce long-range dipole forces [31]. Similar dipole interactions between fermions can be produced by other particles [32–35], for example, by an arion, which is a boson corresponding to a spontaneous breaking of the chiral lepton symmetry [35]. The search for axions and axion-like particles is of considerable interest in relation to the cold dark matter issue. Currently, there is a number of experimental attempts to find axions that encompass the Primakoff effect for the astrophysical axions [36], the polarization measurements for light propagating in a magnetic field, the light shining through wall experiments [37], and the cosmic axion spin precession experiment [38,39].

Another direction in the search for long-range spin-dependent interactions is the prediction of the existence of an unparticle in the context of quantum excitations of scale-invariant interactions [40], along with exotic spin-1 bosons or paraphotons [41], which are currently being actively investigated [42]. Quite generally, the analysis of the behaviors of atomic systems affected by new hypothetical spin-dependent forces gives rise to the constraints on the coupling constants [1] with the sixteen types of potentials characterizing interactions between fermions mediated by the new exotic particles [42,43]. A wide range of relevant laboratory investigations was carried out using the physical methods of atomic and molecular systems, as well as make use of the optical methods: ion capture experiments [44], using nitrogen-vacancy centers in diamonds [45,46], based on molecular and atomic spectroscopies [47].

The gauge symmetry is one of the most powerful physical concepts underlying the description of fundamental interactions. The global gauge $U(1)$ invariance in quantum electrodynamics leads to the electric charge conservation law, whereas the local gauge invariance requires the introduction of a massless vector field that mediates the long-range interaction between charges, with the electric current as a source of the electromagnetic field. In quantum chromodynamics, the invariance under the global group $SU(2)$ yields the isospin conservation law, and the invariance under the local transformations of the group $SU(2)$ leads to the introduction of the Yang–Mills gauge field. The extension of this theory leads to the explanation of the strong interactions in terms of the exchange of gluons.

Recently [48,49], an attempt was made to construct a gauge theory model to describe the weak spin–spin interactions. It was suggested [48] that the invariance of the Lagrangian under the local Lorentz transformation requires the introduction of a massless axial vector gauge field, which gives rise to a super-weak long-range spin–spin force in a vacuum, which is attractive for parallel spins. In this model, the axial vector field couples to the axial vector current of the Dirac fermion field and the photon field or a neutral spin-1 field. The axial vector field was introduced [50] to provide stability of the classical electron and to construct divergence-free quantum electrodynamics. Optical experiments are the most accurate and accessible for measuring the effects of new physical fields. The direct interactions between electromagnetic and gauge fields were considered in [51] in the Poincaré gauge gravity approach, whereas the coupling of the photon to the axial vector gauge field was supported by the local Lorentz symmetry group in the approach [48,49]. Here, we study the possible influence of the axial torsion field on the propagation and polarization of an electromagnetic wave.

The structure of the paper is as follows. In Section 2, we briefly outline the corresponding Lagrange–Noether framework of the Poincaré gauge gravity theory. In Section 3, we study the propagation of the electromagnetic wave under the influence of the uniform homogeneous axial vector torsion field. Finally, in Section 4, we discuss the results obtained and apply them to derive the strong upper limit of the value of the cosmic background torsion field from the astrophysical data. In Appendix A, the structure of the general quadratic Poincaré gauge gravitational field Lagrangian is given, and the effective coupling constants are introduced.

Our basic conventions and notations are as follows. The world indices are labeled by Latin letters $i, j, k, \ldots = 0, 1, 2, 3$ (for example, the local spacetime coordinates x^i), whereas we reserve Greek letters for tetrad indices, $\alpha, \beta, \ldots = 0, 1, 2, 3$ (i.e., for labeling the legs of an anholonomic coframe e_i^α). In order to distinguish separate tetrad indices, we put hats over them. Finally, spatial indices are denoted by Latin letters from the beginning of the alphabet, $a, b, c, \ldots = 1, 2, 3$. The metric of the Minkowski spacetime reads $g_{ij} = \mathrm{diag}(c^2, -1, -1, -1)$, and the totally antisymmetric Levi–Civita tensor η_{ijkl} has the only nontrivial component $\eta_{0123} = c$, so that $\eta_{0abc} = c\varepsilon_{abc}$ with the three-dimensional Levi–Civita tensor ε_{abc}. The spatial components of the tensor objects are raised and lowered with the help of the Euclidean three-dimensional metric δ_{ab}.

2. Gauge Theory of Gravitation

The Poincaré gauge gravity [52–57] is an extension of Einstein's general relativity theory (GR), in which the spin, energy, and momentum are independent sources of the gravitational fields (the metric g_{ij} and connection $\Gamma_{ki}{}^j$), and the spacetime structure is described by the Riemann–Cartan geometry with the curvature and the torsion:

$$R_{kli}{}^j = \partial_k \Gamma_{li}{}^j - \partial_l \Gamma_{ki}{}^j + \Gamma_{kn}{}^j \Gamma_{li}{}^n - \Gamma_{ln}{}^j \Gamma_{ki}{}^n, \tag{1}$$

$$T_{kl}{}^i = \Gamma_{kl}{}^i - \Gamma_{lk}{}^i. \tag{2}$$

The Riemann–Cartan connection can be decomposed into the Riemannian and the post-Riemannian parts,

$$\Gamma_{kj}{}^i = \widetilde{\Gamma}_{kj}{}^i - K_{kj}{}^i, \tag{3}$$

where the Christoffel symbols are determined by the metric

$$\widetilde{\Gamma}_{kj}{}^i = \frac{1}{2} g^{il} (\partial_j g_{kl} + \partial_k g_{lj} - \partial_l g_{kj}), \tag{4}$$

and the contortion tensor is constructed in terms of the torsion

$$K_{kj}{}^i = -\frac{1}{2}(T_{kj}{}^i + T^i{}_{kj} + T^i{}_{jk}). \tag{5}$$

The torsion (2) can be decomposed [57] into three irreducible components,

$$T_{kl}{}^i = {}^{(1)}T_{kl}{}^i + {}^{(2)}T_{kl}{}^i + {}^{(3)}T_{kl}{}^i, \tag{6}$$

where the second irreducible part features the torsion trace vector

$${}^{(2)}T_{kl}{}^i = \frac{1}{3}\Big(\delta^i_k T_l - \delta^i_l T_k\Big), \tag{7}$$

the third irreducible part is constructed in terms of the torsion axial pseudo-vector

$${}^{(3)}T_{kl}{}^i = -\frac{1}{3}\,\eta_{kl}{}^{ij}\overline{T}_j, \tag{8}$$

and the first irreducible purely tensor part has the properties

$$^{(1)}T_{ik}{}^{i} = 0, \qquad ^{(1)}T_{ijk}\eta^{ijkl} = 0. \tag{9}$$

The vector and pseudovector of torsion are defined as

$$T_j := T_{ij}{}^{i}, \qquad \overline{T}^j = \frac{1}{2}T_{kli}\eta^{klij}. \tag{10}$$

Here, we focus on the dynamical realization of the Poincaré gauge theory as a Yang–Mills type model with the most general quadratic in curvature and torsion Lagrangian (A3), see Appendix A for the details. Earlier [25,51], the contributions of the vector and the pseudovector (10) to physical effects at the microscopic level were analyzed in the framework of this theory and the strong constraints were established on the spin–torsion coupling parameters. Following [51], we continue to study the influence of the *axial pseudovector* torsion field $^{(3)}T_{kl}{}^{i}$ on physical matters, and assume that the metric of spacetime is flat, whereas possible post-Riemannian deviations of the spacetime geometry are small.

As a result, the connection (3) reduces to the contortion, $\Gamma_{kj}{}^{i} = -K_{kj}{}^{i} = \frac{1}{2}{}^{(3)}T_{kj}{}^{i}$, and by combining (1) and (8), we find (for the curvature) that $R_{kli}{}^{j} = \frac{1}{3}\eta_i{}^{jn}{}_{[k}\partial_{l]}\overline{T}_n$, for the small post-Riemannian corrections. Then it is straightforward to verify that the Yang–Mills-type gauge gravity Lagrangian (A3) is simplified to

$$L = \hbar\left\{-\frac{1}{4}f_{ij}f^{ij} + \frac{\mu^2}{2}\alpha_i\alpha^i - \frac{\lambda}{2}(\partial_i\alpha^i)^2\right\}. \tag{11}$$

Here, $f_{ij} = \partial_i\alpha_j - \partial_j\alpha_i$ is constructed from the rescaled axial torsion trace vector field

$$\alpha_i = \frac{\ell_\rho}{3}\sqrt{\frac{-\Lambda_5}{2\kappa c\hbar}}\,\overline{T}_i, \tag{12}$$

and the coupling constants (A4)–(A9) of the Poincaré gravity Lagrangian (A3) determine

$$\mu^2 = -\frac{3\mu_1}{\ell_\rho^2\Lambda_5}, \qquad \lambda = \frac{3\Lambda_4}{2\Lambda_5}. \tag{13}$$

It is known that the particle spectrum of the Yang–Mills type Poincaré gauge gravity model (A3) contains, in general, the so-called ghost and tachyon modes that may lead to the loss of stability and unitarity of the theory. These issues were analyzed in References [58–60], and the necessary stability conditions were derived that restrict the choices of the coupling constants. Accordingly, we here specialize to the class of models with $b_3 = 2b_1$, which for the spin 1 sector yields $\lambda = 0$, thus avoiding the stability and unitarity problems. Furthermore, we assume that $a_2 = a_1 - a_0$, which corresponds to the vanishing rest mass $\mu = 0$, in agreement with the estimates derived for the axial vector field [60,61] from the high-energy physics phenomenology.

2.1. Interaction between Fermions and Axial Torsion

In accordance with the minimal coupling principle [29], the interaction between the fermion field ψ and gauge fields is introduced via the spinor covariant derivative $D_i\psi$, where

$$D_i = \partial_i - \frac{iq}{\hbar}A_i + \frac{i}{4}\Gamma_i{}^{\alpha\beta}\sigma_{\alpha\beta}, \tag{14}$$

with the Lorentz group generators $\sigma^{\alpha\beta} = i\gamma^{[\alpha}\gamma^{\beta]}$ constructed from the Dirac matrices γ^α. As a result, the dynamics of the spinor field coupled with the gauge fields on the Minkowski flat metric background is given by the Lagrangian

$$L_D = \frac{i\hbar}{2}\left\{\overline{\psi}\gamma^i\partial_i\psi - (\partial_i\overline{\psi})\gamma^i\psi\right\} - mc\overline{\psi}\psi + qA_i\overline{\psi}\gamma^i\psi + \frac{3}{4}\hbar\chi\alpha_i\overline{\psi}\gamma^i\gamma_5\psi. \tag{15}$$

Here, we used the identity $\gamma^{\mu}\sigma^{\alpha\beta} + \sigma^{\alpha\beta}\gamma^{\mu} = -2\epsilon^{\mu\nu\alpha\beta}\gamma_{\nu}\gamma_5$ and introduced

$$\chi = \frac{1}{\ell_\rho}\sqrt{\frac{2\kappa c\hbar}{-\Lambda_5}}. \tag{16}$$

Thereby, the axial pseudovector torsion field naturally couples to the spinor axial current $j^i_f = \overline{\psi}\gamma^i\gamma_5\psi$ or to the spin and helicity of the Dirac fermion. By recalling the definition of the Planck length $\ell_{\rm Pl}$, we can recast (16) into

$$\chi = \frac{\ell_{\rm Pl}}{\ell_\rho}\sqrt{\frac{16\pi}{-\Lambda_5}}. \tag{17}$$

which demonstrates that the spin–torsion coupling constant χ is *very small*, provided we assume that the characteristic length of the Poincaré gauge gravity is much larger than the Planck scale, $\ell_\rho \gg \ell_{\rm Pl}$.

2.2. Interaction between the Electromagnetic Field and Axial Torsion

Following Pradhan et al. [48–50], the interaction of the axial torsion and the electromagnetic field is derived from the standard Maxwell–Lorentz Lagrangian when the ordinary derivatives are replaced by covariant ones. An apparent breaking of the $U(1)$ gauge invariance can be fixed by the modified Stueckelberg method [62]. Together with the dynamical Lagrangian for the axial vector field (11) and the fermion sector terms (15), the total Lagrangian for the torsion field interacting with the spin of the matter sources then reads [49]

$$\begin{aligned} L = &- \frac{1}{4}\sqrt{\frac{\varepsilon_0}{\mu_0}}F_{ij}F^{ij} + \frac{i\hbar}{2}\left\{\overline{\psi}\gamma^i\partial_i\psi - (\partial_i\overline{\psi})\gamma^i\psi\right\} - mc\overline{\psi}\psi + qA_i\overline{\psi}\gamma^i\psi \\ &+ \hbar\left\{-\frac{1}{4}f_{ij}f^{ij} + \frac{\mu^2}{2}\alpha^2 - \frac{\lambda}{2}(\partial\alpha)^2\right\} + \frac{3}{4}\hbar\chi\alpha_i\overline{\psi}\gamma^i\gamma_5\psi - \chi\sqrt{\frac{\varepsilon_0}{\mu_0}}\alpha_i\eta^{ijkl}A_j\partial_kA_l, \end{aligned} \tag{18}$$

where ε_0 and μ_0 are the electric and magnetic constants of the vacuum. The axial vector torsion is thereby coupled to the axial current density of the electron and photon fields:

$$j^i_f = \overline{\psi}\gamma^i\gamma_5\psi, \qquad j^i_b = \eta^{ijkl}A_j\partial_kA_l, \tag{19}$$

where we explicitly have

$$A_i = \{-\phi, \boldsymbol{A}\}, \qquad \alpha_i = \{-\varphi, \boldsymbol{\alpha}\}. \tag{20}$$

The quantization of the model (18) was analyzed in [51] and the static potential between fermions (due to the exchange of the axial torsion) was computed.

2.3. The Electromagnetic Source of an Axial Vector Field

Let us consider the spin densities of the spinor and electromagnetic fields as the sources of the axial vector field α_i. The components of the axial currents (19) for fermions and photon fields can be obtained by substituting Dirac bispinors $\psi = \begin{pmatrix} u \\ v \end{pmatrix}$ into Equation (19).

$$j^i_f = -\{(u^*v + v^*u)/c,\ (u^*\boldsymbol{\sigma}u + v^*\boldsymbol{\sigma}v)\}, \qquad j^i_b = \frac{1}{c}\{\boldsymbol{A}\cdot\boldsymbol{B},\ (\boldsymbol{E}\times\boldsymbol{A} + \phi\boldsymbol{B})\}. \tag{21}$$

2.4. Field Equations

Neglecting the fermion sector, let us derive the field equations for the Lagrangian (18). Technically, we need to make variations with respect to the axial torsion field α_i and the electromagnetic field potential A_i. The corresponding Euler–Lagrange equation $\delta L/\delta\alpha_i = 0$ for the torsion reads:

$$-\partial_j f^{ij} + \mu^2\alpha^i + \lambda\partial^i(\partial\alpha) + \chi\sqrt{\frac{\varepsilon_0}{\mu_0}}\,\eta^{ijkl}A_j\partial_k A_l = 0. \tag{22}$$

Introducing the "electric" and "magnetic" components of the torsion variable by means of

$$\mathcal{E}_a = f_{a0}, \qquad \mathcal{B}^a = \frac{1}{2}\epsilon^{abc}f_{bc}, \qquad a,b = 1,2,3, \tag{23}$$

along with the usual definitions of the electric and magnetic fields

$$E_a = F_{a0}, \qquad B^a = \frac{1}{2}\epsilon^{abc}F_{bc}, \qquad a,b = 1,2,3, \tag{24}$$

we recast (22) into the three-dimensional form:

$$\boldsymbol{\nabla}\cdot\boldsymbol{\mathcal{E}} - \mu^2\varphi - \lambda\partial_t(\partial\alpha) = -\frac{\chi}{\mu_0}\boldsymbol{A}\cdot\boldsymbol{B}, \tag{25}$$

$$\boldsymbol{\nabla}\times\boldsymbol{\mathcal{B}} - \frac{1}{c^2}\partial_t\boldsymbol{\mathcal{E}} + \mu^2\boldsymbol{\alpha} + \lambda\boldsymbol{\nabla}(\partial\alpha) = -\chi\varepsilon_0(\phi\boldsymbol{B} + \boldsymbol{E}\times\boldsymbol{A}). \tag{26}$$

In a similar way, we derived the modified Maxwell equations as the Euler–Lagrange equation $\delta L/\delta A_i = 0$:

$$-\partial_j F^{ij} + \frac{\chi}{2}\eta^{ijkl}A_j f_{kl} - \chi\,\eta^{ijkl}\alpha_j F_{kl} = 0. \tag{27}$$

An immediate observation is in order. Since $\partial_i\partial_j F^{ij} = 0$ identically (symmetric lower indices contracted with the antisymmetric upper indices), by taking the divergence ∂_i of the field Equation (22) (in the special class of stable and unitary models under consideration with $\lambda = 0$ and $\mu^2 = 0$) and (27), we derive

$$\eta^{ijkl}F_{ij}F_{kl} = 0, \qquad \eta^{ijkl}F_{ij}f_{kl} = 0, \tag{28}$$

respectively. Making use of (23) and (24), we find that only crossed-field configurations are actually allowed:

$$\boldsymbol{E}\cdot\boldsymbol{B} = 0, \qquad \boldsymbol{E}\cdot\boldsymbol{\mathcal{B}} + \boldsymbol{\mathcal{E}}\cdot\boldsymbol{B} = 0. \tag{29}$$

In particular, this includes wave configurations.

By making use of (20), (23), and (24), we rewrite the inhomogeneous Maxwell Equation (27) in the three-dimensional form:

$$\boldsymbol{\nabla}\cdot\boldsymbol{E} = 2\chi c\,\boldsymbol{\alpha}\cdot\boldsymbol{B} - \chi c\,\boldsymbol{A}\cdot\boldsymbol{\mathcal{B}}, \tag{30}$$

$$\boldsymbol{\nabla}\times\boldsymbol{B} - \frac{1}{c^2}\partial_t\boldsymbol{E} = \frac{2\chi}{c}(\varphi\boldsymbol{B} + \boldsymbol{E}\times\boldsymbol{\alpha}) - \frac{\chi}{c}(\phi\boldsymbol{\mathcal{B}} + \boldsymbol{\mathcal{E}}\times\boldsymbol{A}). \tag{31}$$

As usual, we have to add the homogeneous Maxwell system,

$$\boldsymbol{\nabla}\cdot\boldsymbol{B} = 0, \tag{32}$$

$$\boldsymbol{\nabla}\times\boldsymbol{E} + \partial_t\boldsymbol{B} = 0. \tag{33}$$

3. Influence of Axial Torsion on Electromagnetic Wave

Here, we focus on the analysis of the dynamics of the electromagnetic field under the action of the background axial torsion, whereas the full coupled system will be considered elsewhere. Among the possible background configurations, of special interest are the cases of the uniform homogeneous field and the wave configurations. The uniform background field may arise on macroscopic scales, mimicking a distinguished cosmic frame violating the Lorentz symmetry, similar to the mechanisms discussed in [63–66]. It seems natural to turn to the case of the uniform axial torsion background that was extensively considered in earlier literature [28,60,61]

3.1. The Case of the Uniform External Axial Torsion Field

Assuming the uniform axial torsion field, when the components $\alpha_i = \{-\varphi, \boldsymbol{\alpha}\}$ are constant in time and do not change in space, we can solve Maxwell's equations for the plane wave ansatz

$$\boldsymbol{E} = \boldsymbol{E}_0 e^{-i\omega t + i\boldsymbol{k}\cdot\boldsymbol{r}}, \qquad \boldsymbol{B} = \boldsymbol{B}_0 e^{-i\omega t + i\boldsymbol{k}\cdot\boldsymbol{r}}. \tag{34}$$

Substituting this into the homogeneous system, (32) and (33), we derive $\boldsymbol{k}\cdot\boldsymbol{B} = 0$, and

$$\boldsymbol{B} = \frac{\boldsymbol{k}\times\boldsymbol{E}}{\omega}, \tag{35}$$

and making use of this in (31), we obtain the algebraic equation

$$\boldsymbol{k}\times(\boldsymbol{k}\times\boldsymbol{E}) + \frac{\omega^2}{c^2}\boldsymbol{E} + i\frac{2\chi}{c}(\varphi\boldsymbol{k} - \omega\boldsymbol{\alpha})\times\boldsymbol{E} = 0. \tag{36}$$

Evaluating the determinant, we find the dispersion relation

$$\left(\frac{\omega^2}{c^2} - k^2\right)^2\frac{\omega^2}{c^2} - \left(\frac{\omega^2}{c^2} - k^2\right)u^2 - (ku)^2 = 0. \tag{37}$$

Here we denoted

$$\boldsymbol{u} := \frac{2\chi}{c}(\varphi\boldsymbol{k} - \omega\boldsymbol{\alpha}), \tag{38}$$

whereas $k^2 = \boldsymbol{k}\cdot\boldsymbol{k}$, $u^2 = \boldsymbol{u}\cdot\boldsymbol{u}$, and $(ku) = \boldsymbol{k}\cdot\boldsymbol{u}$. It is worthwhile to notice that, similar to the wave in a vacuum, the magnetic field is orthogonal to the electric field and the wave vector,

$$\boldsymbol{B}\cdot\boldsymbol{E} = 0, \qquad \boldsymbol{B}\cdot\boldsymbol{k} = 0, \tag{39}$$

which follows from (35). However, the electric field is not orthogonal to the wave vector, in general,

$$i\boldsymbol{k}\cdot\boldsymbol{E} = 2\chi c\,\boldsymbol{\alpha}\cdot\boldsymbol{B} = \frac{2\chi c}{\omega}\,\boldsymbol{\alpha}\cdot(\boldsymbol{k}\times\boldsymbol{E}), \tag{40}$$

which is derived by substituting the plane wave ansatz (34) into the inhomogeneous Equation (30); this is also a direct consequence of (36).

Special case 1. Assuming $\varphi \neq 0, \boldsymbol{\alpha} = 0$, from (37), we find a simpler dispersion relation

$$\left(\frac{\omega^2}{c^2} - k^2\right)^2 - \left(\frac{2\chi}{c}\right)^2\varphi^2 k^2 = 0. \tag{41}$$

This can be immediately recast into

$$\frac{\omega^2}{c^2} = \boldsymbol{k}\cdot\boldsymbol{k} \pm \frac{2\chi}{c}\,\varphi\sqrt{\boldsymbol{k}\cdot\boldsymbol{k}}. \tag{42}$$

The second term on the right-hand side describes a deformation of the light cone under the influence of the torsion component φ.

Special case 2. Assuming $\varphi = 0, \boldsymbol{\alpha} \neq 0$, the general result (37) reduces to

$$\left(\frac{\omega^2}{c^2} - k^2\right)^2 - (2\chi)^2\left[\left(\frac{\omega^2}{c^2} - k^2\right)\alpha^2 + (k\alpha)^2\right] = 0. \tag{43}$$

Here, $\alpha^2 = \boldsymbol{\alpha}\cdot\boldsymbol{\alpha}$, and $(k\alpha) = \boldsymbol{k}\cdot\boldsymbol{\alpha}$. The resulting dispersion relation can be straightforwardly simplified into

$$\frac{\omega^2}{c^2} = \boldsymbol{k}\cdot\boldsymbol{k} + 2\chi^2\boldsymbol{\alpha}\cdot\boldsymbol{\alpha} \pm 2\chi\sqrt{\chi^2(\boldsymbol{\alpha}\cdot\boldsymbol{\alpha})^2 + (\boldsymbol{k}\cdot\boldsymbol{\alpha})^2} = 0. \tag{44}$$

3.2. Rotation of the Polarization Plane

Without loss of generality, one can assume that the electromagnetic wave propagates along the z-axis, in other words, we take $\boldsymbol{k} = (0, 0, k_z)$. It is more convenient to analyze the two special cases separately. In the first case ($\varphi \neq 0, \boldsymbol{\alpha} = 0$), the dispersion relation (42) yields two values for the wave vector,

$$k_z = k_\pm = \frac{\omega \pm \chi\varphi}{c}, \tag{45}$$

in the leading order of the small coupling constant χ. Hence, there are two independent waves propagating along z with two different phase velocities. As a result, after extracting the corresponding amplitudes $\boldsymbol{E}_\pm$ for the electric field from the algebraic Equation (36), we find the solution

$$\boldsymbol{E} = E_+ e^{-i\omega t + ik_+ z}\,\boldsymbol{e}_+ + E_- e^{-i\omega t + ik_- z}\,\boldsymbol{e}_-\,, \tag{46}$$

where we denote the combinations of the basis vectors

$$\boldsymbol{e}_\pm = \frac{\boldsymbol{e}_x \mp i\boldsymbol{e}_y}{2}. \tag{47}$$

Therefore, from the point of view of physics, the solution (46) describes the superposition of the right-hand (counter-clockwise) circularly polarized and the left-hand (clockwise) circularly polarized waves.

Recalling that the linearly polarized wave arises as the sum of the right-hand and left-hand circular waves with equal amplitudes, $E_+ = E_-$, we then obtain the real solution

$$\boldsymbol{E} = E_0 \cos[\omega(t - z/c)]\{\cos(\chi\varphi z/c)\,\boldsymbol{e}_x - \sin(\chi\varphi z/c)\,\boldsymbol{e}_y\}. \tag{48}$$

Thus, we recover the *Faraday effect* when the polarization vector continuously rotates with the propagation of the plane wave. The polarization rotation angle γ, see Figure 1, from the initial point $z = 0$ until the point $z = h$ is determined by

$$\gamma = \frac{\chi\varphi h}{c}. \tag{49}$$

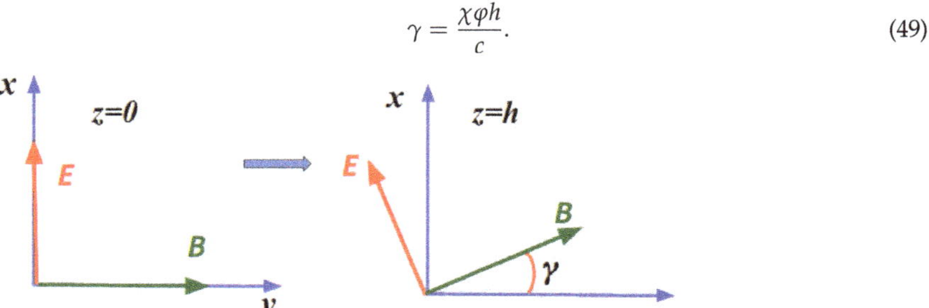

Figure 1. Faraday effect of polarization rotation under the action of uniform axial torsion.

For the second case of the axial torsion field configuration ($\varphi = 0, \boldsymbol{\alpha} \neq 0$), the form of the solution depends on the relative orientation of the $\boldsymbol{\alpha}$ with respect to the wave vector of the electromagnetic wave $\boldsymbol{k}$. Quite generally, given the direction of the wave propagation, we can decompose $\boldsymbol{\alpha} = \boldsymbol{\alpha}_{||} + \boldsymbol{\alpha}_\perp$ into the longitudinal and transversal projections on the wave vector $\boldsymbol{k}$ and the plane orthogonal to it. Accordingly, we derive the solution for the real electric field of the linearly polarized electromagnetic plane wave, to the first order in the interaction constant χ:

$$
\begin{aligned}
E = E_0 \cos[\omega(t - z/c)]\Big\{\cos(\chi\alpha_{||}z)\, e_x - \sin(\chi\alpha_{||}z)\, e_y\Big\} \\
- \frac{2\chi\alpha_\perp c}{\omega} E_0 \sin[\omega(t - z/c)] \sin(\chi\alpha_{||}z + \phi_0)\, e_z\,.
\end{aligned} \tag{50}
$$

Since the wave propagates along the z-axis, we have $\boldsymbol{k} \cdot \boldsymbol{\alpha} = k_z\alpha_{||}$, and

$$
\alpha_{||} = \boldsymbol{e}_z \cdot \boldsymbol{\alpha} = \alpha \cos\theta, \qquad \alpha_\perp = \alpha \sin\theta, \tag{51}
$$

and $\cos\phi_0 = \boldsymbol{e}_x \cdot \boldsymbol{\alpha}_\perp$ measures the angle between the projection $\boldsymbol{\alpha}_\perp$ and the basis vector $\boldsymbol{e}_x$. Obviously, by choosing the coordinate frame appropriately, we can always make $\phi_0 = 0$. It is worthwhile to note that, in accordance with (40), the electric field has a nontrivial component along the wave vector and the third term in (50) vanishes only when the wave propagates along the axial torsion $\boldsymbol{\alpha}$. In that case, the polarization rotation angle

$$
\gamma = \chi\alpha_{||}h = \chi\alpha \cos\theta h \tag{52}
$$

from the initial point $z = 0$ to the endpoint $z = h$ is maximal. However, when the axial vector field is orthogonal to the direction of propagation of the electromagnetic wave $\boldsymbol{\alpha} \perp \boldsymbol{k}$, there will be no rotation of the polarization plane. This generalized Faraday effect is supported by the solution of the dispersion Equation (43),

$$
k_z = k_\pm = \frac{\omega}{c} \pm \chi\alpha_{||} = \frac{\omega}{c} \pm \chi\alpha \cos\theta, \tag{53}
$$

in the leading order of the small coupling constant χ, which gives rise to the two waves traveling with two different phase velocities.

4. Discussion and Conclusions

The search for exotic new forces and interactions generated by the spin of matter particles, fields, and continuous media has a long history. Since the corresponding spin–torsion coupling constant χ is very small, the detection of such new forces becomes a challenging issue and requires high-precision measurements. On the other hand, the new fields should be truly highly penetrating.

Taking into account that optical experiments are among the most accurate ones, the analysis of possible optical effects of the axial vector torsion field appears to be quite promising. Here, the classical dynamics of the axial vector field were studied in the framework of the Yang–Mills type Poincaré gravity model with a focus on the interaction with the electromagnetic field. We derived the dynamical equations for the axial vector torsion field, (25) and (26), and identified the source of the "electric" component of the torsion variable with the helicity density of the electromagnetic field $\sim \boldsymbol{A} \cdot \boldsymbol{B}$, which characterizes non-trivial topological properties of the field configuration, whereas the "magnetic" component of the torsion variable is generated by the spin density of the electromagnetic field.

Continuing the earlier studies of the spin–torsion effects in the fermion sector [24–26], we here turn to the boson sector. Maxwell's equations are modified (30)–(33) in the presence of the axial torsion field. The analysis of the propagation of electromagnetic waves under the action of the axial vector torsion field reveals the Faraday effect of the rotation of the wave's polarization and the angle of rotation is determined by the coupling constant, the magnitude of the axial vector torsion field and the travel distance: $\gamma = \frac{\chi\varphi h}{c}$ or $\gamma = \chi\alpha_{||}h$. This is consistent with the similar effect arising due to the Lorentz symmetry violation [63,64] or due to the action of the pseudoscalar axion [65,66].

Compared to earlier literature that focused on the evaluation of the spin–spin interaction potential [20,30,34,42,43,48,49,51], the results obtained provide a qualitatively new approach to the search for possible manifestations of the spin–torsion coupling with the help of the optical polarization methods. When discussing the most general model of the interacting electromagnetic field and propagating axial torsion field [67,68], one has to pay special attention to the acausal (i.e., superluminal propagation) anomalies. In agreement with the conclusions of [67,68], the class of models under consideration with $b_3 = 2b_1$ yields $\lambda = 0$ for the axial pseudovector field, and so the stability, unitarity, and causality issues are safely avoided.

Assuming the cosmic nature of the background axial torsion, we can apply the results obtained in the astrophysical situation, and analyze the distribution over the sky sphere of the polarization of radiation, coming from distant radio galaxies. Then taking the observational data collected in Table I of Reference [65], and repeating verbatim the computations by replacing the Lorentz-violating parameter with the axial torsion pseudovector, we establish the strong bounds on the torsion's magnitude: $|\overline{T}| \lesssim 8.7 \times 10^{-27}\,\mathrm{m}^{-1}$. This turns out to be significantly lower than the limits found from the analysis of the spin–torsion coupling effects in the fermion sector [12,24–29].

Author Contributions: All authors discussed the results and contributed to the final manuscript. All authors have read and agreed to the published version of the manuscript.

Funding: The work of MIT was supported by the Russian Science Foundation under grant 22-72-00036.

Data Availability Statement: No data is associated with this theoretical work; accordingly no data will be deposited.

Acknowledgments: We express our gratitude to I. A. Kudryashov (Nuclear Physics Research Institute of Moscow State University) for discussing the problem.

Conflicts of Interest: The authors declare no conflict of interest.

Appendix A. Poincaré Gauge Gravity Dynamics

In the literature, the quadratic Poincaré gravity theories are often formulated in terms of the standard tensor objects, which are not decomposed into irreducible parts. As usual, we introduce the *Ricci tensor* and the *co-Ricci tensor* as

$$R_{ij} := R_{kij}{}^{k}, \qquad \overline{R}^{ij} := \frac{1}{2}\, R_{klm}{}^{i}\, \eta^{klmj}, \tag{A1}$$

from which the curvature scalar and pseudoscalar arise naturally as the traces

$$R = g^{ij}R_{ij} = R_{ij}{}^{ji}, \qquad \overline{R} = g_{ij}\overline{R}^{ij} = \frac{1}{2}\, R_{ijkl}\, \eta^{ijkl}. \tag{A2}$$

We consider the general quadratic model with the Yang–Mills type Lagrangian that contains all possible quadratic invariants of the torsion and the curvature:

$$\begin{aligned} L = -\frac{1}{2\kappa c}\Big\{ & a_0 R + \overline{a}_0 \overline{R} + 2\lambda_0 \\ & + a_1\, T_{kl}{}^{i}\, T^{kl}{}_{i} + a_2\, T_i\, T^i + a_3\, T_{kl}{}^{i}\, T_i{}^{kl} \\ & + \overline{a}_1\, \eta^{klmn}\, T_{kli}\, T_{mn}{}^{i} + \overline{a}_2\, \eta^{klmn}\, T_{klm}\, T_n \\ + \ell_\rho^2 \Big(& b_1\, R_{ijkl}R^{ijkl} + b_2\, R_{ijkl}R^{klij} + b_3\, R_{ijkl}R^{ikjl} \\ & + b_4\, R_{ij}R^{ij} + b_5\, R_{ij}R^{ji} + b_6\, R^2 \\ & + \overline{b}_1\, \eta^{klmn}\, R_{klij}R_{mn}{}^{ij} + \overline{b}_2\, \eta^{klmn}\, R_{kl}R_{mn} \\ & + \overline{b}_3\, \eta^{klmn}\, R_{klm}{}^{i}\, R_{ni} + \overline{b}_4\, \eta^{klmn}\, R_{klmn}\, R\Big)\Big\}. \end{aligned} \tag{A3}$$

Here, $\kappa = \frac{8\pi G}{c^4}$ is Einstein's gravitational constant with the dimension of $[\kappa c] = \mathrm{s\,kg}^{-1}$. $G = 6.67 \times 10^{-11}\ \mathrm{m}^3\,\mathrm{kg}^{-1}\,\mathrm{s}^{-2}$ is Newton's gravitational constant. The speed of light: $c = 2.9 \times 10^8\ \mathrm{m/s}$. For completeness, we include the cosmological term λ_0.

Besides the linear "Hilbert type" part characterized by a_0 and $\overline{a}_0$, the Lagrangian (A3) contains several additional coupling constants, which fix the structure of the "Yang–Mills type" part: $a_1, a_2, a_3, \overline{a}_1, \overline{a}_2, b_1, \cdots, b_6, \overline{b}_1, \cdots, \overline{b}_4$, and ℓ_ρ^2. The coupling constants $a_I, \overline{a}_I, b_I$, and $\overline{b}_I$ are dimensionless, whereas the dimension $[\ell_\rho^2] = $[area] so that $[\ell_\rho^2/\kappa c] = [\hbar]$.

The analysis of the particle spectrum for the quadratic model (A3) reveals that the dynamics of gravitational modes in different J^P (spin^{parity}) sectors are determined by the following combinations of the coupling constants: $2^\pm$ sector

$$\Lambda_1 = 4(b_1 + b_2) + 2b_3 + b_4 + b_5, \qquad \Lambda_2 = 4b_1 + b_3, \tag{A4}$$

$0^\pm$ sector

$$\Lambda_3 = 4(b_1 + b_2) + 2b_3 + 4(b_4 + b_5) + 12b_6, \qquad \Lambda_4 = 4b_1 - 2b_3, \tag{A5}$$

and $1^\pm$ sector

$$\Lambda_5 = 4(b_1 - b_2) + b_4 - b_5, \qquad \Lambda_6 = 4b_1 + b_3 + 2b_4, \tag{A6}$$
$$\overline{\Lambda}_5 = 4(\overline{b}_2 - 2\overline{b}_3), \qquad \overline{\Lambda}_6 = -4(\overline{b}_2 + 3\overline{b}_4), \tag{A7}$$

whereas the mass terms are specified by

$$\mu_1 = -a_0 + a_1 - a_2, \qquad \mu_2 = -2a_0 + \frac{2a_1 + a_2 + 3a_3}{4}, \tag{A8}$$
$$\mu_3 = -a_0 - \frac{2a_1 + a_2}{4}, \qquad \overline{\mu}_1 = \frac{8}{3}(4\overline{a}_1 + 3\overline{a}_2 - 2\overline{a}_0). \tag{A9}$$

References

1. Safronova, M.; Budker, D.; DeMille, D.; Kimball, D.F.J.; Derevianko, A.; Clark, C.W. Search for new physics with atoms and molecules. *Rev. Mod. Phys.* **2018**, *90*, 025008. [CrossRef]
2. Cartan, É. On a generalization of the notion of Riemann curvature and spaces with torsion. In *Cosmology and Gravitation: Spin, Torsion, Rotation, and Supergravity*; Bergmann, P.G., De Sabbata, V., Eds.; Plenum: New York, NY, USA, 1980; pp. 489–491. [CrossRef]
3. Sciama, D.W. On the analogy between charge and spin in general relativity. In *Recent Developments in General Relativity, Festschrift for Infeld*; Pergamon Press: Oxford, UK; PWN: Warsaw, Poland, 1962; pp. 415–439.
4. Kibble, T.W.B. Lorentz invariance and the gravitational field. *J. Math. Phys.* **1961**, *2*, 212–221. [CrossRef]
5. Trautman, A. Einstein–Cartan theory. In *Encyclopedia of Mathematical Physics*; Françoise, J.-P., Naber, G.L., Tsou, S.T., Eds.; Elsevier: Oxford, UK, 2006; pp. 189–195.
6. Hehl, F.W.; von der Heyde, P.; Kerlick, G.D. General relativity with spin and torsion and its deviations from Einstein's theory. *Phys. Rev. D* **1974**, *10*, 1066–1069. [CrossRef]
7. Hehl, F.W.; Lemke, J.; Mielke, E.W. Two lectures on fermions and gravity. In *Geometry and Theoretical Physics*; Debrus, J., Hirshfeld, A.C., Eds.; Springer: Heidelberg, Germany, 1991; pp. 56–140.
8. Yasskin, P.B.; Stoeger, W.R. Propagating equations for test bodies with spin and rotation in theories of gravity with torsion. *Phys. Rev. D* **1980**, *21*, 2081–2094. [CrossRef]
9. Hehl, F.W.; Obukhov, Y.N.; Puetzfeld, D. On Poincaré gauge theory of gravity, its equations of motion, and Gravity Probe B. *Phys. Lett. A* **2013**, *377*, 1775–1781. [CrossRef]
10. Obukhov, Y.N.; Puetzfeld, D. Multipolar test body equations of motion in generalized gravity theories. In *Fundamental Theories of Physics*; Springer: Cham, Switzerland, 2015; Volume 179, pp. 67–119. [CrossRef]
11. Hehl, F.W.; von der Heyde, P.; Kerlick, G.D.; Nester, J.M. General relativity with spin and torsion: Foundations and prospects. *Rev. Mod. Phys.* **1976**, *48*, 393–416. [CrossRef]
12. Shapiro, I.L. Physical aspects of the space-time torsion. *Phys. Rept.* **2002**, *357*, 113–213. [CrossRef]
13. Ponomarev, V.N.; Barvinsky, A.O.; Obukhov, Y.N. *Gauge Approach and Quantization Methods in Gravity Theory*; Nauka: Moscow, Russia, 2017.
14. Hehl, F.W. Spin and torsion in general relativity: I. Foundations. *Gen. Relat. Grav.* **1973**, *4*, 333–349. [CrossRef]
15. Hehl, F.W. Spin and torsion in general relativity. II: Geometry and field equations. *Gen. Relat. Grav.* **1974**, *5*, 491–516. [CrossRef]
16. Trautman, A. On the structure of the Einstein–Cartan equations. *Symp. Math.* **1973**, *12*, 139–160.
17. Tiwari, R.N.; Ray, S. Static spherical charged dust electromagnetic mass models in Einstein–Cartan theory. *Gen. Relat. Grav.* **1997**, *29*, 683–690. [CrossRef]

18. Ray, S. Static spherically symmetric electromagnetic mass models with charged dust sources in Einstein–Cartan theory: Lane-Emden models. *Astropys. Space Sci.* **2002**, *280*, 345–355. [CrossRef]
19. Ray, S.; Bhadra, S. Classical electron model with negative energy density in Einstein–Cartan theory of gravitation. *Int. J. Mod. Phys. D* **2004**, *13*, 555–565. [CrossRef]
20. Tseytlin, A.A. Poincaré and de Sitter gauge theories of gravity with propagating torsion. *Phys. Rev. D* **1982**, *26*, 3327–3341. [CrossRef]
21. Obukhov, Y.N.; Ponomariev, V.N.; Zhytnikov, V.V. Quadratic Poincaré gauge theory of gravity: A comparison with the general relativity theory. *Gen. Relat. Grav.* **1989**, *21*, 1107–1142. [CrossRef]
22. Heckel, B.R.; Cramer, C.E.; Cook, T.S.; Adelberger, E.G.; Schlamminger, S.; Schmidt, U. New CP-violation and preferred-frame tests with polarized electrons. *Phys. Rev. Lett.* **2006**, *97*, 021603. [CrossRef] [PubMed]
23. Heckel, B.R.; Adelberger, E.G.; Cramer, C.E.; Cook, T.S.; Schlamminger, S.; Schmidt, U. Preferred-frame and CP-violation tests with polarized electrons. *Phys. Rev. D* **2008**, *78*, 092006. [CrossRef]
24. Obukhov, Y.N.; Yakushin, I.V. Experimental restrictions on spin–spin interactions in gauge gravity. *Int. J. Theor. Phys.* **1992**, *31*, 1993–2001. [CrossRef]
25. Yakushin, I.V. Contribution of torsion to the hyperfine splitting of the hydrogen atom. *Sov. J. Nucl. Phys.* **1992**, *55*, 418–420.
26. Yakushin, I.V. On the contribution of vector torsion component to the effect of interaction of polarized photons. *Moscow Univ. Phys. Bull.* **1992**, *47*, 76–77.
27. Lehnert, R.; Snow, W.M.; Yan, H. A first experimental limit on in-matter torsion from neutron spin rotation in liquid ^{4}He. *Phys. Lett. B* **2014**, *730*, 353–356; Erratum in *Phys. Lett. B* **2015**, *744*, 11. [CrossRef]
28. Lämmerzahl, C. Constraints on space-time torsion from Hughes-Drever experiments. *Phys. Lett. A* **1997**, *228*, 223–231. [CrossRef]
29. Obukhov, Y.N.; Silenko, A.J.; Teryaev, O.V. Spin-torsion coupling and gravitational moments of Dirac fermions: Theory and experimental bounds. *Phys. Rev. D* **2014**, *90*, 124068. [CrossRef]
30. Moody, J.E.; Wilczek, F. New macroscopic forces? *Phys. Rev. D* **1984**, *30*, 130–138. [CrossRef]
31. Peccei, R.; Quinn, H. Constraints imposed by CP conservation in the presence of pseudoparticles. *Phys. Rev. D* **1977**, *16*, 1791–1797. [CrossRef]
32. Gelmini, G.; Nussinov, S.; Yanagida, T. Does nature like Nambu-Goldstone bosons? *Nucl. Phys. B* **1983**, *219*, 31–40. [CrossRef]
33. Carroll, S.M.; Field, G.B. Consequences of propagating torsion in connection-dynamic theories of gravity. *Phys. Rev. D* **1994**, *50*, 3867–3873. [CrossRef]
34. Neville, D.E. Experimental bounds on the coupling strength of torsion potentials. *Phys. Rev. D* **1980**, *21*, 2075–2080. [CrossRef]
35. Ansel'm, A.A. Possible new long-range interaction and methods for detecting it. *JETP Lett.* **1982**, *36*, 55–59.
36. Asztalos, S.J.; Carosi, G.; Hagmann, C.; Kinion, D.; van Bibber, K.; Hotz, M.; Rosenberg, L.J.; Rybka, G.; Hoskins, J.; Hwang, J.; et al. SQUID-based microwave cavity search for dark-matter axions. *Phys. Rev. Lett.* **2010**, *104*, 41301. [CrossRef]
37. Robilliard, C.; Battesti, R.; Fouche, M.; Mauchain, J.; Sautivet, A.M.; Amiranoff, F.; Rizzo, C. No 'light shining through a wall': Results from a photoregeneration experiment. *Phys. Rev. Lett.* **2007**, *99*, 190403. [CrossRef] [PubMed]
38. Flambaum, V.V.; Tran Tan, H.B. Oscillating nuclear electric dipole moment induced by axion dark matter produces atomic and molecular electric dipole moments and nuclear spin rotation. *Phys. Rev. D* **2019**, *100*, 111301. [CrossRef]
39. Garcon, A.; Aybas, D.; Blanchard, J.W.; Centers, G.; Figueroa, N.L.; Graham, P.W.; Kimball, D.F.J.; Rajendran, S.; Sendra, M.G.; Sushkov, A.O.; et al. The cosmic axion spin precession experiment (CASPEr): A dark-matter search with nuclear magnetic resonance. *Quantum Sci. Technol.* **2017**, *3*, 014008. [CrossRef]
40. Georgi, H. Unparticle physics. *Phys. Rev. Lett.* **2007**, *98*, 221601. [CrossRef] [PubMed]
41. Holdom, B. Two $U(1)$'s and ϵ charge shifts. *Phys. Lett. B* **1986**, *166*, 196–198. [CrossRef]
42. Dobrescu, B.A.; Mocioiu, I. Spin-dependent macroscopic forces from new particle exchange. *J. High Energy Phys.* **2006**, *11*, 5. [CrossRef]
43. Dobrescu, B.A. Massless gauge bosons other than the photon. *Phys. Rev. Lett.* **2005**, *94*, 151802. [CrossRef]
44. Kotler, S.; Ozeri, R.; Kimball, D.F.J. Constraints on exotic dipole-dipole couplings between electrons at the micrometer scale. *Phys. Rev. Lett.* **2015**, *115*, 081801. [CrossRef]
45. Rong, X.; Jiao, M.; Geng, J.; Zhang, B.; Xie, T.; Shi, F.; Duan, C.-K.; Cai, Y.-F.; Du, J. Constraints on a spin-dependent exotic interaction between electrons with single electron spin quantum sensors. *Phys. Rev. Lett.* **2018**, *121*, 080402. [CrossRef]
46. Teissier, J.; Barfuss, A.; Appel, P.; Neu, E.; Maletinsky, P. Strain coupling of a nitrogen-vacancy center spin to a diamond mechanical oscillator. *Phys. Rev. Lett.* **2014**, *113*, 020503. [CrossRef]
47. Ficek, F.; Kimball, D.F.J.; Kozlov, M.G.; Leefer, N.; Pustelny, S.; Budker, D. Constraints on exotic spin-dependent interactions between electrons from helium fine-structure spectroscopy. *Phys. Rev. A* **2017**, *95*, 032505. [CrossRef]
48. Naik, P.C.; Pradhan, T. Long-ray interaction between spins. *J. Phys. A Math. Gen.* **1981**, *14*, 2795–2805. [CrossRef]
49. Pradhan, T.; Malik, R.P.; Naik, P.C. The fifth interaction: Universal long range force between spins. *Pramana-J. Phys.* **1985**, *24*, 77–94. [CrossRef]
50. Pradhan, T.; Lahiri, A. Finite quantum electrodynamics. *Phys. Rev. D* **1974**, *10*, 1872–1882. [CrossRef]
51. Obukhov, Y.N.; Yakushin, I.V. On the experimental estimates of the axial torsion mass and coupling constants. In *Modern Problems of Theoretical Physics. Festschrift for Professor D. Ivanenko*; Pronin, P.I., Obukhov, Y.N., Eds.; World Scientific: Singapore, 1991; pp. 197–215.

52. Hehl, F.W.; McCrea, J.D.; Mielke, E.W.; Ne'eman, Y. Metric affine gauge theory of gravity: Field equations, Noether identities, world spinors, and breaking of dilation invariance. *Phys. Rep.* **1995**, *258*, 1–171. [CrossRef]
53. Blagojević, M. *Gravitation and Gauge Symmetries*; Institute of Physics: Bristol, UK, 2002. [CrossRef]
54. Blagojević, M.; Hehl, F.W. (Eds.) *Gauge Theories of Gravitation: A Reader with Commentaries*; Imperial College Press: London, UK, 2013. [CrossRef]
55. Obukhov, Y.N. Poincaré gauge gravity: Selected topics. *Int. J. Geom. Meth. Mod. Phys.* **2006**, *3*, 95–137. [CrossRef]
56. Obukhov, Y.N. Poincaré gauge gravity: An overview. *Int. J. Geom. Meth. Mod. Phys.* **2018**, *15* (Suppl. S1), 1840005. [CrossRef]
57. Obukhov, Y.N. Poincaré gauge gravity primer. In *Modified and Quantum Gravity—From Theory to Experimental Searches on All Scales*; Pfeifer, C., Lämmerzahl, C., Eds.; Springer: Cham, Switzerland, 2022; Chapter 3. [CrossRef]
58. Vasilev, T.B.; Cembranos, J.A.R.; Valcarcel, J.G.; Martín-Moruno, P. Stability in quadratic torsion theories. *Eur. Phys. J. C* **2017**, *77*, 755. [CrossRef]
59. Jiménez, J.B.; Torralba, F.J.M. Revisiting the stability of quadratic Poincaré gauge gravity. *Eur. Phys. J. C* **2020**, *80*, 611. [CrossRef]
60. Belyaev, A.S.; Shapiro, I.L.; do Vale, M.A.B. Torsion phenomenology at the CERN LHC. *Phys. Rev. D* **2007**, *75*, 034014. [CrossRef]
61. Mohanty, S.; Sarkar, U. Constraints on background torsion field from *K*-physics. *Phys. Lett. B* **1998**, *433*, 424–428. [CrossRef]
62. Stueckelberg, E.C.G. Die Wechselwirkungskräfte in der Elektrodynamik und in der Feldtheorie der Kernkräfte. Teil II und III. *Helv. Phys. Acta* **1938**, *11*, 299–328. [CrossRef]
63. Kostelecký, V.A.; Russell, N.; Tasson, J.D. Constraints on torsion from bounds on Lorentz violation. *Phys. Rev. Lett.* **2008**, *100*, 111102. [CrossRef] [PubMed]
64. Kostelecký, V.A.; Li, Z. Searches for beyond-Riemann gravity. *Phys. Rev. D* **2021**, *104*, 044054. [CrossRef]
65. Carroll, S.M.; Field, G.B.; Jackiw, R. Limits on a Lorentz- and parity-violating modification of electrodynamics. *Phys. Rev. D* **1990**, *41*, 1231–1240. [CrossRef] [PubMed]
66. Itin, Y. Carroll-Field-Jackiw electrodynamics in the premetric framework. *Phys. Rev. D* **2004**, *70*, 025012. [CrossRef]
67. Fabbri, L. A discussion on the most general torsion-gravity with electrodynamics for Dirac spinor matter fields. *Int. J. Geom. Meth. Mod. Phys.* **2015**, *12*, 1550099. [CrossRef]
68. Fabbri, L. Fundamental theory of torsion gravity. *Universe* **2021**, *7*, 305. [CrossRef]

Article

On the Role of Einstein–Cartan Gravity in Fundamental Particle Physics

Carl F. Diether III and Joy Christian *

Einstein Centre for Local-Realistic Physics, 15 Thackley End, Oxford OX2 6LB, UK; fred.diether@einstein-physics.org

* Correspondence: jjc@bu.edu

Received: 7 July 2020; Accepted: 3 August 2020; Published: 5 August 2020

Abstract: Two of the major open questions in particle physics are: (1) Why do the elementary fermionic particles that are so far observed have such low mass-energy compared to the Planck energy scale? (2) What mechanical energy may be counterbalancing the divergent electrostatic and strong force energies of point-like charged fermions in the vicinity of the Planck scale? In this paper, using a hitherto unrecognised mechanism derived from the non-linear amelioration of the Dirac equation known as the Hehl–Datta equation within the Einstein–Cartan–Sciama–Kibble (ECSK) extension of general relativity, we present detailed numerical estimates suggesting that the mechanical energy arising from the gravitationally coupled self-interaction in the ECSK theory can address both of these questions in tandem.

Keywords: Einstein–Cartan gravity; Hehl–Datta equation; ultraviolet divergence; renormalisation; modified gravity; quantum electrodynamics; hierarchy problem; torsion; spinors

---

1. Introduction

For over a century, Einstein's theory of gravity has provided remarkably accurate and precise predictions for the behaviour of macroscopic bodies within our cosmos. For the elementary particles in the quantum realm, however, Einstein–Cartan theory of gravity may be more appropriate, because it incorporates spinors and associated torsion within a covariant description [1,2]. For this reason there has been considerable interest in Einstein–Cartan theory, in the light of the field equations proposed by Sciama [3] and Kibble [4]. For example, in a series of papers Poplawski has argued that Einstein–Cartan–Sciama–Kibble (ECSK) theory of gravity [5] solves many longstanding problems in physics [6–9]. His concern has been to avoid singularities endemic in general relativity by proposing that our observed universe is perhaps a black hole within a larger universe [7]. Our concern, on the other hand, is to point out using numerical estimates that ECSK theory also offers solutions to two longstanding problems in particle physics.

The first problem we address here concerns the well known fact that in the limit of vanishing radii (or point limit) the electrostatic and strong force self-energies of point-like fermions become divergent. We will show, however, that torsion contributions within the ECSK theory resolves this difficulty as well, at least numerically, by counterbalancing the divergent electrostatic and strong force energy densities near the Planck scale. In fact, the negative torsion energy associated with the spin angular momentum of elementary fermions may well be the long sought after mechanical energy that counteracts the divergent positive energies stemming from their electrostatic and strong nuclear charges. As a result of this counterbalancing, however, our suggestion does not have anything to do with high energy physics.

The second of these problems can be traced back to the fact that gravity is a considerably weaker "force" compared to the other forces. When Newton's gravitational constant is combined with the speed of light and Planck's constant, one arrives at the energy scale of $\sim 10^{19}$ GeV, which is some 17 orders of magnitude larger than the heaviest known elementary fermion (the top quark) observed at the mass-energy of $\sim$172 GeV. Thus, there is a difference of some 17 orders of magnitude between the electroweak scale and the Planck scale. There have been many attempts to explain this difference, but none is as simple as our explanation based on the torsion contributions within the ECSK theory.

Now one of the reasons why ECSK theory is not widely accepted as a viable theory of gravity is the lack of any experimental evidence for the gravitational torsion. However, as reviewed in [10], gravitational torsion appears to be entirely confined to elementary fermions, and therefore it is not directly detectable. Elementary fermions are at the heart of all matter and can be viewed as defects in spacetime associated with torsion. Thus, as we have argued elsewhere [11], existence of matter itself may be taken as a proof that gravitational torsion exists, albeit only inside of matter, and therefore it does not propagate. In other words, detection of matter itself may be taken as an indirect detection of gravitational torsion. As we will see, our results below lend considerable support to this possibility.

2. Evaluation of the Charged Fermionic Self-Energy within ECSK Theory

The ECSK theory of gravity is an extension of general relativity allowing spacetime to have torsion in addition to curvature, where torsion is determined by the density of intrinsic angular momentum, reminiscent of the quantum-mechanical spin [1–9,12–18]. As in general relativity, the gravitational Lagrangian density in the ECSK theory is proportional to the curvature scalar. Unlike in general relativity, the affine connection

$$\Gamma_{i\,j}^{\,k} = \omega_i{}^{\mu}{}_{\nu} e^k_{\mu} e^{\nu}_j + e^k_{\mu} \partial_i e^{\mu}_j \tag{1}$$

is not restricted to be symmetric, although metric compatibility conditions are retained, defined by the requirements that the covariant derivatives of the tetrad fields e^{μ}_k and the Minkowski metric vanish, implying the antisymmetry $\omega_i{}^{\mu\nu} = -\omega_i{}^{\nu\mu}$ of the spin connection [19]. The antisymmetric part of the connection, namely

$$S^k{}_{ij} = \Gamma_{[i\,j]}^{\,k} = \partial_i e^k_j - \partial_j e^k_i + \omega_i{}^k{}_{\lambda} e^{\lambda}_j - \omega_j{}^k{}_{\lambda} e^{\lambda}_i \tag{2}$$

(i.e., the torsion tensor), is then regarded as a dynamical variable analogous to the metric tensor g_{ij} in general relativity. Consequently, variation of the total action for the gravitational field and matter with respect to the metric tensor gives Einstein-type field equations that relate the curvature to the dynamical energy-momentum tensor $T_{ij} = (2/\sqrt{-g})\delta\mathfrak{L}/\delta g^{ij}$, where $\mathfrak{L}$ is the matter Lagrangian density. On the other hand, variation of the total action with respect to the torsion tensor gives the Cartan equations for the spin tensor of matter [5]:

$$s^{ijk} = \frac{1}{\kappa} S^{[ijk]}, \quad \text{where } \kappa = \frac{8\pi G}{c^4}. \tag{3}$$

Thus, ECSK theory of gravity extends general relativity to include intrinsic spin of matter, with fermionic fields such as those of quarks and leptons providing natural sources of torsion. Torsion, in turn, modifies the Dirac equation for elementary fermions by adding to it a cubic term in the spinor fields, as observed by Kibble, Hehl and Datta [1,4,5].

It is this nonlinear Hehl–Datta equation that provides the theoretical background for our proposal. The cubic term in this equation corresponds to an axial-axial self-interaction in the matter Lagrangian, which, among other things, generates a spinor-dependent vacuum-energy term in the energy-momentum tensor (see, for example, Reference [15]). The torsion tensor $S^k{}_{ij}$ appears in the matter Lagrangian via covariant derivative of a Dirac spinor with respect to the affine connection. The spin tensor for the Dirac spinor ψ can then be derived [5], and turns out to be totally antisymmetric:

$$s^{ijk} = -\frac{i\hbar c}{4} \bar{\psi} \gamma^{[i} \gamma^j \gamma^{k]} \psi, \tag{4}$$

where $\bar{\psi} \equiv \psi^{\dagger}\gamma^{0} := (\psi_1^*,\ \psi_2^*,\ -\psi_3^*,\ -\psi_4^*)$ is the Dirac adjoint of ψ and γ^i are the Dirac matrices: $\gamma^{(i}\gamma^{j)} = 2g^{ij}$. The Cartan Equations (3) render the torsion tensor to be quadratic in spinor fields. Substituting it into the Dirac equation in the Riemann–Cartan spacetime with metric signature $(+,\ -,\ -,\ -)$ gives the cubic Hehl–Datta equation [1,4,5]:

$$i\hbar\,\gamma^k\psi_{:k} = mc\,\psi + \frac{3\kappa\hbar^2 c}{8}\left(\bar{\psi}\gamma^5\gamma_k\psi\right)\gamma^5\gamma^k\psi, \tag{5}$$

where the colon denotes a general-relativistic covariant derivative with respect to the Christoffel symbols, and m is the mass of the spinor. The Hehl–Datta Equation (5) and its adjoint can be obtained by varying the following action with respect to $\bar{\psi}$ and ψ, respectively, without varying it with respect to the metric tensor or the torsion tensor [15]:

$$\mathcal{I} = \int d^4x\,\sqrt{-g}\left\{-\frac{1}{\kappa}R + i\hbar c(\bar{\psi}\gamma^k\psi_{:k} - \bar{\psi}_{:k}\gamma^k\psi) - mc^2\bar{\psi}\psi - \frac{3\kappa\hbar^2c^2}{8}(\bar{\psi}\gamma^5\gamma_k\psi)(\bar{\psi}\gamma^5\gamma^k\psi)\right\}. \tag{6}$$

The last term in this action corresponds to the effective axial-axial, self-interaction mentioned above:

$$\mathfrak{L}_{\mathrm{AA}} = -\sqrt{-g}\,\frac{3\kappa\hbar^2c^2}{8}(\bar{\psi}\gamma^5\gamma_k\psi)(\bar{\psi}\gamma^5\gamma^k\psi). \tag{7}$$

This self-interaction term is not renormalisable. It is an effective Lagrangian density in which only the metric and spinor fields are dynamical variables. The original Lagrangian density for a Dirac field in which the torsion tensor is also a dynamical variable (giving the Hehl–Datta equation), is renormalisable, since it is quadratic in spinor fields. As we will see, renormalisation may not be required if ECSK gravity turns out to be what is realised in Nature, because it gives physical justification for the counter terms.

Before proceeding further we note that the above action is not the most general possible action within the present context. In addition to the axial-axial term, an axial-vector and a vector-vector term can be added to the action, albeit as non-minimal couplings (see, for example, Reference [17]). However, it has been argued in Reference [15] that minimal coupling is the most natural coupling of fermions to gravity because non-minimal couplings are sourced by components of the torsion that do not appear naturally in the models of spinning matter. For this reason we will confine our treatment to the minimal coupling of fermions to gravity and the corresponding Hehl–Datta equation, while recognizing that strictly speaking our neglect of non-minimal couplings amounts to an approximation, albeit a rather good approximation, at least as far as electrodynamics is concerned.

2.1. S-Matrix Evaluation of the Charged Fermionic Self-Energy within QED

It is instructive for our purposes to first review in this subsection the standard treatment of the charged fermionic self-energy within QED, ignoring the ECSK gravity. To this end, recall that the canonical evaluation of the electron self-energy was performed by Weisskopf in 1939 [20], which revealed that in general the self-energy of an electron is logarithmically divergent. An S-matrix evaluation for the electromagnetic mass, δm, confirms Weisskopf's result [21–23]. In the units of $\hbar = c = 1$ and $\alpha = e^2/4\pi$, together with a high-energy cutoff Λ, this electromagnetic mass can be expressed as

$$\delta m = \frac{3\alpha\, m}{2\pi}\ln\left(\frac{\Lambda}{m}\right) \quad (\text{for } \Lambda >> m), \tag{8}$$

where "ln" stands for the natural logarithm. Let us investigate this result using simple numerical analysis. What happens, for example, if we use Planck mass for the cutoff? This should be a reasonable assumption since the Planck scale can be considered a limiting scale in the same fashion as $\hbar$ and c are considered. Then, for an electron we have

$$\delta m_e = \frac{3\alpha\, m_e}{2\pi} \ln\left(\frac{m_P}{m_e}\right) \approx 0.1795\, m_e. \tag{9}$$

This suggests that the electromagnetic mass-energy of an electron in this case would be about 18 percent of the total rest mass-energy. That does not seem unreasonable, but there is no way to confirm that it is correct. Moreover, according to some views, if the electromagnetic mass is truly infinite then there should be a compensating negative infinite mechanical mass that produces the observed positive rest mass. On the other hand, Weinberg in Reference [22] gives a renormalised expression for the "complete self-energy function" as

$$\begin{aligned}
\Sigma^*_{order\, e^2}(p) &= \Sigma^*_{1\, loop}(p) - (Z_2 - 1)(i\not{p} + m_x) + Z_2\, \delta m_x \\
&= \frac{-2\pi^2 e^2}{(2\pi)^4} \int_0^1 dx \left\{ [i(1-x)\not{p} + 2m_x] \ln\left(\frac{m_x^2(1-x)}{p^2 x(1-x) + m_x^2 x}\right) \right. \\
&\qquad - m_x[1+x] \ln\left(\frac{1-x}{x^2}\right) \\
&\qquad \left. - (i\not{p} + m_x)\left[(1-x)\ln\left(\frac{1-x}{x^2}\right) - \frac{2(1-x^2)}{x}\right] \right\}.
\end{aligned} \tag{10}$$

The middle term in this integrand represents the renormalised electromagnetic mass so that for an electron it gives

$$\delta m_{e\, renorm} = \frac{-2\pi^2 e^2}{(2\pi)^4} \int_0^1 dx \left\{ -m_x[1+x] \ln\left(\frac{1-x}{x^2}\right) \right\} = \frac{3\alpha\, m_e}{8\pi} \approx 0.0008711\, m_e. \tag{11}$$

In this case the electromagnetic mass is considerably less than 1 percent of the total rest mass. That indicates that the cutoff should be very close to the rest mass of the electron. Thus we seem to have two conflicting results. The conflict can be resolved by realizing that the mass in the equations leading to these results was plugged in by hand in an ad hoc manner. Moreover, from a quantum field theory and Feynman diagram perspective, it is easy to see that the mass comes in via the fermion propagator. Therefore by definition it is off-mass-shell, since the propagator represents an internal virtual line. It can thus have any value. In other words, it is really a variable. For the time being we will treat it as a variable and use the relation $m = 1/r$. Now that the apparent conflict is resolved, in what follows we focus on the renormalisation process.

It is easy to see that, in the rest frame of the particle in question with $i\not{p} = -m$, the renormalised complete self-energy function in Equation (10) is identically zero. That may seem odd, since one would expect that in the rest frame the self-energy would be equal to the rest mass-energy. It really just means that the propagator is zero, as it should be for the rest frame. Thanks to the Hehl–Datta equation, we have an additional term in the Lagrangian to consider in the renormalisation process. In our view, this additional term represents the mechanical (bare) mass, since it is negative with respects to the observed mass. Thus, in the renormalised self-energy function in Equation (10), the counter terms are superfluous since the torsion term from the Hehl–Data equation is the counter term! Moreover, we will see later that the rest mass is produced in a natural way from energy considerations generated only from the physical constants and geometry.

Using $m = 1/r$, we can now put the renormalised electromagnetic mass equation in a more familiar form for electromagnetic energy,

$$\delta m_{renorm} = \frac{3\alpha m}{8\pi} = \frac{3\alpha}{8\pi r} = \frac{3\, e^2}{32\pi^2\, r}. \tag{12}$$

However, δm in this expression still goes to infinity as $r \to 0$. Perhaps a Planck length cutoff may be used to tame this infinity. We will soon see that r does take on a finite value very close to Planck length. Now we know from experiments that the radius of an electron is likely to be less than 10^{-22} m [24]. Substituting that bound for r gives

$$\delta m = \frac{3\,e^2}{32\pi^2\,(10^{-22}\mathrm{m})} \gtrapprox 1.719 \times 10^3\,\mathrm{GeV}, \tag{13}$$

which, although finite, is still a very large electromagnetic energy contribution, with the actual value likely to be even greater. Thus, according to this estimate, the electromagnetic mass is going to be very large near the Planck length and will have to be compensated for in order to recover the observed rest mass of a charged fermion. The compensation will have to be negative mass-energy relative to the positive electromagnetic energy:

$$m_{obs} = -X + \delta m_R = -X + \frac{3\,\alpha}{8\pi r_B} \longrightarrow -\frac{X}{r} + \frac{3\,\alpha}{8\pi r_B}. \tag{14}$$

We suspect that the unknown variable X might be related to the Hehl–Datta self-interaction term (7) because that term varies as $1/r_x$. Our goal then is to investigate this possibility.

To that end, note that the full second order S-matrix calculation within QED worked out by Milonni [21] gives

$$S^{(2)}_{fi}(E) = -i(2\pi)^4\delta^4(p_f - p_i)\sqrt{\frac{m^2}{E_i E_f}\frac{1}{r^3}}\bar{u}(p_f, s_f)\Sigma(p_i)u(p_i, s_i), \tag{15}$$

$$\text{where}\quad \Sigma(p_i) = -ie^2\int \frac{d^4k}{(2\pi)^4}\frac{g_{\mu\nu}}{k^2 + i\epsilon}\gamma^\mu \frac{1}{\not{p}_i - \not{k} - m + i\epsilon}\gamma^\nu. \tag{16}$$

Milonni's evaluation of $\Sigma(p_i)$ produces the result for δm given in Equation (8). However, we doubt this evaluation is appropriate since, as we noted previously, the mass comes in via the propagator and therefore it is off mass shell. We will not use it in our evaluation.

2.2. S-Matrix Evaluation for the Charged Fermionic Self-Energy within Einstein–Cartan Gravity

Building on the results of the previous subsection, we now evaluate the self-energy S-matrix process including the contribution from ECSK gravity. Our calculation will be different from the old way of calculating self-energy in the manner of Weisskopf because in our approach we have the gravitational torsion term as a built-in counter-term. The following calculations will be somewhat unconventional as there is no Feynman diagram for evaluation in the rest frame. We begin with our complete QED Lagrangian for free particles:

$$\frac{\mathfrak{L}_{\mathrm{QED}}}{\sqrt{-g}} = i\hbar c\,\bar{\psi}\gamma^\mu\partial_\mu\psi + e\,\bar{\psi}\gamma^\mu A_\mu\psi - \frac{1}{4}F_{\mu\nu}F^{\mu\nu} - mc^2\bar{\psi}\psi - \frac{3\kappa\hbar^2c^2}{8}(\bar{\psi}\gamma^5\gamma_\mu\psi)(\bar{\psi}\gamma^5\gamma^\mu\psi). \tag{17}$$

Here the last term is from the Hehl–Datta equation for gravitational torsion via spin density squared. We take it to be a self-interaction involving intrinsic spin. In other words, the spin interacts with itself much like the charge interacts with the field it creates. For our analysis of a charged lepton in the rest frame, only three terms are applicable because the derivative term is zero and normal gravity is negligible. The self-interaction Hamiltonian density for the first term with $\hbar = c = 1$ and $(x) = (\mathbf{x}, t)$ is then

$$\mathrm{h}_I(x)_1 = e\,\bar{\psi}(x)\gamma^\mu A_\mu\psi(x). \tag{18}$$

The next self-interaction term is

$$h_I(x)_2 = -\frac{3\kappa}{8}(\bar{\psi}(x)\gamma^5\gamma_k\psi(x))(\bar{\psi}(x)\gamma^5\gamma^k\psi(x)). \tag{19}$$

For our purposes it is sufficient to evaluate the S-matrix in the rest frame, where the quantities $\bar{\psi}(x)\gamma^5\gamma_k\psi(x)$ and $\bar{\psi}(x)\gamma^5\gamma^k\psi(x)$ appearing on the RHS are nonzero even in the rest frame because the coupling between fermions and anti-fermions is an intermediate step in the self-interaction, as we have shown in Equations (A14) and (A15) of Appendix B. Furthermore, as shown in Appendix B, the first order S-matrix terms in the rest frame are now simplified to

$$S_{fi}^{(1)}(E1) = -it\frac{\alpha}{r}\begin{pmatrix} 1 & 0 & 0 & 0 \\ 0 & 1 & 0 & 0 \\ 0 & 0 & 0 & 0 \\ 0 & 0 & 0 & 0 \end{pmatrix}. \tag{20}$$

$$S_{fi}^{(1)}(E2) = it\frac{3\kappa}{8r^3}\begin{pmatrix} 1 & 0 & 0 & 0 \\ 0 & 1 & 0 & 0 \\ 0 & 0 & 0 & 0 \\ 0 & 0 & 0 & 0 \end{pmatrix}. \tag{21}$$

On the other hand, for $m\bar{\psi}\psi$ in the rest frame, applying Equation (A8) from Appendix B we have

$$S_{fi}^{(1)}(E3) = im\int d^4x\,\langle f \mid \bar{\psi}(x)\psi(x) \mid i\rangle = i\frac{m\,r^3\,t}{r^3}\,\bar{u}^f(m)u^i(m) = i\,mt\begin{pmatrix} 1 & 0 & 0 & 0 \\ 0 & 1 & 0 & 0 \\ 0 & 0 & 0 & 0 \\ 0 & 0 & 0 & 0 \end{pmatrix}. \tag{22}$$

Equation (5) suggests that the sum of these three terms is equal to zero:

$$-it\frac{\alpha}{r}\begin{pmatrix} 1 & 0 & 0 & 0 \\ 0 & 1 & 0 & 0 \\ 0 & 0 & 0 & 0 \\ 0 & 0 & 0 & 0 \end{pmatrix} + it\frac{3\kappa}{8r^3}\begin{pmatrix} 1 & 0 & 0 & 0 \\ 0 & 1 & 0 & 0 \\ 0 & 0 & 0 & 0 \\ 0 & 0 & 0 & 0 \end{pmatrix} + i\,mt\begin{pmatrix} 1 & 0 & 0 & 0 \\ 0 & 1 & 0 & 0 \\ 0 & 0 & 0 & 0 \\ 0 & 0 & 0 & 0 \end{pmatrix} = 0. \tag{23}$$

Consequently, we have

$$\frac{\alpha}{r} - \frac{3\kappa}{8r^3} = m. \tag{24}$$

Reinstating $\hbar$ and c we thus arrive at

$$\frac{\alpha\hbar c}{r} - \frac{3\kappa(\hbar c)^2}{8r^3} = mc^2. \tag{25}$$

An identical result can be obtained also for the anti-fermion spinor, $v(m)$, so that this equation remains the same for both fermion and anti-fermion. In what follows we will use this S-matrix result for our numerical approximations. From these results, after solving for r, it is evident that the complete process is finite, without divergences. This suggests that the above is the correct S-matrix solution for the fermion self-energy problem, with all other orders of the S-matrix expansion vanishing because the genuine fermionic self-energy must naturally be evaluated only in the rest frame, with all other contributions summing to zero. Any higher loop corrections in squared charge will be automatically compensated by higher loop corrections from the squared spin.

It is also worth noting that without ameliorating the Dirac equation with a cubic term, the Dirac equation would reduce for an electron to $\alpha\hbar/r_e = m_e c$, giving $r_e = \alpha\hbar/(m_e c) \sim 10^{-15}$ m, where $\alpha = e^2/(4\pi\hbar c)$ is the fine structure constant. This is the classical electron radius. Experimental evidence, however, suggests that electron radius is much smaller [24]. As we shall see, our calculations with the cubic term included predicts the electron radius to be of the order of 10^{-34} m, which is closer to the Planck length. This may turn out to be the correct value of the electron radius.

Needless to say, what we have presented above is a derivation of Equation (25) within a theory that may be viewed as a quantum field theory of Dirac fields in a Riemann–Cartan spacetime [4,5]. It can be interpreted also as a theory of gravity-coupled self-interaction within standard general relativity [1,5]. However, any such generalization must necessarily reproduce the Hehl–Datta Equation (5) for single fermions even at reasonably high energies, just as Dirac equation remains valid for single fermions at high energies [18].

Finally, it is important to note here that, despite the appearance of four spinors in the interaction term of Equation (6), it describes the self-interaction of a single fermion, of range $\sim 10^{-34}$ m, not mutual interactions among the spins of four distinct fermions. That is to say, it does not describe a "spin field" of some sort as a carrier of a new interaction [5]. If, however, one insists on interpreting the interaction term in Equation (6) as describing interactions among four distinct fermions, then the mass of the corresponding exchange boson would have to exceed 10^{15} GeV, which is evidently quite unreasonable. Moreover, as we shall see in Section 3, this energy is a fictitious quantity and therefore there is no justification for assuming some kind of a new torsion interaction between different fermions. What is more, as we shall soon see, within our scheme any corrections due to vacuum polarization are automatically compensated for in the production of electroweak mass-energy, dictated by Equation (25) above.

2.3. *Evaluation of Charged Fermionic Self-Energy by Dimensional Analysis*

It is instructive to compare the results of the previous two subsections with the evaluation of the charged fermionic self-energy using only dimensional analysis. To this end, we begin with the following physically reasonable assumptions:

1. ECSK theory of gravity is the correct theory of spacetime for addressing the fermionic self-energy problem since it allows the dimension-full gravitational constant, G, to enter elementary particle physics in a natural manner.
2. Since experiments to date indicate that an electron is a point-like particle without any substructure and put an upper bound of 10^{-22} meters on its radius, we assume that the radius of electron is much less than that value.
3. We assume that the radial distance, r, on which the electromagnetic self-interaction depends is the same radial distance on which the self-interaction arising from the ECSK gravity-induced torsion spin density also depends.

Given these assumptions, we ask: What physical mechanism is responsible for the observed rest mass m_x of the elementary charged fermions? To answer this question we express the rest mass energy in CGS units, and assume that it is at least partially[1] electromagnetic in nature, so that it satisfies a relation like

$$m_x c^2 \sim \frac{e^2}{r} + X, \tag{26}$$

where the dimensionality of X is necessarily that of energy. However we already know that the value of $r < 10^{-15}\, m$ produces an energy greater than $m_x c^2$. Therefore X must be negative energy, giving

$$m_x c^2 \sim \frac{e^2}{r} + (-X). \tag{27}$$

[1] As is well known, assuming that the rest mass energy is entirely electromagnetic in nature leads to the classical radius of the electron, which has been ruled out by experiments. Our assumption of it being at least partially electromagnetic in nature is quite reasonable.

Now, since fermions have spin $\hbar/2$, it is reasonable to assume that it is involved in a mechanical-like energy resulting from the spin interacting with itself analogous to charge, so that we may have a relation like

$$-X \sim -\left(\frac{\hbar}{2}\right)^2 \frac{1}{r}. \tag{28}$$

It is evident from this expression that what we have on its RHS is energy $\times$ mass $\times$ length $= EML$, so we will have to divide out by mass and length to get the dimensions of energy, giving

$$-X \sim -\left(\frac{\hbar}{2r}\right)^2 \frac{1}{M}, \tag{29}$$

which in terms of the gravitational constant $G \sim L^3/MT^2$ can be written as

$$-X \sim -G\left(\frac{\hbar}{2r}\right)^2. \tag{30}$$

The dimensions on the RHS of this expression now give us $E \times L^3/T^2$, so we will have to cancel out L^3/T^2. A natural candidate to accomplish that is the speed of light, c, giving

$$-X \sim -\frac{G}{c^2 r}\left(\frac{\hbar}{2r}\right)^2 = -\frac{G}{r^3}\left(\frac{\hbar}{2c}\right)^2, \tag{31}$$

provided we cancel out the extra L with $1/r$. Thus, we now have the dimensions of energy on the RHS, so that we finally have

$$m_x c^2 \sim \frac{e^2}{r_x} - \frac{G}{r_x^3}\left(\frac{\hbar}{2c}\right)^2. \tag{32}$$

This is the basic form of our central equation (apart from some numerical constants) which we have derived also using two other methods in the previous subsections. Solving this equation for r_x with electron mass for m_x gives a value of the order of 10^{-34} m. However, rather surprisingly, we also found a solution for r_x for the classical electron radius.

3. Particle Masses via Torsion Energy Contribution

We start with our S-matrix result for a charged fermion:

$$\frac{\alpha\hbar c}{r} - \frac{3\pi G\hbar^2}{c^2\, r^3} = mc^2. \tag{33}$$

The first term in Equation (33) diverges as $r \to 0$. If we set r to Planck length we obtain

$$\frac{\alpha\hbar c}{l_P} \approx 8.909 \times 10^{16}\,\text{GeV}, \tag{34}$$

which is close to Planck energy. Although finite, this is still an extremely large energy. Such a large energy for charged leptons is never realised in Nature. A natural question then is: Is there a negative mechanical energy that cancels out most of this energy to produce the observed rest mass-energy of leptons? We believe the answer lies in the second term of Equation (33), which—as we saw above—arises from the non-linear amelioration of the Dirac equation within ECSK theory. Indeed, if we again set the Planck length for r in the second term of Equation (33), then we obtain

$$-\frac{3\pi\, G\hbar^2}{2\, c^2\, (l_P)^3} \approx -5.773 \times 10^{19}\ \text{GeV}. \tag{35}$$

Comparing this value with the electrostatic energy at the Planck length estimated in Equation (34) we see at once that the torsion-induced mechanical energy in Equation (35) can indeed counterbalance the huge electrostatic energy. This is a surprising observation, considering the widespread belief that "the numerical differences which arise [between GR and ECSK theories] are normally very small, so that the advantages of including torsion are entirely theoretical" [18].

Moving forward to our goal of numerical estimates, let us note that whenever terms quadratic in spin happen to be negligible, then the ECSK theory is observationally indistinguishable from general relativity. Therefore, for post-general-relativistic effects, the density of spin-squared has to be comparable to the density of mass. The corresponding characteristic length scale, say for a nucleon, is referred to as the Cartan or Einstein–Cartan radius [2,18], defined as

$$r_{Cart} \approx (l_P^2\,\lambda_C)^{\frac{1}{3}}, \tag{36}$$

where λ_C is the Compton wavelength of the nucleon. Now it has been noted by Poplawski [6–9,25,26] that quantum field theory based on the Hehl–Datta equation may avert divergent integrals normally encountered in calculating radiative corrections, by self-regulating propagators. Moreover, the multipole expansion applied to Dirac fields within the ECSK theory shows that such fields cannot form singular, point-like configurations because these configurations would violate the conservation law for the spin density, and thus the Bianchi identities. These fields in fact describe non-singular particles whose spatial dimensions are at least of the order of their Cartan radii, defined by the condition

$$\epsilon \sim \kappa s^2, \tag{37}$$

where $\sqrt{s^2} \sim \hbar c\,|\psi|^2$ is the spin density, $\epsilon \sim mc^2|\psi|^2$ is the rest energy density and $|\psi|^2 \sim 1/r^3$ is the probability density, giving the radius in Equation (36). Consequently, at the least the de Broglie energy associated with the Cartan radius of a fermion (which is approximately 10^{-27} m for an electron) may introduce an effective ultraviolet cutoff for it in quantum field theory in the ECKS spacetime. The avoidance of divergences in radiative corrections in quantum fields may thus come from spacetime torsion originating from intrinsic spin. Poplawski and others, however, took ϵ to be the mass-energy density of the fermion to arrive at the Cartan radius Equation (36). It is easy to work out from the first term of our Equation (33) that at the Cartan radius the electrostatic energy density for an electron is still extremely large:

$$\frac{\alpha\hbar c}{(10^{-27}m)^4} \approx 1.440 \times 10^{90}\,\mathrm{GeV\,m^{-3}}. \tag{38}$$

For this reason it is not correct to identify ϵ with the rest mass-energy density, which is $\approx 5.1099 \times 10^{77}$ GeV m^{-3} for an electron at the Cartan radius. The electrostatic energy density of an electron is thus about thirteen orders of magnitude higher. Therefore ϵ is better identified with the electrostatic energy density provided most of it is cancelled out.

If in Equation (33) we set the electrostatic energy appearing in its first term to be equal to the spin squared energy induced by the self-interaction appearing in its second term and solve for r, then we obtain

$$\boxed{r_t = \sqrt{\frac{3\pi}{\alpha}}\;l_P}\,, \tag{39}$$

the value of which works out to be

$$r_t \approx 5.808 \times 10^{-34}\,\mathrm{m}. \tag{40}$$

Thus, our r_t is about 36 times larger than the Planck length, and, as we can see, it is a remarkably simple constant in terms of the Planck length, l_P, and the fine structure constant, α. According to Equation (33), r_t is the effective radius at which energy density due to spin density should completely compensate the huge electrostatic energy seen in Equation (34). In our view, this is the correct Cartan radius, at least for the charged leptons, and that may still provide a plausible mechanism for averting singularities, since it is still larger than the Planck length. It is important to note, however, that these huge energy densities never actually occur in Nature, because according to our Equation (33) they are automatically compensated. The physical mechanism described above is simply to enable extraction of the radius r_x for different charged fermions.

We can now use Equation (33) to solve for r_x for the different charged leptons and anti-leptons which leads us to the following formula for our numerical estimates:

$$\frac{\alpha\hbar c}{r_x} - \frac{3\,\pi G\hbar^2}{c^2\, r_x^3} = +m_x c^2 \,. \tag{41}$$

As shown in the Appendix A below, we were able to find solutions for r_x for the charged leptons using arbitrary precision in *Mathematica*. The first in our results listed below is the solution for r_t up to 24 significant figures. Then, using the same precision for comparison, we list the results for r_e for an electron, r_μ for a muon and r_τ for a tauon, along with the anti-fermions:

$$\begin{aligned} r_t &= 5.80838808109165274355010 \times 10^{-34}\ \mathrm{m} \longrightarrow 0.0\ \mathrm{MeV}\,, \\ r_{e-} = r_{e+} &= 5.80838808109165274414872 \times 10^{-34}\ \mathrm{m} \longrightarrow 0.511\ \mathrm{MeV}\,, \\ r_{\mu-} = r_{\mu+} &= 5.80838808109165286732523 \times 10^{-34}\ \mathrm{m} \longrightarrow 106\ \mathrm{MeV}\,, \\ r_{\tau-} = r_{\tau+} &= 5.80838808109165482503366 \times 10^{-34}\ \mathrm{m} \longrightarrow 1777\ \mathrm{MeV}\,. \end{aligned}$$

We should note that there is also a positive solution obtained for the reduced Compton wavelength for these fermions in the form of

$$r_x = \alpha\frac{\hbar}{m_x c}. \tag{42}$$

This is so because the spin density squared term for an electron becomes very small of order 10^{-38} MeV when r_x is equal to that result so it can be considered effectively as zero. This solution for fermion radius appears to have been ruled out by scattering experiments. However, while it is true that no structure has been found via scattering experiments, there does seem to be some structure involving the magnetic moment and zitterbewegung. The apparent conflict with the scattering experiments can be resolved if there is a very small object for electric charge near Planck length, as our solution indicates, that is "circulating" about the Compton wavelength. Then the scattering at high energies can be understood as from that point-like object and the scattering at low energies can be understood as from the Compton wavelength "size" due to the Coulomb field, with the Coulomb field "barrier" near the Compton wavelength penetrated in the high energy scatterings.

Evidently, very minute changes in the radii are seen to cause large changes in the observed rest mass-energies of the fermions. As the differences in the radii go larger, the resultant mass-energies go higher, as one would expect. It seems extraordinary that Nature would subscribe to such tiny differences resulting from a large number of significant figures, but that might explain why the underlying relationship between the observed values of the masses of the elementary particles has remained elusive so far. In addition to the possible reasons for this mentioned above, it is not inconceivable that the difference between the spin energy density and the electrostatic energy density radii with respect to r_t arises due to purely geometrical factors. We also suspect that there may possibly be some kind of symmetry breaking mechanism at work similar to the Higgs mechanism, and this symmetry breaking results in the observed mass-energy generation.

As a consistency check, let us verify that the tiny length differences seen above vanish, $\Delta r \to 0$, as the corresponding rest mass-energy differences tend to zero: $\Delta E \to 0$. To this end, we recast Equation (41) for arbitrary r_x in a form involving only rest mass-energy on the RHS as:

$$\frac{\alpha\hbar c}{r_x} - \frac{3\pi\, l_P^2 \hbar c}{r_x^3} = m_x c^2. \tag{43}$$

If we now set

$$A \equiv \alpha\hbar c \quad \text{and} \quad B \equiv 3\pi\, l_P^2 \hbar c\,, \tag{44}$$

then, with $\Delta E = m_x c^2$ and setting $r_x = r_t$ as the cancellation radius for which $\Delta E = 0$, we obtain

$$r_t = \sqrt{B/A}\,. \tag{45}$$

This allows us to derive a general expression for r_x when $\Delta E \neq 0$:

$$\frac{A}{r_x} - \frac{B}{r_x^3} = \Delta E. \tag{46}$$

From this expression it is now easy to see that

$$\lim_{\Delta E \to 0} \left\{ \left(\frac{A}{r_x} - \frac{B}{r_x^3} \right) = \Delta E \right\} = \left(\frac{\sqrt{A}^3}{\sqrt{B}} - \frac{\sqrt{A}^3}{\sqrt{B}} \right) \ln \frac{\sqrt{B/A}}{l_P} = 0 \implies r_x = \sqrt{B/A} = r_t\,, \tag{47}$$

and conversely, using Equation (45),

$$\lim_{r_x \to r_t} \left\{ \left(\frac{A}{r_x} - \frac{B}{r_x^3} \right) = \Delta E \right\} \implies \Delta E = 0\,. \tag{48}$$

Consequently, with $\Delta r \equiv |\, r_t - r_x|$, we see from the above limits that $\Delta r \to 0$ as $\Delta E \to 0$, and vice versa.

As a rough estimate, the calculation for the radius r_q of elementary quarks can be performed in a similar manner as that for charged leptons, since at such short distances the strong force reduces to a Coulomb-like force. One must also factor-in the electrostatic energy, so that a relationship like the following must be calculated, say, for the top quark:

$$\frac{9\,\alpha\hbar c}{4\,r_{qx}} + \frac{\alpha_s \hbar c}{3\,r_{qx}} - \frac{3\pi G\hbar^2}{c^2\, r_{qx}^3} = m_t c^2\,. \tag{49}$$

Here α_s is the appropriate strong force coupling (we use 0.1). Needless to say, a cancellation radius different from that of the charged leptons should be calculated for comparison, by setting

$$\frac{9\,\alpha\hbar c}{4\,r_{qt}} + \frac{\alpha_s \hbar c}{3\,r_{qt}} - \frac{3\pi G\hbar^2}{c^2\, r_{qt}^3} = 0. \tag{50}$$

A calculation of the radius for the top quark based on Equation (49) can be found in the Appendix A and is $\approx 2.594 \times 10^{-34}$m. We expect it to be only a very rough estimate of the actual value of the radius. Since only one spin density is involved, the above calculation might be able to approximate the behaviour of the quarks. The calculation of the radii r_{qx} for the up and down quark will probably be problematic, since their masses are not well known.

With regard to neutrinos, the self-energy function is very similar to that for charged leptons with the Z boson replacing the photon and weak coupling replacing α. A rough calculation gives the same order for the radius at $\approx 10^{-34}$ m. However, if this does not hold for the neutrino self-energy, then the story is quite different because they would not have self-energy due to electric or colour charge. That means that their rest mass-energy comes entirely from torsion energy due to intrinsic spin. Solving that gives us a radius for electron neutrinos of the order of 10^{-26} m. However, the torsion (spin-squared) self-energy is negative relative to the positive rest mass. We suspect that this means that neutrinos could have anti-gravity properties [27]. The anti-gravity effect with normal matter for single neutrinos is practically negligible but the cosmological implications could be large.

4. Possible Solution of the Hierarchy Problem

As alluded to in the introduction, the Hierarchy Problem refers to the fact that gravitational interaction is extremely weak compared to the other known interactions in Nature. One way to appreciate this difference is by combining the Newton's gravitational constant G with the reduced Planck's constant $\hbar$ and the speed of light c. The resulting mass scale is the Planck mass, m_P, which some have speculated to be associated with the existence of smallest possible black holes [7]. If we compare the Plank mass with the mass of the top quark (the heaviest known elementary particle),

$$m_P = \sqrt{\frac{\hbar c}{G}} \approx 2.1765 \times 10^{-8}\,\text{kg},$$
$$m_t = \frac{173.21\ \text{GeV}}{c^2} \approx 3.1197 \times 10^{-25}\,\text{kg},$$

then we see that there is some 17 orders of magnitude difference between them. This illustrates the enormous difference between the Planck scale and the electroweak scale. Many solutions have been proposed to explain this difference, such as supersymmetry and large extra dimensions, but none has been universally accepted, for one reason or another. Furthermore, recent experiments performed with the Large Hadron Collider are gradually ruling out some of these proposals. Regardless of the nature of any specific proposal, it is clear from the above values that predictions of numbers with at least 17 significant figures are necessary to successfully explain the difference between m_P and m_t.

We saw from our numerical demonstration in the previous section that within the ECSK theory minute changes in length can induce sizable changes in the observed masses of elementary particles, and that we do have numbers at our disposal with more than 17 significant figures for producing those masses. Moreover, all length changes occurring in our demonstration are taking place close to the Planck length. Thus, since we are "cancelling out" near the Planck length to obtain masses down to the electroweak scale, ours is clearly a possible mechanism for resolving the Hierarchy Problem. We can appreciate this fact by simplifying our central equation by setting $\hbar = c = 1$, which then reduces to

$$\frac{\alpha}{r_x} - \frac{3\pi G}{r_x^3} = m_x\,. \tag{51}$$

It is now easy to see from this equation that the observed mass-energy only depends on the coupling constants and the radii (geometry). Moreover, it is confined entirely within a volume close to the Planck volume, as we saw in our calculations in the previous section. Thus, we are led to

$$\text{Planck Scale} \Longrightarrow \text{Electroweak Scale.}$$

In other words, there is no hierarchy problem in the ECSK theory, because Planck scale physics is producing the electroweak scale physics in the form of the mass-energy of fermions as a byproduct of the very geometry of spacetime.

Within the ECSK theory, which extends general relativity to include spin-induced torsion, gravitational effects near micro scales are not necessarily weak. On the other hand, since torsion is produced in the ECSK theory by the spin density of matter, it is confined to that matter, and thus is a very short range effect, unlike the infinite range effect of Einstein's gravity produced by mass-energy. In fact, the torsion field falls off as $1/r^6$, as shown in the calculations of Section 3, since it is produced by spin density squared, confined to the matter distribution [9].

To compare the strengths of gravitational and torsion effects at various scales, we may define a mass-dependent dimensionless gravitational coupling constant, $Gm^2/(\hbar c)$, and evaluate it for the electron, top quark and Planck masses:

$$\alpha_{G_e} = \frac{Gm_e^2}{\hbar c} \approx 1.7517 \times 10^{-45},$$
$$\alpha_{G_t} = \frac{Gm_t^2}{\hbar c} \approx 1.1620 \times 10^{-36},$$
$$\alpha_{G_P} = \frac{Gm_P^2}{\hbar c} = 1,$$
$$\alpha_e = \frac{e^2}{4\pi\hbar c} \simeq 7.2973 \times 10^{-3}.$$

Here α_e is the electromagnetic coupling constant, or the fine structure constant. From these values we see that near the Planck scale the gravitational coupling is very strong compared to the electromagnetic coupling. However, as we noted above and in Section 3, near the Planck scale torsion effects due to spin density are also very strong, albeit with opposite polarity compared to that of Einstein's gravity, akin to a kind of "anti-gravity" effect of a very short range.

For our demonstration above we have used electrostatic energy density and spin density for matter in a static approximation, for which the field equation within the ECSK theory reduces to $G^{00} = T^{00}$. A numerical estimate for G^{00} from the contributions of the electrostatic energy and spin density parts of T^{00} at our cancellation-radius gives

$$G^{00}_{stat} = \frac{8\pi G}{c^4}\frac{\alpha\hbar c}{r_t^4} \approx +4.209 \times 10^{62}\ \mathrm{m}^{-2} \tag{52}$$

and

$$G^{00}_{spin} = -\frac{8\pi G}{c^4}\frac{3\,\pi G\hbar^2}{c^2 r_t^6} \approx -4.209 \times 10^{62}\ \mathrm{m}^{-2}. \tag{53}$$

Evidently, these field strengths at the cancellation radius are quite large even for a single electron. Fortunately they are never realised in Nature, because, as we can see, they cancel each other out to produce $G^{00}_{net} = 0$. On the other hand, if we use only the mass-energy density for electron at the cancellation radius, then we obtain $G^{00}_{mass} \approx 3.0674 \times 10^{43}\,\mathrm{m}^{-2}$, which is again some 19 orders of magnitude off the mark. What is more, the latter field strength does not fall off as fast as that due to the spin-induced torsion field. Thus, it is reasonable to conclude that without the cancellation of divergent energies due to the spin self-interaction we have explored here, our universe would be highly improbable.

While there have been other approaches to the hierarchy problem from the viewpoint of the ECSK theory [28–32], our partial solution to the problem is simpler. We have fermions with near Planck scale radii (size) producing rest mass energy in the electro-weak scale. While we have not explained why higher generation fermions do not exist, given that as an assumption, our solution is more complete.

5. Concluding Remarks

In this paper we have addressed two longstanding questions in particle physics: (1) Why do the elementary fermionic particles that are so far observed have such low mass-energy compared to the Planck scale? (2) What mechanical energy may be counterbalancing the divergent electrostatic and strong force energies of point-like charged fermions in the vicinity of the Planck scale? Using a hitherto unrecognised mechanism extracted from the well known Hehl–Datta equation, we have presented numerical estimates suggesting that the torsion contributions within the Einstein–Cartan–Sciama–Kibble extension of general relativity can address both of these questions in conjunction.

The first of these problems, the Hierarchy Problem, can be traced back to the extreme weakness of gravity compared to the other forces, inducing a difference of some 17 orders of magnitude between the electroweak scale and the Planck scale. There have been many attempts to explain this huge difference, but none is simpler than our explanation based on the spin induced torsion contributions within the ECSK theory of gravity. The second problem we addressed here concerns the well known divergences of the electrostatic and strong force self-energies of point-like fermions at short distances. We have demonstrated above, numerically, that torsion contributions within the ECSK theory resolves this difficulty as well, by counterbalancing the divergent electrostatic and strong force energies close to the Planck scale.

It is widely accepted that in the standard model of particle physics charged elementary fermions acquire masses via the Higgs mechanism. Within this mechanism, however, there is no satisfactory explanation for how the different couplings required for the fermions are produced to give the correct values of their masses. While the Higgs mechanism does bestow masses correctly to the heavy gauge bosons and a massless photon, and while our demonstration above does not furnish a fundamental explanation for the fermion masses either, we believe that what we have proposed in this paper is worthy of further research, since our proposal also offers a possible resolution of the Hierarchy Problem.

In Reference [33], Singh points out that there appears to be a symmetry between small and large masses for spin-torsion coupling and energy-curvature coupling. We have noted that there also appears to be a symmetry in that the energy-curvature coupling is effectively infinite while the spin-torsion coupling is very short-ranged near the Planck length.

One may wonder why a gravitational coupling would be involved in the torsion term involving spin-squared, but we suspect it has more to do with Planck length than gravity. The torsion term with our r_t cancellation length can be simplified to

$$-\frac{3\,\pi G\hbar^2}{c^2\,r_t^3} = -\frac{3\pi(l_P)^2\hbar c}{(\sqrt{\frac{3\pi}{\alpha}}l_P)^3} = -\frac{\alpha\hbar c}{\sqrt{\frac{3\pi}{\alpha}}l_P} \approx -2.479\times 10^{15}\,\text{GeV}. \tag{54}$$

Instead of gravitational coupling, now the term has become simple and involves only constants and the Planck length.

Needless to say, the geometrical cancellation mechanism for divergent energies we have proposed here also dispels the need for mass-renormalisation, since we have obtained finite solutions for r_x taming the infinities. Thus, both classical and quantum electrodynamics appear to be more complete with torsion contributions included.

Author Contributions: The authors C.F.D.III and J.C. contributed to the article equally. All authors have read and agreed to the published version of the manuscript.

Funding: This research received no external funding.

Acknowledgments: The authors wish to thank Tejinder P. Singh for encouragement and discussions concerning the significance of torsion.

Conflicts of Interest: The authors declare no conflict of interest.

Appendix A. Calculations of Fermion Radii using Wolfram Mathematica

In this appendix we explain how we used the arbitrary-precision in *Mathematica* to solve the numerical equations out to 24 significant figures. Each equation displayed below—derived from our central Equation (33)—is simplified so that only the numerical factors have to be used, since the dimensional units cancel out, leaving lengths in meters. For decimal factors, the numbers must be padded out to 26 digits with zeros. Then the numerical part of electrostatic energy density is defined as A and the numerical part of spin energy density is defined as B, just as in Equation (44) above. These are then used throughout to perform the calculations. For the values of various physical constants involved in the calculations we have used the 2014 CODATA values, Reference [34] and values from the Particle Data Group, Reference [35].

Calculation of the Cancellation Radius for Charged Leptons using Formula Equation (41):

$$\alpha\hbar c\, r_t^2 - \frac{3\,\pi G\hbar^2}{c^2} = 0 \tag{A1}$$

$A{:=}N[(7.2973525664000000000000000 \times 10^{-3})(1.0545718000000000000000000 \times 10^{-34}) \times (2.9979245800000000000000000 \times 10^{8}), 26];$
$B{:=}N[(3\pi(6.6740800000000000000000000 \times 10^{-11})(1.0545718001391130000000000 \times 10^{-34})^2)/ (2.9979245800000000000000000 \times 10^{8})^2, 26];$
$N[\text{Solve}[A * r_t^2 - B == 0, r_t], 26]\ //\text{Last}$
$\{r_t \to 5.8083880810916527435501011 5 \times 10^{-34}\}$

Calculation of Radius r_e of Electron and Positron

$$\alpha\hbar c - \frac{3\pi G\hbar^2}{c^2 r_e^2} - m_e c^2 r_e = 0 \tag{A2}$$

$C1{:=}N[(9.1093835600000000000000000 \times 10^{-31})((2.9979245800000000000000000 \times 10^{8})^2), 26];$
$N\left[\text{Solve}\left[\left(A - B/\left(r_e^2\right)\right) - C1 * r_e == 0, r_e\right], 26\right]\ //\text{Last}$
$\{r_e \to 5.8083880810916527441487187 3 \times 10^{-34}\}$

Calculation of Radius r_μ of Muon and Anti-Muon

$$\alpha\hbar c - \frac{3\pi G\hbar^2}{c^2 r_\mu^2} - m_\mu c^2 r_\mu = 0 \tag{A3}$$

$C2{:=}N[(1.8835315940000000000000000 * 10^{-28})((2.9979245800000000000000000 * 10^{8})^2), 26];$
$N\left[\text{Solve}\left[\left(A - B/\left(r_\mu^2\right)\right) - C2 * r_\mu == 0, r_\mu\right], 26\right]\ //\text{Last}$
$\{r_\mu \to 5.8083880810916528673252292 8 \times 10^{-34}\}$

Calculation of Radius r_τ of Tau and Anti-Tau

$$\alpha\hbar c - \frac{3\,\pi G\hbar^2}{c^2 r_\tau^2} - m_\tau c^2 r_\tau = 0 \tag{A4}$$

$C3{:=}N[(3.1674700000000000000000000 \times 10^{-27})((2.9979245800000000000000000 * 10^{8})^2), 26];$
$N\left[\text{Solve}\left[\left(A - B/\left(r_\tau^2\right)\right) - C3 * r_\tau == 0, r_\tau\right], 26\right]\ //\text{Last}$
$\{r_\tau \to 5.8083880810916548250336629 5 \times 10^{-34}\}$

Calculation of the Cancellation Radius for (2e/3) Quarks using Formula (50):

$$\frac{9\alpha\hbar c}{4} + \frac{\alpha_s\hbar c}{3} - \frac{3\pi G\hbar^2}{c^2 r_{qt}^2} = 0. \tag{A5}$$

$D{:=}N[(1/3)(1/10)(1.0545718000000000000000000 \times 10^{-34})(2.9979245800000000000000000 \times 10^{8}), 26];$
$N\left[\text{Solve}[((4/9)A + D) * r_{qt}^2 - B == 0, r_{qt}], 26\right]\ //\text{Last}$
$\{r_{qt} \to 2.5943980965877948941460173 3 \times 10^{-34}\}$

Calculation of Radius r_{tq} of Top Quark:

$$\frac{\alpha\hbar c}{6\pi} + \frac{\alpha_s\hbar c}{3} - \frac{3\pi G\hbar^2}{2c^2 r_{tq}^2} - m_t c^2 r_{tq} = 0 \tag{A6}$$

$E{:=}N[(3.0877000000000000000000000 \times 10^{-25})((2.9979245800000000000000000 \times 10^{8})^2), 26];$
$N[\text{Solve}[\left((4/9)A + D - B/\left(r_{tq}^2\right)\right) - E * r_{tq} == 0, r_{tq}], 26]\ //\text{Last}$
$\{r_{tq} \to 2.5943980965878029705779804 9 \times 10^{-34}\}$

Appendix B. Miscellaneous Derivations for S-matrix Evaluation

For use in deriving the following, we begin with the fermion anticommutator,

$$\{b_i(p), b_j^\dagger(p')\} = \{d_i(p), d_j^\dagger(p')\} = (2\pi)^3 \frac{E}{m} \delta_{ij} \delta^3(\mathbf{p} - \mathbf{p}'), \tag{A7}$$

and since there is only one electron and no positron in both the initial and final states for the rest frame and $E_i = E_f = p_i = p_f = m = \omega = 2\pi/t$ with the spins being summed over, we have

$$\begin{aligned}
\psi(x) \mid i\rangle &= \sqrt{\frac{m}{E_i r^3}} \psi(x) b_i^\dagger(p_i) \mid 0\rangle \\
&= \sqrt{\frac{1}{r^3}} \int \frac{d^3p}{(2\pi)^3} \sum_{j=1}^{2} b_j(m) u^j(m) e^{-iEt} b_i^\dagger(m) \mid 0\rangle \\
&= \sqrt{\frac{1}{r^3}} \int \frac{d^3p}{(2\pi)^3} \sum_{j=1}^{2} u^j(m) e^{-iEt} \{b_j(m), b_i^\dagger(m)\} \mid 0\rangle \\
&= \sqrt{\frac{1}{r^3}} u^i(m) e^{-iEt},
\end{aligned}$$

and similarly

$$\begin{aligned}
\langle f \mid \bar{\psi}(x) &= \sqrt{\frac{1}{r^3}} \bar{u}^f(m) e^{+iEt}, \\
\psi(x) \mid i\rangle &= \sqrt{\frac{1}{r^3}} v^i(m) e^{+iEt}, \\
\langle f \mid \bar{\psi}(x) &= \sqrt{\frac{1}{r^3}} \bar{v}^f(m) e^{-iEt},
\end{aligned} \tag{A8}$$

in which the fermion anticommutator Equation (A7) is used for simplification [21]. Note that the spatial components, $\sqrt{1/r^3}$, of the spin vector are factored out in these plane-wave equations.

For the derivation of Equation (20), we begin with Equation (18) for the electrostatic self-interaction contribution,

$$\mathrm{h}_I = e\,\bar{\psi}(x)\gamma^\mu A_\mu \psi(x), \tag{A9}$$

and substitute it into the first order S-Matrix,

$$S_{fi}^{(1)}(E1) = -i \int d^4x \, \langle f \mid e\,\bar{\psi}(x)\gamma^\mu A_\mu \psi(x) \mid i\,\rangle. \tag{A10}$$

Making substitutions using Equation (A8) from above, and then going to the rest frame, along with using $A_0 = e/4\pi r$, $\alpha = e^2/4\pi$ and taking r to be constant, the S-matrix expression works out to give

$$
\begin{aligned}
S_{fi}^{(1)}(E1) &= -i\frac{e^2}{4\pi r}\int d^4x\left(\sqrt{\frac{1}{r^3}}\bar{u}^f(m)e^{+imt}\gamma^0\sqrt{\frac{1}{r^3}}u^i(m)e^{-imt}\right)\\
&= -i\frac{\alpha}{r^4}\left(\bar{u}^f(m)\gamma^0 u^i(m)\right)\int d^4x\\
&= -i\frac{\alpha\, r^3 t}{r^4}\begin{pmatrix}1&0&0&0\\0&1&0&0\\0&0&0&0\\0&0&0&0\end{pmatrix}\\
&= -i\,t\,\frac{\alpha}{r}\begin{pmatrix}1&0&0&0\\0&1&0&0\\0&0&0&0\\0&0&0&0\end{pmatrix},
\end{aligned}
\tag{A11}
$$

where we have taken $\int d^4x = r^3\,t$.

The S-matrix evaluation for Equation (21) in the rest frame is as follows. We begin with Equation (19),

$$
\mathrm{h}_I(x) = -\frac{3\kappa}{8}(\bar{\psi}(x)\gamma^5\gamma_k\psi(x))(\bar{\psi}(x)\gamma^5\gamma^k\psi(x)),
\tag{A12}
$$

and substitute it into the first order S-Matrix:

$$
S_{fi}^{(1)}(E2) = i\frac{3\kappa}{8}\int d^4x\,\langle f\mid(\bar{\psi}(x)\gamma^5\gamma_k\psi(x))(\bar{\psi}(x)\gamma^5\gamma^k\psi(x))\mid i\rangle.
\tag{A13}
$$

Then, because the spatial components, $\sqrt{1/r^3}$, of the spin vectors are factored out in the plane-wave equations that we derived in Equation (A8), by substituting Equation (A8) into Equation (A13) and then going to the rest frame, the S-matrix expression works out to give

$$
\begin{aligned}
S_{fi}^{(1)}(E2) &= i\frac{3\kappa}{8}\int d^4x\,\frac{1}{r^3}\left(\bar{u}^f(m)e^{+imt}\gamma^5\gamma_0 v^i(m)e^{+imt}\right)\frac{1}{r^3}\left(\bar{v}^f(m)e^{-imt}\gamma^5\gamma^0 u^i(m)e^{-imt}\right)\\
&= i\frac{3\kappa}{8r^6}\int d^4x\left(\bar{u}^f(m)\gamma^5\gamma_0 v^i(m)\,\bar{v}^f(m)\gamma^5\gamma^0 u^i(m)\right)\\
&= i\frac{3\kappa}{8r^6}\left(\bar{u}^f(m)\gamma^5\gamma_0(-1)\begin{pmatrix}0&0&0&0\\0&0&0&0\\0&0&1&0\\0&0&0&1\end{pmatrix}\gamma^5\gamma^0 u^i(m)\right)\int d^4x\\
&= i\frac{3\kappa\, r^3 t}{8r^6}\begin{pmatrix}1&0&0&0\\0&1&0&0\\0&0&0&0\\0&0&0&0\end{pmatrix}\\
&= i\,t\,\frac{3\kappa}{8r^3}\begin{pmatrix}1&0&0&0\\0&1&0&0\\0&0&0&0\\0&0&0&0\end{pmatrix},
\end{aligned}
\tag{A14}
$$

where we have again taken $\int d^4x = r^3\,t$. Similarly, the S-matrix expression for the anti-fermion works out to give

$$
\begin{aligned}
S_{fi}^{(1)}(E4) &= i\frac{3\kappa}{8}\int d^4\,x\,\frac{1}{r^3}\left(\bar{v}^f(m)e^{-imt}\,\gamma^5\gamma_0\,u^i(m)e^{-imt}\right)\frac{1}{r^3}\left(\bar{u}^f(m)e^{+imt}\,\gamma^5\gamma^0\,v^i(m)e^{+imt}\right)\\
&= i\frac{3\kappa}{8\,r^6}\left(\bar{v}^f(m)\,\gamma^5\gamma_0\begin{pmatrix}1&0&0&0\\0&1&0&0\\0&0&0&0\\0&0&0&0\end{pmatrix}\gamma^5\gamma^0\,v^i(m)\right)\int d^4\,x\\
&= i\,\frac{3\kappa\,r^3\,t}{8\,r^6}(-1)\begin{pmatrix}0&0&0&0\\0&0&0&0\\0&0&1&0\\0&0&0&1\end{pmatrix}\\
&= -i\,t\,\frac{3\kappa}{8\,r^3}\begin{pmatrix}0&0&0&0\\0&0&0&0\\0&0&1&0\\0&0&0&1\end{pmatrix}.
\end{aligned}
\tag{A15}
$$

From the last two equations it is easy to see that the fermion and anti-fermion S-matrix equations are coupled. In other words, one might initially think that in the rest frame $\bar{\psi}(t)\gamma^5\gamma^0\psi(t) = 0$, which would result in the entire term being zero, but that is not the case because of the coupling of mixed states between fermions and anti-fermions is an intermediate step in the self-interaction.

References

1. Hehl, F.W.; von der Heyde, P.; Kerlick, G.D.; Nester, J.M. General relativity with spin and torsion: Foundations and prospects. *Rev. Mod. Phys.* **1976**, *48*, 393. [CrossRef]
2. Trautman, A. Einstein-Cartan Theory. In *Encyclopedia of Mathematical Physics*; Francoise, J-P., Naber, G.L., Tsun, T.S., Eds.; Elsevier: Oxford, UK, 2006; Volume 2, pp. 189–195.
3. Sciama, D.W. The physical structure of general relativity. *Rev. Mod. Phys.* **1964**, *36*, 463. [CrossRef]
4. Kibble, T.W.B. Lorentz invariance and the gravitational field. *J. Math. Phys.* **1961**, *2*, 212. [CrossRef]
5. Hehl, F.W.; Datta, B.K. Nonlinear spinor equation and asymmetric connection in general relativity. *J. Math. Phys.* **1971**, *12*, 1334. [CrossRef]
6. Poplawski, N.J. Matter-antimatter symmetry and dark matter from torsion. *Phys. Rev. D* **2011**, *83*, 084033. [CrossRef]
7. Poplawski, N.J. Nonsingular, big-bounce cosmology from spinor-torsion coupling. *Phys. Rev. D* **2012**, *85*, 107502. [CrossRef]
8. Poplawski, N.J. Cosmological consequences of gravity with spin and torsion. *Astron. Rev.* **2013**, *8*, 108. [CrossRef]
9. Poplawski, N.J. Universe in a Black Hole in Einstein–Cartan Gravity. *Astrophys. J.* **2016**, *832*, 96. [CrossRef]
10. Tecchiolli, M. On the Mathematics of Coframe Formalism and Einstein–Cartan Theory—A Brief Review. *Universe* **2019**, *5*, 206. [CrossRef]
11. Diether, C.F., III; Christian, J. Existence of Matter as a Proof of the Existence of Gravitational Torsion. *Prespacetime J.* **2019**, *10*, 610.
12. Ortin, T. *Gravity and Strings: Cambridge Monographs on Mathematical Physics*; Cambridge University Press: Cambridge, UK, 2004.
13. Rohrlich, F. *Classical Charged Particles*, 3rd ed.; World Scientific: Singapore, 2007.
14. Blagojević, M.; Hehl, F.W. (Eds.) *Gauge Theories of Gravitation*; World Scientific: Singapore, 2013.
15. De Berredo-Peixoto, G.; Freidel, L.; Shapiro, I.L.; de Souza, C.A. Dirac fields, torsion and Barbero-Immirzi parameter in cosmology. *J. Cosmol. Astropart. Phys.* **2012**, *6*, 017. [CrossRef]
16. Magueijo, J.; Zlosnik, T.G.; Kibble, T.W.B. Cosmology with a Spin. *Phys. Rev. D* **2013**, *87*, 063504. [CrossRef]
17. Rudenko, A.S.; Khriplovich, I.B. Gravitational four-fermion interaction in the early Universe. *Phys. Uspekhi* **2014**, *57*, 167. [CrossRef]

18. Boos, J.; Hehl, F.W. Gravity-induced four-fermion contact interaction: Liberating the intermediate W and Z type gauge bosons., *Int. J. Theor. Phys.* **2017**, *56*, 751. [CrossRef]
19. Fabbri, L.; Vignolo, S. Dirac fields in f(R)-gravity with torsion. *Classical Quantum Gravity* **2011**, *28*, 12. [CrossRef]
20. Weisskopf, V.F. On the Self-Energy and the Electromagnetic Field of the Electron. *Phys. Rev.* **1939**, *56*, 72. [CrossRef]
21. Milonni, P.W. *The Quantum Vacuum: An Introduction to Quantum Electrodynamics*; Academic Press: San Diego, CA, USA, 1994; Chapter 12.
22. Weinberg, S. *The Quantum Theory of Fields, Vol. I*; Cambridge University Press: New York, NY, USA, 2005; Chapter 11.
23. Halzen, F.; Martin, A. *Quarks and Leptons: An Introductory Course in Modern Particle Physics*; Wiley, Inc.: Hoboken, NJ, USA, 1984; Chapter 7.
24. Dehmelt, H. A Single Atomic Particle Forever Floating at Rest in Free Space: New Value for Electron Radius. *Phys. Scr.* **1988**, *1988-T22*, 102. [CrossRef]
25. Poplawski, N.J. Uncertainty principle for momentum, torsional regularization, and bare charge. *arXiv* **2017**, arXiv:1712.09997.
26. Poplawski, N.J. Torsional regularization of vertex function. *arXiv* **2018**, arXiv:1807.07068.
27. Yablon, J. Grand Unified SU(8) Gauge Theory Based on Baryons which Are Yang-Mills Magnetic Monopoles. *J. Mod. Phys.* **2013**, *4*, 94. [CrossRef]
28. Zubkhov, M.A. Torsion Instead of Technicolor. *Mod. Phys. Lett. A* **2010**, *25*, 2885. [CrossRef]
29. Zubkhov, M.A. Dynamical torsion as the microscopic origin of the neutrino seesaw. *Mod. Phys. Lett. A* **2014**, *29*, 1450111. [CrossRef]
30. Zubkhov, M.A. Gauge theory of Lorentz group as a source of the dynamical electroweak symmetry breaking. *J. High Energy Phys.* **2013**, *44*. [CrossRef]
31. Fabbri, L. A simple assessment on the hierarchy problem. *Int. J. Geom. Meth. Mod. Phys.* **2016**, *13*, 1650068. [CrossRef]
32. Castillo-Felisola, O.; Corral, C.; Villavicencio, C.; Zerwekh, A.R. Fermion masses through condensation in spacetimes with torsion. *Phys. Rev. D* **2013**, *88*, 124022. [CrossRef]
33. Singh, T.P. A new length scale, and modified Einstein-Cartan-Dirac equations for a point mass. *Int. J. Mod. Phys.* **2018**, *27*, 1850077. [CrossRef]
34. Mohr, P.J.; Newell, D.B.; Taylor, B.N. CODATA Recommended Values of the Fundamental Physical Constants: 2014. *Rev. Mod. Phys.* **2016**, *88*, 035009. [CrossRef]
35. Olive, K.A.; Agashe, K.; Amsler, C.; Antonelli, M.; Arguin, J.-F.; Asner, D.M.; Baer, H.; Band, H.R.; Barnett, R.M.; Basaglia, T.; et al. Particle Data Group. *Chin. Phys. C* **2014**, *38*, 090001; see also 2015 update. [CrossRef]

Article

A Possible Explanation of Dark Matter and Dark Energy Involving a Vector Torsion Field

Graeme W. Milton

Department of Mathematics, The University of Utah, Salt Lake City, UT 84112 USA; milton@math.utah.edu

Abstract: A simple gravitational model with torsion is studied, and it is suggested that it could explain the dark matter and dark energy in the universe. It can be reinterpreted as a model using the Einstein gravitational equations where spacetime has regions filled with a perfect fluid with negative energy (pressure) and positive mass density, other regions containing an anisotropic substance that in the rest frame (where the momentum is zero) has negative mass density and a uniaxial stress tensor, and possibly other "luminal" regions where there is no rest frame. The torsion vector field is inhomogeneous throughout spacetime, and possibly turbulent. Numerical simulations should reveal whether or not the equations are consistent with cosmological observations of dark matter and dark energy.

Keywords: general relativity with torsion; dark matter; dark energy

Citation: Milton, G.W. A Possible Explanation of Dark Matter and Dark Energy Involving a Vector Torsion Field. *Universe* **2022**, *8*, 298. https://doi.org/10.3390/universe8060298

Academic Editors: Fabbri Luca and Matteo Luca Ruggiero

Received: 26 March 2022
Accepted: 16 May 2022
Published: 25 May 2022

Publisher's Note: MDPI stays neutral with regard to jurisdictional claims in published maps and institutional affiliations.

1. Introduction

One of the outstanding problems in physics is to account for the apparent dark energy and dark matter in the universe since it accounts for roughly 95% of the total matter in the universe. Reviews of the dark matter and dark energy cosmological problem, and the models that have been introduced to account for it, include those of Peebles and Ratra [1], Sahni [2], Copeland, Sami, and Tsujikawa [3], Frieman, Turner, and Huterer [4], Amendola and Tsujikawa [5], Li, Li, Wang, and Wang [6], and Arun, Gudennavar, and Sivaram [7]. We will not survey the literature here as these reviews do an excellent job of that. As is often the case, we use dimensions where the speed of light c is 1; we use the Einstein summation convention where sums over repeated indices are assumed, and a comma in front of a lower index such as $f_{,i}$ denotes differentiation of f with respect to x^i.

Maybe the most favored model is the ΛCDM model. Here, Λ is Einstein's cosmological constant, giving rise to dark energy with $p = -\mu_0$, and CDM is cold dark matter introduced to give the observed ratio of pressure to total mass density, which is about -0.8. Constraints on dark matter and dark energy properties are imposed by the results of the DES collaboration [8,9]. Gravitational lensing measurements [10] give a Hubble constant that is consistent with long-period Cepheid measurements in the large Magellanic cloud [11], but both strongly indicate significant discrepancies with the ΛCDM model. Experimental tests of the strong equivalence principle [12] provide further evidence casting doubt on the model in favor of modified gravity theories. Alternatively, there may be late time dark matter creation [13].

The relativistic model we introduce here has no adjustable parameters and incorporates a torsion vector field. It is perhaps the simplest gravitational model involving torsion; yet, we believe it could explain the dark energy and dark mass in the universe. If the simplicity of the underlying equations is to be a guiding principle in physics, then these equations surely meet that principle. Of course, our equations still need to be compatible with both existing and future experimental observations, both qualitatively and quantitatively, and this remains to be seen. It is to be stressed that our equations govern the curvature of empty space and do not fully determine the interaction between matter and the curvature. We believe the simpler problem of obtaining the equations for empty space should be addressed first, as a stepping stone towards a more general theory where matter is included. The main demands that drive our formulation of the equations are:

- That the new equations should be as simple as possible, involving as few assumptions as possible
- That, correspondingly, the new equations should be linear constraints on the curvature tensor.
- That, clearly, the number of unknowns in the torsion field and in the metric, modulo coordinate transformations, should be equal to the number of independent scalar constraints imposed by the new equations.
- Any solution to Einstein's equations is also a solution to the new equations.

It may be argued that these should not be assumed a priori, but that convincing physical arguments should be presented as well. On the other hand, Einstein's equations for empty space can be obtained from the first three of these requirements without any necessity to introduce physical considerations. Indeed, as is well known, it is natural that the Ricci tensor (with the possible addition of the cosmological constant term) is zero in empty space as this provides 10 equations for the 10 metric elements, with the 4 functions associated with freedom in the choice of coordinates being compensated by the 4 Bianchi identities. Only when matter is present is physics needed to determine the full Einstein equations, as embodied in the constraints that the equations reduce to Newton's gravitational equations when the spacetime curvature is small and that small test particles follow geodesics. Since we do not consider the full interaction of matter and curvature, we cannot claim that small test particles will still follow geodesics: that would be a natural demand to be required of a more general theory.

Despite the simplicity of our underlying equations the resultant dynamics of the torsion vector field, even in the weak field approximation, is enormously complicated, suggesting the torsion vector field has some sort of turbulent behavior. This is the main novel feature of our theory: the suggestion that torsion may induce intrinsic inhomogeneity on many length scales, even in the absence of matter. This goes further than the idea that space is inhomogeneous on the Planck length scale and is also a feature of anti-de Sitter spacetime [14]. Other work shows that the inhomogeneities of matter in the universe may account for the perceived acceleration of the universe without any need to introduce a negative cosmological constant ($\Lambda < 0$) (see [15] and the references therein).

Numerical simulations of the torsion field behavior will almost certainly be necessary to test the theory and assess its compatibility with astronomical and cosmological observations. The equations can be reinterpreted as a model using the Einstein gravitational equations where spacetime has regions filled with a perfect fluid with negative energy (pressure) and positive mass density, other regions containing an anisotropic substance, which in the local rest frame (where the momentum is zero) has negative mass density and a uniaxial stress tensor, and possibly other "luminal" regions where there is no natural local "rest frame". We emphasize, though, that all three regions are manifestations of the torsion vector field, and the three regions accordingly correspond to regions where the vector field points inside, outside, or on the boundary of the light cone. Our theory predicts that dark energy and dark matter (which are both manifestations of the torsion field) interact and exchange energy. Other models where dark energy and dark matter interact were reviewed by Wang, Abdalla, Atrio-Barandela, and Pavón [16] (see also the more recent work of Borges and Wands [17]).

It has been noted before by De Sabbata and Sivaram [18] that torsion provides a natural framework for negative mass, as has been suggested to occur in the early universe. Cosmological models with negative mass have been studied by Ray, Khlopov, Ghosh, and Mukhopadhyay [19] and by Famaey and McGaugh [20] and yield promising explanations for the acceleration of the expansion rate of the universe.

In addition to the cosmological dark mass problem, there is also the dark mass problem, which is associated with the observations of higher-than-expected rotational velocities of stars far from the galactic center. One empirically motivated model that successfully accounts for this is Modified Newtonian Dynamics (MOND), first introduced by Milgrom [21]. He suggested that Newton's law, where the gravitational force is proportional to the acceleration, be replaced at low accelerations, below a critical acceleration $a_0 = \approx 1.2 \times 10^{-10}$ ms^{-2}, by one where the force is proportional to the square of the acceleration; see Figure 1. Later, this idea motivated a relativistic theory developed by Bekenstein [22] and generalized by Skordis [23]. One prediction of MOND, later verified, was that there should be a universal relation between the rotation speeds of stars in the outermost parts of a galaxy and the total mass, not dark mass, of the galaxy; see the book of Merritt [24] for further discussion on this point. In particular, on the basis of this, it seems unlikely that unseen particles will provide the explanation for the galactic missing mass problem. Other reviews of MOND, including these and other relativistic extensions and their implications for cosmology, have been given by Famaey and McGaugh [20], Merritt [24], and Milgrom [25]. It is not yet clear whether the torsion field model developed here will be successful in explaining the galactic dark mass problem, though the success of Farnes [26] in explaining the flattening of rotation curves by introducing negative mass suggests that it might meet with success on this front.

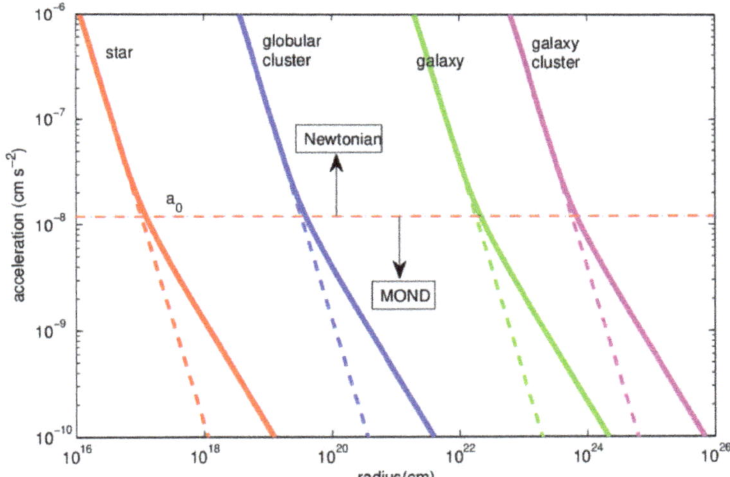

Figure 1. Figure, courtesy of M. Milgrom, taken from http://www.scholarpedia.org/article/The_MOND_paradigm_of_modified_dynamics, (accessed on 31 March 2020) showing its predictions, which are consistent with experimental observations. Plotted is the acceleration as a function of the distance from an isolated mass M, for a star with $M = M_\odot$ (red), a globular cluster with $M = 10^5 M_\odot$ (blue), a galaxy with $M = 3 \times 10^{10} M_\odot$ (green), and a galaxy cluster with $M = 3 \times 10^{13} M_\odot$ (magenta), in which $M_\odot$ represents one solar mass.

Torsion is the antisymmetric part of the affine connection. The affine connection determines how vectors change under parallel displacements. Cartan introduced torsion and applied it to develop generalizations of Einstein's gravitational equations. His work dates back to the early 1920s; see [27] and the references therein (translated in [28]). A brief introduction to torsion is in the classic book on gravitation by Misner, Thorne, and Wheeler [29]. More extensive reviews of general relativistic models that include torsion, with further developments, include those of Hehl, von der Heyde, and Kerlick [30], De Sabbata and Sivaram [18], Hehl, McCrea, Mielke, and Ne'eman [31], Shapiro [32], Ortín [33], Trautman [34], Poplawski [35], and Fabbri [36]. Interestingly, Jose Beltrán Jiméneza, Lavinia Heisenberg, and Tomi S. Koivisto have recently shown [37] that Einstein's gravitational equations can be reformulated in terms of the torsion alone, eliminating the metric.

Typically, general relativistic models with torsion have been introduced to allow for the intrinsic spin of matter and are quite complicated. By contrast, our focus here is on developing a simple model that may account for the dark mass and dark energy in the universe.

Ivanov and Wellenzohn suggested that the Einstein–Cartan theory may account for dark energy [38]. Another gravitational model that incorporates the same torsion vector field we use, as well as additional fields and a fifth dimension, was developed by Sengupta [39], who suggests it may solve both the cosmological and galactic dark matter problem. Other models incorporating torsion, quite different from the one explored here, that may explain the accelerated expansion of the universe have been developed by Watanabe and Hayashi [40], Minkevich [41], de Berredo-Peixoto and de Freitas [42], Belyaev, Thomas, and Shapiro [43], and Vasak, Kirsch, and Struckmeier [44].

The analysis in the following sections is more or less standard, though equivalent formulations are clearly possible according to one's mathematical taste. The key step to arriving at our equations is simply to postulate that geodesics and autoparallels coincide. There is nothing difficult in the analysis leading to our equations governing the spacetime curvature.

2. Metric and Affinities

The functions g_{uv} of the metric field describe with respect to the arbitrarily chosen system of co-ordinates the metrical relations of the spacetime continuum:

$$ds^2 = g_{uv}dx^u dx^v. \tag{1}$$

Here, we will assume that the g_{uv} are real and symmetric in the indices u and v, and thus, (1) provides the defining equation for g_{uv} with respect to a given coordinate system.

Now, consider the affinity Γ^i_{st}, which determines a vector after parallel displacement. To a real contravariant vector $\mathbf{A}$ with components A^i at a point P with coordinates x^t, we correlate a vector $\mathbf{A} + \delta\mathbf{A}$ with components $A^i + \delta A^i$ at the infinitesimally close point with coordinates $x^t + \delta x^t$ by

$$\delta A^i = -\Gamma^i_{st} A^s \delta x^t. \tag{2}$$

Since the magnitude of $\mathbf{A}$ in parallel displacement does not change to first order in that displacement, we obtain

$$0 = \delta[g_{uv}A^u A^v] = \frac{dg_{uv}}{dx^\alpha} A^u A^v dx^\alpha + g_{uv}A^u(\delta A^v) + g_{uv}A^v(\delta A^u), \tag{3}$$

and so, using (2), we obtain

$$g_{uv,\alpha} - g_{u\beta}\Gamma^\beta_{v\alpha} - g_{v\beta}\Gamma^\beta_{u\alpha} = 0, \tag{4}$$

where the comma denotes partial differentiation. Now, by considering this equation together with the two equations

$$g_{v\alpha,u} - g_{v\beta}\Gamma^\beta_{\alpha u} - g_{\alpha\beta}\Gamma^\beta_{vu} = 0, \tag{5}$$

$$g_{\alpha u,v} - g_{\alpha\beta}\Gamma^\beta_{uv} - g_{u\beta}\Gamma^\beta_{\alpha v} = 0, \tag{6}$$

that are obtained by a cyclic interchange of indices and by subtracting (4) from the sum of (5) and (6), we obtain

$$[u\ v, \alpha] + g_{v\beta}\widehat{\Gamma}^\beta_{u\alpha} + g_{u\beta}\widehat{\Gamma}^\beta_{v\alpha} - g_{\alpha\beta}\widehat{\Gamma}^\beta_{uv} = g_{\alpha\beta}\Gamma^\beta_{uv}, \tag{7}$$

where $[u\ v, \alpha]$ is the Christoffel symbol of the first kind, given by

$$[u\ v, \alpha] = \tfrac{1}{2}(g_{\alpha u,v} + g_{\alpha v,u} - g_{uv,\alpha}), \quad \widehat{\Gamma}^\beta_{ij} = \tfrac{1}{2}(\Gamma^\beta_{ij} - \Gamma^\beta_{ji}). \tag{8}$$

The antisymmetric part of the affinity $\widehat{\Gamma}^{\beta}_{ij}$, in contrast to Γ^{β}_{ji}, is a tensor—Cartan's torsion tensor.

3. Equating Geodesics with Autoparallels

Geodesics are trajectories $\mathbf{x}(s)$, which we chose to parametrize by the distance s along them, that have an extremal distance between two points. Since they clearly only depend on the metric, they satisfy the standard formula:

$$\frac{d^2x^\mu}{ds^2} + g^{\mu r}[\alpha\ \beta, r]\frac{dx^\alpha}{ds}\frac{dx^\beta}{ds} = 0. \tag{9}$$

Alternatively, we may consider an autoparallel constructed in such a way that successive elements arise from each other by parallel displacements. An element is the vector $d\mathbf{x}/ds$, and under parallel displacement, its components transform as

$$\delta\left(\frac{dx^u}{ds}\right) = -\Gamma^\mu_{\alpha\beta}\frac{dx^\alpha}{ds}\delta x^\beta. \tag{10}$$

The left-hand side is to be replaced by $(d^2x^\mu/ds^2)\delta s$, giving

$$\frac{d^2x^\mu}{ds^2} + \Gamma^\mu_{\alpha\beta}\frac{dx^\alpha}{ds}\frac{dx^\beta}{ds} = 0. \tag{11}$$

We postulate that geodesics coincide with autoparallels, thus giving

$$\left\{\Gamma^\mu_{\alpha\beta} - g^{\mu r}[\alpha\ \beta, r]\right\}\frac{dx^\alpha}{ds}\frac{dx^\beta}{ds} = 0, \tag{12}$$

or equivalently,

$$\left\{\tfrac{1}{2}(\Gamma^\mu_{\alpha\beta} + \Gamma^\mu_{\beta\alpha}) - g^{\mu r}[\alpha\ \beta, r]\right\}\frac{dx^\alpha}{ds}\frac{dx^\beta}{ds} = 0. \tag{13}$$

This postulate is fundamental to the theory. While it is absent of any physical justification, aside from removing possible ambiguity in the path that test particles are required to follow in a more general theory, it is essential to keep the governing equations as simple as possible. This is our motivation for this constraint.

As (13) holds for all dx^α/ds and dx^β/ds, we obtain

$$\overline{\Gamma}^\mu_{\alpha\beta} \equiv \tfrac{1}{2}(\Gamma^\mu_{\alpha\beta} + \Gamma^\mu_{\beta\alpha}) = g^{\mu r}[\alpha\ \beta, r]. \tag{14}$$

Multiplying both sides by $g_{\mu s}$ and summing over μ gives

$$g_{\mu s}\Gamma^\mu_{\alpha\beta} + g_{\mu s}\Gamma^\mu_{\beta\alpha} = 2[\alpha\ \beta, s]. \tag{15}$$

Combining this with (7) then yields

$$S_{\alpha\beta\mu} \equiv g_{\alpha r}\widehat{\Gamma}^r_{\beta\mu} = -g_{\beta r}\widehat{\Gamma}^r_{\alpha\mu} = -S_{\beta\alpha\mu} = S_{\beta\mu\alpha}. \tag{16}$$

Therefore, $S_{\alpha\beta\mu}$ is antisymmetric with respect to the interchange of any pair of its three indices, and this implies (see, for example, the text below Equation (2.16) in [30]) that

$$\widehat{\Gamma}^i_{jk} = g^{ir}e_{rjk\ell}U^\ell, \tag{17}$$

for some contravariant vector density $\mathbf{U}$, where, as is standard, $e_{rjk\ell}$ is the Levi-Civita tensor density, with $e_{1234} = 1$ and antisymmetric with respect to the interchange of any pair of indices. $\mathbf{U}$ is known as the axial part of the torsion [30]. A parallel derivation of the complete antisymmetry of torsion is in the review of Fabbri [36]. Combining (17) with (14) gives

$$\Gamma^{\mu}_{\alpha\beta} = \overline{\Gamma}^{\mu}_{\alpha\beta} + g^{\mu r} e_{r\alpha\beta\ell} U^{\ell}. \tag{18}$$

4. The Ricci Tensor

Let us express the Ricci tensor:

$$R_{jk} = \Gamma^{i}_{ri}\Gamma^{r}_{jk} - \Gamma^{i}_{rk}\Gamma^{r}_{ji} + \Gamma^{i}_{jk,i} - \Gamma^{i}_{ji,k}, \tag{19}$$

which is associated with the local curvature of spacetime, in terms of the symmetric and antisymmetric parts of the affinity:

$$\begin{aligned} R_{jk} &= (\overline{\Gamma}^{i}_{ri} + \widehat{\Gamma}^{i}_{ri})(\overline{\Gamma}^{r}_{jk} + \widehat{\Gamma}^{r}_{jk}) - (\overline{\Gamma}^{i}_{rk} + \widehat{\Gamma}^{i}_{rk})(\overline{\Gamma}^{r}_{ji} + \widehat{\Gamma}^{r}_{ji}) + (\overline{\Gamma}^{i}_{jk} + \widehat{\Gamma}^{i}_{jk})_{,i} - (\overline{\Gamma}^{i}_{ji} + \widehat{\Gamma}^{i}_{ji})_{,k}, \\ &= \overline{\Gamma}^{i}_{ri}(\overline{\Gamma}^{r}_{jk} + \widehat{\Gamma}^{r}_{jk}) - (\overline{\Gamma}^{i}_{rk} + \widehat{\Gamma}^{i}_{rk})(\overline{\Gamma}^{r}_{ji} + \widehat{\Gamma}^{r}_{ji}) + (\overline{\Gamma}^{i}_{jk} + \widehat{\Gamma}^{i}_{jk})_{,i} - \overline{\Gamma}^{i}_{ji,k}, \end{aligned} \tag{20}$$

where we used the fact that $\widehat{\Gamma}^{i}_{ri} = 0$, as follows from (17). Therefore, now, we have

$$R_{jk} = R^{0}_{jk} - \widehat{\Gamma}^{i}_{rk}\widehat{\Gamma}^{r}_{ji} + \widehat{\Gamma}^{i}_{kr}\overline{\Gamma}^{r}_{ji} + \overline{\Gamma}^{i}_{rk}\widehat{\Gamma}^{r}_{ij} - \overline{\Gamma}^{i}_{ri}\widehat{\Gamma}^{r}_{kj} - \widehat{\Gamma}^{i}_{kj,i}, \tag{21}$$

where

$$R^{0}_{jk} = \overline{\Gamma}^{i}_{ri}\overline{\Gamma}^{r}_{jk} - \overline{\Gamma}^{i}_{rk}\overline{\Gamma}^{r}_{ji} + \overline{\Gamma}^{i}_{jk,i} - \overline{\Gamma}^{i}_{ji,k} \tag{22}$$

is the usual Ricci curvature tensor associated just with the metric. We now consider the symmetric part of R_{jk} as it is central to our equations:

$$\overline{R}_{jk} \equiv \tfrac{1}{2}(R_{jk} + R_{kj}) = R^{0}_{jk} - \widehat{\Gamma}^{i}_{rk}\widehat{\Gamma}^{r}_{ji} = R^{0}_{jk} - g^{si} e_{srk\ell} U^{\ell} g^{tr} e_{tjih} U^{h}. \tag{23}$$

Given an arbitrary point, we can always find a new coordinate system such that the metric is orthogonal at that point. In this new coordinate system at this one point

$$\begin{aligned} \overline{R}_{11} &= R^{0}_{11} - g^{ii} e_{ir1\ell} U^{\ell} g^{rr} e_{r1ih} U^{h}, \\ \overline{R}_{12} &= R^{0}_{11} - g^{ii} e_{ir1\ell} U^{\ell} g^{rr} e_{r2ih} U^{h}, \end{aligned} \tag{24}$$

where a sum over i and r is implied. For $e_{ir1\ell} e_{r1ih}$ to be nonzero, it is necessary that $r \neq i$ and $ir\ell$ must be a permutation of rih (and a permutation of 234), implying $\ell = h$. Therefore, we obtain

$$\overline{R}_{11} = R^{0}_{11} - 2g^{22}g^{33}(U^{4})^{2} - 2g^{44}g^{22}(U^{3})^{2} - 2g^{33}g^{44}(U^{2})^{2}. \tag{25}$$

Furthermore, for $e_{ir1\ell} e_{r2ih}$ to be nonzero, ℓ must be 2 and h must be 1, implying

$$\overline{R}_{12} = R^{0}_{12} + 2g^{33}g^{44}U^{3}U^{4}. \tag{26}$$

Of course, similar formulas hold for the other elements of $\overline{R}_{jk}$. Hence, at this point, in this coordinate system,

$$\overline{R}_{jk} = R^{0}_{jk} + 2g^{-1} g_{jn} g_{km} U^{m} U^{n} - 2g^{-1} g_{jk} g_{mn} U^{m} U^{n}, \tag{27}$$

where $g = g_{11}g_{22}g_{33}g_{44}$ is the determinant of the metric tensor, or, introducing a contravariant vector N^{k} such that $N^{k} = U^{k}/\sqrt{-g}$, we obtain

$$\overline{R}_{jk} = R^{0}_{jk} + 2g_{jk}g_{mn}N^{m}N^{n} - 2g_{jn}g_{km}N^{m}N^{n}. \tag{28}$$

This equation being a tensor equation will be true in any coordinate system, as well as at any point since the original point was arbitrarily chosen. Raising indices gives

$$\overline{R}^{j}_{k} = (R^{0})^{j}_{k} + 2\delta^{j}_{k} g_{mn} N^{m} N^{n} - 2g_{km} N^{m} N^{j}. \tag{29}$$

Finally, contracting indices, we obtain

$$\overline{R} \equiv \overline{R}^j_j = R^0 + 6g_{mn}N^m N^n, \tag{30}$$

where $R^0 = (R^0)^j_j$. We will call **N** the torsion field.

5. The Proposed New Gravitational Equations

In this section, we investigate how torsion affects the geometry of empty space. For our purposes, it is to be observed that in the Einstein–Cartan–Sciama–Kibble theory, as reviewed in [30,33,34], one has

$$N_j = -2\kappa S_j, \tag{31}$$

where S_j is the spin axial-vector field, and so, there would be no torsion in empty space. Nevertheless, more general theories of torsion-gravity, in which torsion propagates, do not need to verify such a constraint, and therefore, **N** can still be nonzero even if $\mathbf{S} = 0$ identically; see Fabbri [36] and the references therein. Having emphasized that there exist theories in which torsion can still be nonzero, even in a vacuum, we will however not be specifying any particular Lagrangian. Instead, we will work from a very general perspective. The new gravitational field equations are

$$\overline{R}_{jk} - \tfrac{1}{2}g_{jk}\overline{R} = \kappa T'_{jk}. \tag{32}$$

where the T'_{jk} are the elements of the symmetric stress–energy–momentum tensor $\mathbf{T}'$ and $\kappa \approx 2 \times 10^{-43}\mathrm{s}^2\mathrm{m}^{-1}\mathrm{kg}^{-1}$ is the gravitational constant. This then has the equivalent form:

$$R^0_{jk} - \tfrac{1}{2}g_{jk}R^0 - g_{jk}g_{mn}N^m N^n - 2g_{jn}g_{km}N^m N^n = \kappa T'_{jk}, \tag{33}$$

or

$$R^0_{jk} - \tfrac{1}{2}g_{jk}R^0 = \kappa T_{jk}, \tag{34}$$

with

$$T_{jk} = T'_{jk} + [g_{jk}g_{mn}N^m N^n + 2g_{jn}g_{km}N^m N^n]/\kappa. \tag{35}$$

Thus, **T** is the equivalent stress–energy–momentum tensor if we were to reinterpret our equations in the format of Einstein's original gravitational equation. Therefore, if the torsion field **N** is small enough, we recover Einstein's original equations to a good approximation and, hence, those of Newtonian gravity. From here onwards, until the last section, we assume that $\mathbf{T}' = 0$, i.e., that no ordinary matter is present in the region of spacetime being studied. By multiplying (32) by g^{kj} and summing over indices, we see that $\overline{R} = 0$, and hence, (33) can be rewritten as

$$\overline{R}_{jk} = R^0_{jk} + 2g_{jk}g_{mn}N^m N^n - 2g_{jn}g_{km}N^m N^n = 0, \tag{36}$$

or, raising indices,

$$\overline{R}^{jk} = \{R^0\}^{jk} + 2g^{jk}g_{mn}N^m N^n - 2N^j N^k = 0. \tag{37}$$

These equations are consistent, for example, with those of Sengupta [39] (see his Equation (27)), which, however, are not the same as they include an extra dimension and incorporate additional fields.

The well-known Bianchi identities between the components of the contracted curvature tensor imply

$$[\{R^0\}^{jk} - \tfrac{1}{2}g^{jk}R^0]_{,k} = 0, \tag{38}$$

and as is well known, this implies $T^{ij}_{,j} = 0$, reflecting conservation of energy and momentum. Together with (37) and (30), we obtain

$$[g^{jk}g_{mn}N^m N^n + 2N^j N^k]_{,k} = 0. \tag{39}$$

We can view these as the extra four equations needed to determine the four components of **N** in empty space. One slightly unsatisfactory feature of the equations is that **N** is only determined up to a sign change. In other words, given a solution in a spacetime region, another solution can be obtained by reversing the sign of **N** within a subregion. Thus, we do not consider our theory to be complete. At the quantum Planck length scale, it likely needs modification, and the modified theory could prevent abrupt changes in the sign of **N**. Alternatively, one could take the view that there is no torsion, but rather, $\mathbf{N}(\mathbf{x})$ is just a vector field pervading all space. Then, the sign of $\mathbf{N}(\mathbf{x})$ is immaterial, but still, one would expect modifications at the Planck length scale to provide a lower limit to the length scales of "turbulence" in the vector field $\mathbf{N}(\mathbf{x})$.

6. The Weak Field Approximation

Now, consider the weak field approximation where $g_{\alpha\beta} = g^0_{\alpha\beta} + \kappa h_{\alpha\beta}$ and $N^i = \sqrt{\kappa} n_i$ where κ is a small parameter, and the $g^0_{\alpha\beta}$ correspond to the Minkowski metric:

$$g^0_{aa} = \{g^0\}^{aa} = 1, \quad g^0_{ab} = \{g^0\}^{ab} = 0, \quad g^0_{a4} = \{g^0\}^{a4} = 0, \quad g^0_{44} = \{g^0\}^{44} = -1 \tag{40}$$

in which a, b are indices taking the values 1, 2, or 3 with $a \neq b$. There is some freedom in the choice of the $h_{\alpha\beta}$ due to the coordinate shifts that we can make to first order in κ. This freedom can be eliminated by imposing the harmonic gauge that

$$h^{jk}_{,k} = \tfrac{1}{2}\{g^0\}^{jk} h_{,k}, \tag{41}$$

in which $h = \{g^0\}^{st} h_{st}$ and $h^{jk} = \{g^0\}^{js}\{g^0\}^{kt} h_{st}$. To first order in κ (37) implies

$$0 = \overline{R}^{jk}/\kappa = -\frac{1}{2} g^0_{mn} \frac{\partial h^{jk}}{\partial x_m \partial x_n} + 2\{g^0\}^{jk} g^0_{mn} n^m n^n - 2n^j n^k. \tag{42}$$

Furthermore, to first order in κ, (39) implies

$$[\{g^0\}^{jk} g^0_{mn} n^m n^n + 2n^j n^k]_{,k} = 0. \tag{43}$$

Not all 10 equations in (42) are independent, as a consequence of the Bianchi identities (38). To see this directly, multiply (42) by g^0_{hj} and contract indices to give

$$0 = \overline{R}/\kappa = -\frac{1}{2} g^0_{mn} \frac{\partial h}{\partial x_m \partial x_n} + 6 g^0_{mn} n^m n^n, \tag{44}$$

which is also implied by taking the first-order approximation to (30). Thus, we have

$$0 = (\overline{R}^{jk} - \tfrac{1}{2} g^{jk}\overline{R})/\kappa = -\frac{\partial}{\partial x_m \partial x_n}(h^{jk} - \tfrac{1}{2}\{g^0\}^{jk} h) - [\{g^0\}^{jk} g^0_{mn} n^m n^n + 2n^j n^k]. \tag{45}$$

With (41), we recover (43). In summary, we should first use the four equations (43) to determine the $n^i(\mathbf{x})$, $i = 1, 2, 3, 4$. Then, we should use the 16 Equations (41) and (42), of which only 10 are independent, to determine the 10 functions $h_{ij}(\mathbf{x})$. Writing out Equation (42) explicitly, we obtain

$$\begin{aligned} \nabla^2 h^{ab} - \frac{\partial^2}{\partial t^2} h^{ab} &= 4[\delta_{ab}(n^2 - n_4^2) - n_a n_b], \\ \nabla^2 h^{a4} - \frac{\partial^2}{\partial t^2} h^{a4} &= 4 n_a n_4, \\ \nabla^2 h^{44} - \frac{\partial^2}{\partial t^2} h^{44} &= -4n^2, \end{aligned} \tag{46}$$

where the indices a and b take values from 1 to 3, $n^2 = n_1^2 + n_2^2 + n_3^2$, and $n_i = g^0_{ij} n^j$. As we used the harmonic gauge, there is the additional restriction that the h^{jk} satisfy (41), i.e., that

$$\begin{aligned} h^{a1}_{,1} + h^{a2}_{,2} + h^{a3}_{,3} + h^{a4}_{,4} &= \tfrac{1}{2}(h^{11} + h^{22} + h^{33} - h^{44})_{,a}, \quad a = 1,2,3, \\ h^{41}_{,1} + h^{42}_{,2} + h^{43}_{,3} + h^{44}_{,4} &= -\tfrac{1}{2}(h^{11} + h^{22} + h^{33} - h^{44})_{,4}. \end{aligned} \tag{47}$$

The identities (43) imply $T^{ij}_{,j} = 0$ with, to zeroth order in κ,

$$\begin{aligned} T^{aa} &= 2n_a^2 + n^2 - n_4^2, \quad T^{ab} = 2n_a n_b, \\ T^{44} &= 3n_4^2 - n^2, \quad T^{a4} = -2n_a n_4. \end{aligned} \tag{48}$$

Equivalently, the matrix $\mathbf{T}$ with elements T^{ij} takes the block form:

$$\mathbf{T} = \begin{pmatrix} 2\mathbf{n} \otimes \mathbf{n} + (n^2 - n_4^2)\mathbf{I} & -2n_4 \mathbf{n} \\ -2n_4 \mathbf{n}^T & 3n_4^2 - n^2 \end{pmatrix}, \tag{49}$$

where $\mathbf{n}^T$ is the row vector, which is the transpose of $\mathbf{n}$, defined as $\mathbf{n} = (n_1, n_2, n_3)$.

7. Subluminal, Luminal, and Superluminal Regions of Spacetime

In this section, we do not make the weak field approximation, but we consider any point P in spacetime and choose the Minkowski metric (40) at that point.

7.1. Subluminal Regions and the Equivalent Perfect Fluid with Negative Energy That Occupies Them

Consider a region where $k = n_4^2 - n^2 > 0$. We call such a region a subluminal region. Define the 4-velocity $\mathbf{V}$ with components

$$V_a = n_a/\sqrt{k}, \quad V_4 = n_4/\sqrt{k} \tag{50}$$

satisfying $V_1^2 + V_2^2 + V_3^2 - V_4^2 = -1$. In terms of this velocity, (48) implies

$$\begin{aligned} T^{aa} &= (2V_a^2 - 1)k, \quad T^{ab} = 2V_a V_b k, \\ T^{44} &= (2V_4^2 + 1)k, \quad T^{a4} = 2V_a V_4 k. \end{aligned} \tag{51}$$

By comparison, a perfect fluid moving with 4-velocity $\mathbf{V}$ has

$$\begin{aligned} T^{aa} &= (\mu_0 + p)V_a^2 + p, \quad T^{ab} = (\mu_0 + p)V_a V_b, \\ T^{44} &= (\mu_0 + p)V_4^2 - p, \quad T^{a4} = (\mu_0 + p)V_a V_4, \end{aligned} \tag{52}$$

where $p = p$ is the pressure and μ_0 is the rest density (in the frame with the same velocity as the fluid). Thus, T corresponds to a fluid with

$$p = -\mu_0/3, \quad \mu_0 = 3k. \tag{53}$$

Note that, in this case, it always possible to choose a moving frame of reference with respect to which the fluid is not locally moving, i.e., $n^2 = 0$.

7.2. Superluminal Regions and the Equivalent Substance with Negative Mass That Occupies Them

Consider those regions where $k = n_4^2 - n^2 < 0$, which we call superluminal. Then, it is impossible to move to a reference frame such that $n^2 = 0$ at a given point. Rather, we can move to a frame where $n_4 = 0$ at this point. In this frame,

$$\begin{aligned} T^{aa} &= 2n_a^2 + n^2, \quad T^{ab} = 2n_a n_b, \\ T^{44} &= -n^2, \quad T^{a4} = 0. \end{aligned} \tag{54}$$

This corresponds to some sort of substance, which, in this frame, has no momentum, a negative mass density $-n^2$, and a stress:

$$\sigma = -n^2\mathbf{I} - 2\mathbf{n} \otimes \mathbf{n}, \tag{55}$$

corresponding to a pressure of n^2 and an additional uniaxial compression in the direction $\mathbf{n}$.

7.3. Luminal Regions

Finally, consider the regions where $k = n_4^2 - n^2 = 0$, which we call luminal. Then,

$$T^{aa} = 2n_a^2, \quad T^{ab} = 2n_a n_b, \quad T^{44} = 2n_4^2, \quad T^{a4} = -2n_a n_4. \tag{56}$$

Clearly, a luminal boundary or luminal region must separate regions that are subluminal or superluminal. In a luminal region, one cannot move to a frame where $n^2 = 0$, nor where $n_4^2 = 0$, unless both are zero. The momentum density, mass density, and stress are nonzero everywhere, except where the torsion field vanishes.

8. Some Solutions and Perturbative Solutions for the Torsion Field in the Weak Field Approximation

Let us consider solutions of $T^{ij}_{\ ,j} = 0$ in a flat metric given by (40). Using (49), we obtain

$$\begin{aligned} 0 &= \frac{\partial}{\partial t}[3n_4^2 - n^2] - 2\nabla \cdot (n_4\mathbf{n}), \\ 0 &= \nabla \cdot (\mathbf{n} \otimes \mathbf{n}) - \frac{\partial(n_4\mathbf{n})}{\partial t} + \tfrac{1}{2}\nabla(n^2 - n_4^2) \\ &= (\mathbf{n} \cdot \nabla)\mathbf{n} + \mathbf{n}\nabla \cdot \mathbf{n} - \frac{\partial(n_4\mathbf{n})}{\partial t} + \tfrac{1}{2}\nabla(n^2 - n_4^2), \end{aligned} \tag{57}$$

where the first equation represents the conservation of energy and the second the balance of forces.

In the subluminal regions, if we look for solutions where $\mathbf{n} = 0$ globally and not just at one point, the conservation of energy implies $\partial n_4^2/\partial t = 0$, while the balance of forces implies $\nabla(n_4^2) = 0$. Thus, n_4^2 must be a constant in spacetime. On the other hand, if we allow for small values of $\mathbf{n}$, with $n^2 \ll n_4^2$, then to first order in the perturbation with $\mathbf{e} = n_4\mathbf{n}$ and $f = n_4^2$, we obtain

$$\frac{\partial \mathbf{e}}{\partial t} = -\frac{1}{2}\nabla f, \quad 3\frac{\partial f}{\partial t} = 2\nabla \cdot \mathbf{e}, \tag{58}$$

giving

$$3\frac{\partial^2 f}{\partial t^2} = -\nabla^2 f. \tag{59}$$

This has exponentially growing solutions such as

$$f = 1 + \delta e^{\tau} \cos \mathbf{k} \cdot \mathbf{x}, \quad \text{with } k^2 \equiv \mathbf{k} \cdot \mathbf{k} = 3\tau^2, \tag{60}$$

where δ is a small parameter. After a finite time, this solution for f reaches negative values, but before which, our assumption that $n^2 \ll n_4^2$ is violated. Thus, the solution with $\mathbf{n} = 0$ is unstable to perturbations.

In the superluminal regions, if we look for solutions where $n_4 = 0$ globally and not just at one point, then the conservation of energy implies that n^2 must not vary with time, and the balance of forces implies

$$\nabla(n^2) + 2\nabla \cdot (\mathbf{n} \otimes \mathbf{n}) = 0. \tag{61}$$

This provides three equations to be satisfied by the three functions $n_a(x_1, x_2, x_3, t)$, $a = 1, 2, 3$. There is a manifold of functions satisfying (61), and we can choose any trajectory $\mathbf{n}(x_1, x_2, x_3, t)$ that lies on this manifold and is such that $n^2(x_1, x_2, x_3) = \mathbf{n}(x_1, x_2, x_3, t) \cdot \mathbf{n}(x_1, x_2, x_3, t)$ is independent of time. Unless $\mathbf{n}(x_1, x_2, x_3, t)$ only depends on t, it seems likely that this second condition will generally force $\mathbf{n}(x_1, x_2, x_3, t)$ to be independent of time (up to a sign change in $\mathbf{n}$). If we investigate the effect of perturbations, with $\mathbf{n}_4^2 \ll n^2$ and both n_1 and ϵ depending only on x_1 and t and say $n_2 = n_3 = 0$, we obtain

$$\frac{\partial e_1}{\partial t} = \frac{3}{2}\frac{\partial d}{\partial x_1}, \quad \frac{\partial d}{\partial t} = -2\frac{\partial e_1}{\partial x_1}, \tag{62}$$

where $e_1 = n_4 n_1$ and $d = n_1^2$. This gives

$$\frac{\partial^2 d}{\partial t^2} = -3\frac{\partial^2 d}{\partial x_1^2}, \tag{63}$$

which has exponentially growing solutions such as

$$d = 1 + e^{\tau} \cos \mathbf{k} \cdot \mathbf{x}, \quad \text{with } k^2 = \tau^2/3. \tag{64}$$

While, after a finite time, our assumption that $\mathbf{n}_4^2 \ll n^2$ becomes violated, the calculation shows that the solution with $n_4 = 0$ is unstable to perturbations.

In luminal regions where $n^2 - n_4^2 = 0$, we can use this identity to eliminate n_4 from (57) and obtain

$$\begin{aligned} 0 &= \frac{\partial n^2}{\partial t} \pm \nabla \cdot (|\mathbf{n}|\mathbf{n}), \\ 0 &= \nabla \cdot (\mathbf{n} \otimes \mathbf{n}) \pm \frac{\partial(|\mathbf{n}|\mathbf{n})}{\partial t}, \\ &= (\mathbf{n} \cdot \nabla)\mathbf{n} + \mathbf{n}\nabla \cdot \mathbf{n} \pm \frac{\partial(|\mathbf{n}|\mathbf{n})}{\partial t}, \end{aligned} \tag{65}$$

where the plus or minus sign is taken according to whether $n_4 = \pm|\mathbf{n}|$. In the special case where $n_2 = n_3 = 0$ (after making a spatial rotation if necessary), we obtain $n_4 = n_1$ (or $n_4 = -n_1$), and (65) reduces to the single equation:

$$\frac{\partial n_1}{\partial t} = \frac{\partial n_1}{\partial x_1} \tag{66}$$

to be satisfied by the function $n_1(x_1, x_2, x_3, t)$, describing a wave propagating at the speed of light in the direction of the x_1-axis. We call them localized longitudinal torsion waves, longitudinal because $\mathbf{n}$ is aligned with the direction of propagation. We now look for perturbation solutions with $n_4 = n_1 - \epsilon$, where ϵ is a small parameter and both n_1 and ϵ depend only on x_1 and t while $n_2 = n_3 = 0$. Letting $\epsilon n_1 = \eta$ and $d = n_1^2$, we obtain

$$\frac{\partial(d-\eta)}{\partial t} = -\frac{\partial(d-\eta)}{\partial x_1}, \quad \frac{\partial(d-3\eta)}{\partial t} = -\frac{\partial(d-\eta)}{\partial x_1}. \tag{67}$$

The first wave equation has the solution $\eta = d + h(x_1 - t)$, where $h(y)$ is an arbitrary function, and substituting in the second gives $\partial d/\partial t = h'(x_1 - t)$, where $h'(y)$ is the derivative of $h(y)$. We conclude that

$$\eta = \sigma(x_1), \quad d = \sigma(x_1) - h(x_1 - t), \tag{68}$$

where $\sigma(x_1)$ satisfies $h(y) \leq |\sigma(y)| \ll |h(y)|$ for all y, to ensure that d is non-negative and that the perturbation is small ($|\epsilon(x_1,t)| \ll |n_1(x_1,t)|$ for all (x_1,t), but otherwise is an arbitrary function. Thus, $h(y)$ can only take negative values, and the perturbation travels at the speed of light in the direction of $\mathbf{n}$.

We now present various other solutions of the equations, without investigating their stability.

8.1. Plane Wave Solutions

Here, we consider plane wave solutions to the equations in the weak field approximation. It is to be emphasized that since the equations are non-linear, specifically quadratic in $\mathbf{n}$, one cannot generally superimpose our plane wave solutions to obtain another solution.

The simplest case is when the fields only depend on, say, x_1. Then, we deduce that T^{1j} is a constant, i.e.,

$$3n_1^2 + n_2^2 + n_3^2 - n_4^2 = k_1, \quad n_1 n_2 = k_2, \quad n_1 n_3 = k_3, \quad n_1 n_4 = k_4, \tag{69}$$

where the k_i are constants. Multiplying the first equation by n_1^2, we obtain

$$n_1^4 = (k_1 n_1^2 - k_2^2 - k_3^2 + k_4^2)/3, \tag{70}$$

which requires the constants k_i to be such that the right-hand side is non-negative. Thus, n_1^2 is constant, and the last three equations in (69) imply that n_2^2, n_3^2, and n_4^2 are constants as well, unless $n_1^2 = 0$. Therefore, the only interesting case is when $n_1^2 = 0$, implying that $k_2 = k_3 = k_4 = 0$. Then, according to whether $-k_1 = n_4^2 - n^2$ is positive, zero, or negative, the solution will be subluminal, superluminal, or luminal. Thus, subject to the constraint that $n^2 \geq k_1$ (relevant only when $k_1 > 0$), $n_2(x_1)$ and $n_3(x_1)$ can be chosen arbitrarily and determine $n_4^2 = n^2 - k_1$. In particular, if $k_1 = 0$, one may choose $n_2(x_1)$ and $n_3(x_1)$ to be zero outside an interval of values of x_1. In a frame of reference moving with velocity $-v_1$ in direction x_1, this will look like a wave pulse traveling a velocity v_1, as all the field components will be functions of $x_1 - v_1 t$. We call them localized transverse torsion waves, transverse because $\mathbf{n}$ is perpendicular to the wave front. Unlike longitudinal torsion waves, which can only travel at the speed of light, these can have any velocity less than c.

Similarly, when the fields only depend on $t = x_4$, we deduce that T^{4j} is a constant, i.e.,

$$n_4 \mathbf{n} = \mathbf{k}', \quad 3n_4^2 - n^2 = k_4', \tag{71}$$

in which $\mathbf{n} = (n_1, n_2, n_3)$ and $n^2 = \mathbf{n} \cdot \mathbf{n}$ and where k_4' and $\mathbf{k}' = (k_1', k_2', k_3')$ are constants. Multiplying the last formula by n_4^2 shows that

$$n_4^4 = (k_4' n_4^2 + \mathbf{k}' \cdot \mathbf{k}')/3 \tag{72}$$

is constant, implying n_4^2 is constant and, through the first equation in (71), that n_1^2, n_2^2, and n_3^2 are constant as well, unless $n_4^2 = 0$. When $n_4^2 = 0$, then $\mathbf{k}' = 0$. The last formula in (71) then forces n^2 to be constant. Subject to this constraint, $\mathbf{n}$ can have an arbitrary dependence on time (with $\mathbf{n}(t)$ being independent of x_1, x_2, and x_3). However, note that the inevitable spatial variation of $\mathbf{n}$ may eliminate this arbitrariness, as may going beyond the weak field approximation.

8.2. Solutions with Cylindrical Symmetry, Including Torsion-Rolls

Consider cylindrical coordinates (r, θ, z, t) taking r to be the radial distance from the z-axis, θ to be the angular variable, and t to be the time. We seek solutions where $\mathbf{n} = (n_r, n_\theta, n_z)$ and n_4 only depend on r, so that

$$\begin{aligned}
\nabla \cdot (n_4 \mathbf{n}) &= \frac{1}{r}\frac{d(rn_4n_r)}{dr}\hat{\mathbf{r}}, \\
(\mathbf{n} \cdot \nabla)\mathbf{n} &= \left(n_r\frac{dn_r}{dr} - \frac{n_r^2}{r}\right)\hat{\mathbf{r}} + \left(n_r\frac{dn_\theta}{dr} + \frac{n_rn_\theta}{r}\right)\hat{\theta} + \left(n_r\frac{dn_z}{dr}\right)\hat{\mathbf{z}}, \\
\mathbf{n}(\nabla \cdot \mathbf{n}) &= \frac{1}{r}\frac{d(rn_r)}{dr}(n_r\hat{\mathbf{r}} + n_\theta\hat{\theta} + n_z\hat{\mathbf{z}}), \\
\tfrac{1}{2}\nabla(n^2 - n_4^2) &= \tfrac{1}{2}\left[\frac{d}{dr}(n^2 - n_4^2)\right]\hat{\mathbf{r}},
\end{aligned} \tag{73}$$

where we used the standard formulas for the gradient, divergence, and $\mathbf{n} \cdot \nabla$ in cylindrical coordinates. Then, the conservation laws (57) take the form

$$\begin{aligned}
0 &= \frac{1}{r}\frac{d(rn_rn_4)}{dr}, \\
0 &= \frac{n_r^2 - n_\theta^2}{r} + \tfrac{1}{2}\frac{d}{dr}\left[2n_r^2 + n^2 - n_4^2\right], \\
0 &= \frac{2n_rn_\theta}{r} + \frac{d(n_rn_\theta)}{dr}, \\
0 &= \frac{n_rn_z}{r} + \frac{d(n_rn_z)}{dr}.
\end{aligned} \tag{74}$$

If we consider an interface at a constant radius $r = r_0$, with outwards unit normal $\hat{\mathbf{r}}$, then the weak form of the equations $T^{ij}_{\ ,j} = 0$ imply the jump conditions on the elements T^{ij} that

$$\mathbf{T}\begin{pmatrix}\hat{\mathbf{r}} \\ 0\end{pmatrix}$$

must be continuous across the interface, where $\mathbf{T}$ is given by (49). This implies that the quantities

$$c_4 = n_4n_r, \quad c_\theta = n_\theta n_r, \quad c_z = n_zn_r, \quad c_r = 3n_r^2 + n_\theta^2 + n_z^2 - n_4^2 \tag{75}$$

must all be continuous across the interface $r = r_0$. Multiplying the last equation by n_r^2, we see that

$$n_r^4 = (c_rn_r^2 - c_\theta^2 - c_z^2 + c_4^2)/3 \tag{76}$$

must be continuous as well, and the first three equations imply that all components of $(\mathbf{n}, n_4)$ are continuous across the interface, up to a change of sign, unless $n_r^2 = 0$ at the interface. If n_r^2 is zero at the interface, it follows that $c_4 = c_\theta = c_z = 0$ at the interface. Therefore, across $r = r_0$, any jumps in $n_\theta(r,t)$, $n_z(r,t)$, and $n_4(r,t)$ that maintain the continuity of $n^2 - n_4^2$ are possible provided $n_r(r,t)$ is continuous and $n_r(r_0,t) = 0$.

The first, third, and last equations in (74) imply

$$rn_rn_4 = k_4, \quad rn_rn_z = k_z, \quad r^2n_rn_\theta = k_\theta \tag{77}$$

where k_4, k_z, and k_θ are constants. In the case $n_r = 0$, all are satisfied with $k_4 = k_z = k_\theta = 0$. The remaining second equation in (74) becomes

$$\frac{d}{dr}\left[n^2 - n_4^2\right] = \frac{2n_\theta^2}{r}. \tag{78}$$

Thus, there is only one constraint among the three functions $n_\theta(r)$, $n_z(r)$, and $n_4(r)$. We see that $n^2 - n_4^2$ must monotonically increase with r, in a manner controlled by $n_\theta^2(r)$, and if it tends to zero at infinity, then $n^2 - n_4^2$ must be negative for all r, corresponding to a subluminal region. If $\mathbf{n}$ and n_4 vanish outside a certain radius, then we call this solution a torsion-roll. Physically, the pressure increases to larger negative values as the radius decreases, and its gradient provides the centripetal force that holds the "fluid" circulating around the z-axis with a velocity governed by n_θ. In a moving frame of reference, which is not moving in the z-direction, the torsion-roll will appear to be moving.

Of course, if $n^2 - n_4^2$ is constant and positive outside a certain radius (corresponding, for example, to a superluminal region where, say, n_z is constant and $n_\theta = n_4 = 0$), then $n^2 - n_4^2$ can remain positive for all r, or can transition from positive to negative values at a particular radius. This example demonstrates that transitions between subluminal and superluminal regions are possible.

Alternatively, if n_r is nonzero, then (77) implies

$$n_4 = k_4/(rn_r), \quad n_z = k_z/(rn_r), \quad n_\theta = k_\theta/(r^2 n_r). \tag{79}$$

Substituting these in the second equation in (74) yields

$$\frac{ds}{dr} = \frac{2s(3k_\theta^2 - r^4 s^2 - 2kr^2)}{r(3r^4 s^2 - k_\theta^2 + kr^2)}, \quad \text{where} \quad s = n_r^2, \quad k = k_4^2 - k_z^2. \tag{80}$$

This gives us a flow-field in the (r, s) phase plane. Note that (80) remains invariant under the transformation

$$r \to \lambda_1 r, \quad s \to \lambda_2 s, \quad k_\theta^2 \to \lambda_1^4 \lambda_2^2 k_\theta^2, \quad k \to \lambda_1^2 \lambda_2^2 k. \tag{81}$$

Thus, without loss of generality, we may, by rescaling any solution, take k_θ to be 0 or 1 and k to be -1, 0, or 1. If $k = 0$, then there is essentially just one solution: $s_0(r)$ satisfying $s_0(1) = 1$ with all other solutions (with $k_\theta = 1$) taking the form $s(r) = \lambda^2 s_0(\lambda r)$, parametrized by λ. The solutions for $s_0(r) = n_r^2(r)$ and $n_\theta^2(r) = 1/(r^4 n_r^2)$ are shown in Figure 2 along with the flow field. One can see that the solution does not exist below a critical value of r, which looks unsatisfactory. This critical radius is associated with the vanishing of the denominator in (80).

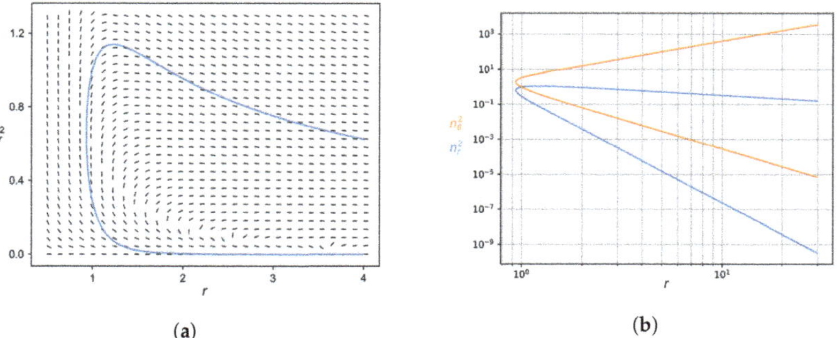

(a) (b)

Figure 2. Solution for the torsion field with cylindrical symmetry with $n_r \neq 0$, $k = 0$, and $k_\theta = 1$. (**a**) The flow field when $k = 0$ and $k_\theta = 1$ and the particular solution satisfying $n_r^2 = 1$ when $r = 1$. (**b**) The same solution for n_r^2 on a log–log plot and the accompanying function $n_\theta^2 = 1/(r^4 n_r^2)$.

To obtain satisfactory solutions that exist for all $r \neq 0$, one may take $k_\theta = 0$ and $k = 1$ to avoid the denominator in (80) vanishing, except at $r = 0$. Then, (80) reduces to

$$\frac{ds}{dr} = -\frac{2s(r^2 s^2 + 2)}{r(3r^2 s^2 + k)}, \quad \text{where} \quad s = n_r^2. \tag{82}$$

There is again essentially just one solution: $s_0(r)$ satisfying $s_0(1) = 1$ with all other solutions (with $k = 1$) taking the form $s(r) = \lambda s_0(\lambda r)$, parametrized by λ. The solution is graphed in Figure 3. There is a singularity at $r = 0$, and while $n_r^2(r)$ goes rapidly to zero as $r \to \infty$, $n_4^2(r)$ and $n_z^2(r)$ (unless it is zero) diverge to ∞ as $r \to \infty$. This solution is satisfactory once one takes into account that the weak field approximation is not valid near the singularity at $r = 0$, nor as $r \to \infty$, and one should use the full equations (37) there. For this example with $k = 1$ and $k_\theta = 0$, it is interesting that there is a transition from a superluminal region inside to a subluminal region outside according to the sign of

$$n^2 - n_4^2 = n_r^2 + \frac{k_z^2}{r^2 n_r^2} - \frac{k_4^2}{r^2 n_r^2} = n_r^2 - \frac{1}{r^2 n_r^2}, \tag{83}$$

which is also plotted in Figure 3.

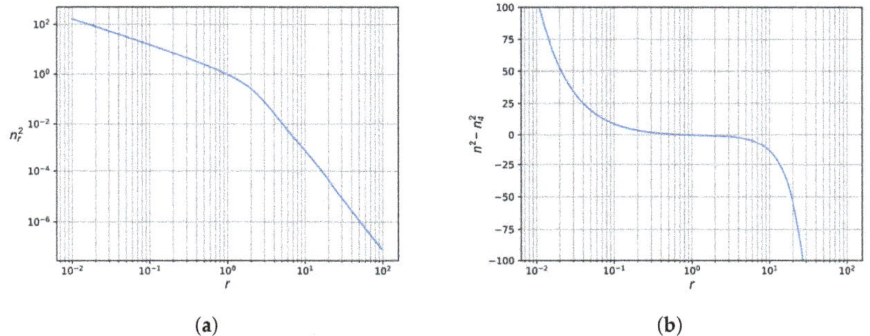

(a) (b)

Figure 3. Solution for the torsion field with cylindrical symmetry with $n_r \neq 0$, $k = 1$, and $k_\theta = 0$. (**a**) The graph of $n_r^2 = 1$ showing its divergence as $r \to 0$. (**b**) The plot of $n^2 - n_4^2 = n_r^2 - 1/(r^4 n_r^2)$ showing a transition from superluminal to subluminal as r increases.

9. Extension of the Schwarzschild Solutions With Spherical Symmetry

Here, we generalize Schwarzschild's solution for a spherically symmetric metric solving Einstein's equations in the absence of matter. The important point is that in appropriate limits, some of the solutions here approach the Schwarzschild solution. Consequently, existing experimental results of black holes do not invalidate our theory, but rather place constraints on the magnitude of the torsion field. This magnitude should be tied to the radius of the universe, and, hence, to the critical acceleration in MOND. Thus, experiments in the near vicinity of a star or black hole would not typically reveal the difference with Schwarzschild's solution. We have not explored the situation regarding rotating black holes.

As shown by Schwarzschild, the metric in "polar" coordinates spherically symmetric about the origin must be of the form

$$ds^2 = a\,dr^2 + r^2(d\theta^2 + \sin^2\theta\,d\phi^2) - b\,dt^2, \tag{84}$$

in which a and b are functions of r and t. Here, we look for solutions where they are functions of r only. Setting $x_1 = r$, $x_2 = \theta$, $x_3 = \phi$, $x_4 = t$ allows us to use (84) to identify the coefficients:

$$g_{11} = a, \quad g_{22} = r^2, \quad g_{33} = r^2 \sin^2\theta, \quad g_{44} = -b. \tag{85}$$

From (36), we obtain the ten equations:

$$
\begin{aligned}
0 &= \overline{R}_{11} = \frac{a'}{ar} + \frac{a'b'}{4ab} + \frac{(b')^2}{4b^2} - \frac{b''}{2b} + 2a[r^2(N^2)^2 + r^2\sin^2\theta(N^3)^2 - b(N^4)^2], \\
0 &= \overline{R}_{22} = 1 - \frac{1}{a} + \frac{ra'}{2a^2} - \frac{rb'}{2ab} + 2r^2[a(N^1)^2 + r^2\sin^2\theta(N^3)^2 - b(N^4)^2], \\
0 &= \overline{R}_{33} = [1 - \frac{1}{a} + \frac{ra'}{2a^2} - \frac{rb'}{2ab}]\sin^2\theta + 2r^2\sin^2\theta[a(N^1)^2 + r^2(N^2)^2 - b(N^4)], \\
0 &= \overline{R}_{44} = \left(\frac{b'}{ar} + \frac{b''}{2a} - \frac{(b')^2}{4ab} - \frac{a'b'}{4a^2}\right) - 2b[a(N^1)^2 + r^2(N^2)^2 + r^2\sin^2\theta(N^3)^2], \\
0 &= \overline{R}_{mn} = -2g_{mm}g_{nn}N^mN^n \quad \text{for all } m,n \quad \text{with } m \neq n, \quad \text{no sum on } m,n, \qquad (86)
\end{aligned}
$$

where the terms not involving $\mathbf{N}$ can be identified with the standard formulas for the elements R^0_{ij} that are zero when $i \neq j$. Here, differentiation with respect to $x_1 = r$ is denoted by the prime, with the double prime denoting the second derivative. The second and third equations and the last equation force $N^2 = N^3 = 0$, which is not surprising considering the symmetry of the problem. Two possibilities remain: either $N^1 = 0$ or $N^4 = 0$. The first case corresponds to a subluminal solution and the second to a superluminal solution.

Let us consider first the case where $N^1 = N^2 = N^3 = 0$. Multiplying the second last equation in (86) by a/b and adding it to the first gives

$$
\frac{a'}{a} + \frac{b'}{b} - 2q = 0 \quad \text{where } q = rab(N^4)^2 \geq 0. \qquad (87)
$$

The second equation in (86) implies

$$
\frac{a'}{a} - \frac{b'}{b} + 2(a-1)/r - 4q = 0. \qquad (88)
$$

Adding and subtracting these equations gives

$$
\begin{aligned}
a'/a &= \frac{1}{r} - \frac{a}{r} + 3q, \\
b'/b &= \frac{a}{r} - \frac{1}{r} - q. \qquad (89)
\end{aligned}
$$

Multiplying the last by br, differentiating it, and using the result to eliminate b'' from the first equation in (86) yield

$$
q' = 2q^2 + \frac{q}{r}. \qquad (90)
$$

This has the solution

$$
q = \frac{\alpha^2 r}{1 - \alpha^2 r^2}, \qquad (91)
$$

where α is a constant. Furthermore, by replacing q with $rab(N^4)^2$, one obtains

$$
\begin{aligned}
2q^2 + \frac{q}{r} = q' &= ab(N^4)^2 + (ra'/a)ab(N^4)^2 + (rb'/b)ab(N^4)^2 + rab\frac{d(N^4)^2}{dr}, \\
&= \frac{q}{r}[1 + (1 - a + 3qr) + (a - 1 - qr)] + q\frac{d(N^4)^2}{dr} = \frac{q}{r} + 2q^2 + q\frac{d(N^4)^2}{dr}. \qquad (92)
\end{aligned}
$$

This implies that $(N^4)^2$ is a constant that we call β^2, giving

$$
\frac{a}{r} = \frac{q}{br^2(N^4)^2} = \frac{\alpha^2}{br\beta^2(1 - \alpha^2 r^2)}. \qquad (93)
$$

Substituting this back in the second equation in (89) gives the linear first-order differential equation:

$$\frac{db}{dr} + b\left[\frac{1}{r} + \frac{\alpha^2}{1-\alpha^2 r^2}\right] = \frac{\alpha^2}{\beta^2 r(1-\alpha^2 r^2)}. \tag{94}$$

Multiplying both sides by the integrating factor of $r/\sqrt{1-\alpha^2 r^2}$ gives

$$\frac{d}{dr}\left[br/\sqrt{1-\alpha^2 r^2}\right] = \frac{\alpha^2}{\beta^2(1-\alpha^2 r^2)\sqrt{1-\alpha^2 r^2}}. \tag{95}$$

Integrating both sides and recalling (93), we obtain

$$\begin{aligned} b &= \frac{\alpha^2}{\beta^2} - 2m\frac{\sqrt{1-\alpha^2 r^2}}{r}, \\ a &= \frac{\alpha^2}{b\beta^2(1-\alpha^2 r^2)}, \end{aligned} \tag{96}$$

where m is a constant of integration. In particular, with $\alpha^2 = \beta^2$, this becomes

$$\begin{aligned} b &= 1 - 2m\frac{\sqrt{1-\alpha^2 r^2}}{r}, \\ a &= \frac{1}{b(1-\alpha^2 r^2)}, \end{aligned} \tag{97}$$

which in the limit $\alpha \to 0$ reduces to the familiar Schwarzschild solution:

$$a = \frac{1}{1-2m/r}, \quad b = 1 - 2m/r, \tag{98}$$

which becomes Euclidean at large r. Once we allow nonzero α, the space is no longer Euclidean at large r, but it still has a black hole at the center, with a diverging when $r = 2m\sqrt{1-\alpha^2 r^2}$ and at $r = 1/\alpha^2$, the latter corresponding to the closed universe studied in the next section.

Now, consider the second possibility that $N^2 = N^3 = N^4 = 0$. Again, multiplying the second last equation in (86) by $a/(b)$ and adding it to the first give

$$\frac{a'}{a} + \frac{b'}{b} - 2w = 0 \quad \text{where } w = ra^2(N^1)^2 \geq 0. \tag{99}$$

Furthermore, the second equation in (86) implies

$$\frac{a'}{a} + \frac{b'}{b} + 2(a-1)/r + 4w = 0. \tag{100}$$

Adding and subtracting these equations give

$$\begin{aligned} a'/a &= \frac{1}{r} - \frac{a}{r} - w, \\ b'/b &= \frac{a}{r} - \frac{1}{r} + 3w. \end{aligned} \tag{101}$$

Multiplying the last by br, differentiating it, and using the result to eliminate b'' from the first equation in (86) yield

$$w' = -2w^2 + w(1 - \tfrac{4}{3}a)/r. \tag{102}$$

The equations (101) and (102) appear to have no simple analytic solution. One may eliminate $a(r)$ from the two equations that do not involve $b(r)$ to obtain

$$\frac{w''}{w} = \frac{7(w')^2}{4w^2} - \frac{3w'}{2wr} - \frac{2w}{r} + w^2 - \frac{1}{4r^2}, \tag{103}$$

and from a solution $w(r)$, (102) easily gives $a(r)$. Alternatively, one may eliminate $w(r)$ from these equations tc obtain

$$\frac{v''}{v} = \frac{3(v')^2}{v^2} + \frac{v'}{vr} + (5v' + 2v^2)/3 + v/r. \tag{104}$$

where $v = a/r$, and given a solution $a(r) = rv(r)$, the first equation in (101) yields $w(r)$. In either case, $b(r)$ is found by integrating the last equation in (101). Note that if $b(r)$ is a solution, then so will be $\lambda^2 b(r)$ for any constant λ, i.e., $b(r)$ is only determined up to a multiplicative constant. This reflects the fact that we are free to rescale the time coordinate, replacing t by t/λ in (84).

Rather than dealing with these second-order equations for $w(r)$ and $v(r)$, one can numerically solve (101) and (102) directly. Figure 4 shows some typical solutions, excluding unphysical examples where, say, $a(r)$ or $w(r)$ remain negative for all r

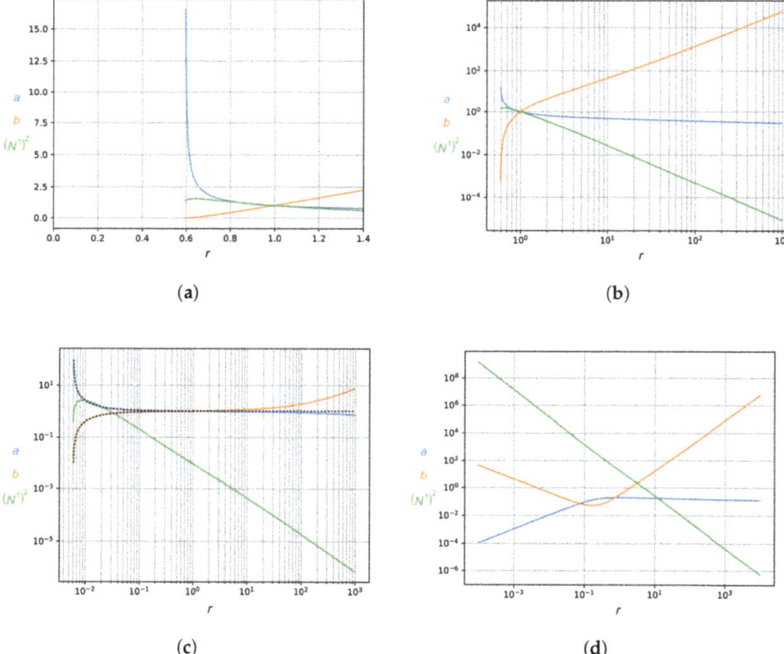

Figure 4. Numerical solutions of Equations (101) and (102). (**a**) Graph with $w(1) = a(1) = b(1) = 1$ showing a "black hole"-type singularity at $r = 0.5959$. (**b**) Same as for (**a**), but on a log–log plot. Note the blow up of $b(r)$ as $r \to \infty$. (**c**) Graph with $w(1) = 0.01$ and $a(1) = b(1) = 1$. Comparing this with (**b**) and taking note of the different vertical scales, one can see the approach to the usual Schwarzschild solution as $w(1) \to 0$. (**d**) Graph with $w(1) = 0.8$, $a(1) = 0.2$, and $b(1) = 0.3$ showing a different type of solution with no critical "black hole" radius, but rather, a singularity at $r = 0$. The solution for $b(r)$ still clearly blows up as $r \to \infty$.

10. Homogeneous Expanding Universe

It should be emphasized that the solution given here, which is incompatible with the observations, is for a homogeneous universe devoid of ordinary matter. It does not apply to a universe where spacetime itself has fluctuations that are not due to ordinary matter. Even ignoring ordinary gravitational effects, the perturbation results at the beginning of Section 8 imply such fluctuations occur, and so, one should expect that its cosmological predictions deviate from those presented in this section. This is explained further in the next section.

We take the Friedmann–Lemaître–Robertson–Walker metric in reduced-circumference polar coordinates:

$$ds^2 = a^2\left[\frac{dr^2}{1-kr^2} + r^2(d\theta^2 + \sin^2\theta\, d\phi^2)\right] - dt^2 \tag{105}$$

where a, the reduced circumference, can be a function of time, while r, θ, and ϕ are time independent, and $k = 1, 0, -1$, according to whether the universe is spatially closed, flat, or open with negative curvature. With $x_1 = r$, $x_2 = \theta$, $x_3 = \phi$, and $x_4 = t$, the corresponding metric coefficients are

$$g_{11} = a^2/(1-kr^2), \quad g_{22} = a^2r^2, \quad g_{33} = a^2r^2\sin^2\theta, \quad g_{44} = -1. \tag{106}$$

Assuming $N^1 = N^2 = N^3 = 0$ and defining

$$P = 2k + (a\ddot{a} + 2\dot{a}^2), \tag{107}$$

where the dot and double dot denote first and second derivatives with respect to time, the equations become

$$\begin{aligned}
0 &= \overline{R}_{11} = -2(N^4)^2a^2/(1-kr^2) - P/(1-kr^2),\\
0 &= \overline{R}_{22} = -2(N^4)^2a^2r^2 - Pr^2,\\
0 &= \overline{R}_{33} = -2(N^4)^2a^2r^2\sin^2\theta - Pr^2\sin^2\theta,\\
0 &= \overline{R}_{44} = 3\ddot{a}/a,
\end{aligned} \tag{108}$$

where the terms not involving N^4 can be identified with the standard formulas for R^0_{ij}. The last equation in (108) implies $\dot{a}$ is a constant that we define to be β. We obtain

$$P = 2k + 2\dot{a}^2 = 2(k+\beta^2), \quad a = \beta t + \gamma, \tag{109}$$

where γ is an integration constant that we can choose to be zero by redefining our origin of time appropriately (except in the trivial case of a spatially flat universe independent of time with $N_4 = 0$). From the remaining three equations in (107), which are all equivalent, we obtain

$$(N^4)^2 = -\frac{P}{2a^2} = -\frac{k+\beta^2}{\beta^2 t^2}, \tag{110}$$

which implies that $k = -1$ (an open universe, like in anti-de Sitter spacetime) and $\beta^2 < 1$.

11. Addressing the Dark Matter and Dark Energy Problem

The result of the previous section giving an expansion rate $\dot{a}$ independent of time agrees with the well-known result that $\ddot{a} = 0$ for a model with $p = -\mu_0/3$. However, this is based on the premise that spacetime is homogeneous. The expansion of the universe appears to be accelerating with measurements indicating $p = -0.8\mu_0$ [8], and this could be a consequence of our theory, as we now explain.

Dark matter itself is known to be inhomogeneous; see, for example, [45] and the references therein. Spacetime is also inhomogeneous in our model. As the analysis at the beginning of Section 8 shows, if there is a small fluctuation in the torsion vector field in subluminal or superluminal regions of spacetime, then that perturbation will grow. Moreover, ordinary gravitational effects might add to the inhomogeneity: if there is a higher equivalent mass density in two different regions, then there could be gravitational attraction between these regions, leading to accretion. At the same time, "collisions" between accreting regions should tend to disperse the torsion vector field density. Thus, there will be a certain amount of equivalent kinetic energy associated with the torsion field accounting for some additional "dark energy". More importantly, there could be substructures in the torsion field containing differing ratios of "dark energy" to "dark mass". The structures could collide and give rise to different structures. In particular, there might be "negative mass structures", by which we mean structures in the torsion vector field incorporating superluminal regions. Accounting for these effects should reduce the total mass density, providing a higher p/μ_0 ratio, which may be consistent with the experimental value of -0.8.

It is to be emphasized that both our full equations (37) and their weak field approximations (46), (47), and (57) have no intrinsic length scale. There is a length scale associated with the overall density of the torsion vector field (connected with the mass density of the apparent dark matter and dark energy in our theory), but this is of the order of the radius of the universe. It seems likely that the torsion vector field could be quite turbulent with structures on many length scales, down to some lower cutoff length scale where the current theory breaks down. This cutoff could be the Planck length scale.

To provide quantitative predictions, one needs a better idea of the behavior of the torsion vector field within spacetime, and this will almost certainly require sophisticated numerical simulations to obtain an approximation to the "macroscopic equation of state". Simulations are needed to provide a better understanding of torsion fluid behavior in intergalactic and interstellar regions, as well as around stars, globular clusters, galaxies, and galaxy clusters. These may require the introduction of some parameter that provides a lower length scale to the "turbulence" in the torsion vector field, which ultimately could be taken to be very small. Simulating the dynamics of the torsion vector field over the continuum of length scales may also require a sort of numerical renormalization group approach. While we have not investigated the stability of the torsion waves and torsion-rolls, it is not important that they are stable, even in the weak field approximation. The purpose of our exact solutions in the weak field approximation was mainly to illustrate the rich dynamics of the torsion vector field, to give some insight into possible dynamics and to show that one can have transitions between subluminal and superluminal regions, as noted at the end of Section 8.2.

Regarding the question as to whether our model can account for the galactic dark mass problem, an encouraging sign is the apparent cosmological connection between the critical acceleration $a_0 \approx 1.2 \times 10^{-10}$ m/s^2 in MOND, the radius of the universe, and the density of dark matter or energy in the universe, as reviewed in [46]. Thus, the density of dark matter or energy, roughly $\varrho \approx 10^{-27}$ kg/m^3, which in our theory is related to the strength of the torsion field $\mathbf{N}$, has an associated length scale $1/\sqrt{c^2\varrho\kappa} \approx 6 \times 10^{26}$ meters (approximately the radius of the universe), which agrees with the length scale $c^2/a_0 \approx 7.5 \times 10^{26}$ meters associated with the critical acceleration a_0 in MOND.

12. Discussion

The theory presented here is largely aimed at providing equations governing the behavior of spacetime and the torsion field in regions devoid of matter. An initial test of the theory would entail numerical simulations of the torsion field in the weak field approximation with ordinary gravitational effects neglected, as governed by Equation (57), and allowing for field fluctuations. These fluctuations in the torsion field should be truncated at a small length scale, perhaps at the Planck length scale. This should give an approximation to the effective equation of state. The next step would be to determine the evolution of a homogeneous closed universe with this equation of state. Then, perturbations could be introduced and the evolution studied. For the theory to be viable, without modification, the results need to be consistent with cosmological observations.

Beyond the need for a lower cutoff, the equations are still incomplete. As remarked already, one can change the sign of $\mathbf{N}(\mathbf{x})$ in any region and still satisfy the equations, indicating that there is a deeper theory that prevents such discontinuous solutions for $\mathbf{N}(\mathbf{x})$. Perhaps this also enters at the Planck length scale, and both it and the truncation of fluctuations in the torsion field are accounted for by appropriate quantum equations. Assuming there is only weak coupling between the torsion fluid with matter, aside from the coupling due to gravitation (spacetime curvature), then one might think there is conservation of momentum and energy both for the stress–energy–momentum tensor of the torsion vector field and for the stress–energy–momentum tensor of matter. On the other hand, if one regards the conservation of momentum and energy as a consequence of the Bianchi identities, then there appears to be no reason why they should be separately conserved. For this reason, our current theory, while it describes the curvature of spacetime and the accompanying torsion vector field in regions devoid of matter, is incomplete in regions containing matter.

One appealing feature of Cartan's equations, and which is absent in our current theory, is that they allow for the incorporation of intrinsic spin—something that was discovered in 1925–1926 after Cartan first arrived at his remarkable equations. Cartan was originally motivated by the work of the Cosserat brothers [47], who, like his equations, allowed for a non-symmetric stress field. His focus was on deriving equations where the source (matter) field automatically satisfied energy and momentum conservation. Sciama [48] and Kibble [49] independently developed the same generalization of Cartan's theory, known as U_4 or the Einstein–Cartan–Sciama–Kibble theory. Their theory and the original Cartan theory imply that the torsion field is zero in empty space and, so, reduce to the Einstein equations when matter is not present. An advantage of these theories, not yet incorporated in our theory as there is no coupling with matter, is that they account for the conservation of angular momentum [30].

As others have also realized, departing from Cartan's approach has the potential for explaining dark energy and dark matter as manifestations of a revised gravitational theory. Our theory is perhaps the simplest theory with that potential. As stressed already, conservation of energy and momentum still hold provided one reinterprets the equations as Einstein's equation with an energy–momentum–stress tensor associated with "empty space", i.e., associated with the torsion field. It could be that more complicated equations involving torsion will provide the final answer (and, as observed in the introduction, many candidates, besides Cartan's and those of Sciama and Kibble, have been proposed, and undoubtedly, others will be put forward in the future). In that case, it could be that the ultimate theory only slightly perturbs the results in our theory in the intergalactic and interstellar regions, yet provides some lower limit to the likely "turbulence" in the torsion field. Thus, if successful, the theory proposed here may provide a guide in the search for the ultimate theory. It may be that the most important "take home" message of this paper is highlighting the importance of considering torsion theories that allow for dynamics in empty space on multiple length scales of the torsion field (and hence, of the accompanying metric). Interestingly, even in the absence of any torsion, anti-de Sitter space has a weakly turbulent instability [14].

If warranted by experimental observations, a natural modification of our theory would be to add a term involving Einstein's cosmological constant Λ. However, it would be far more satisfying if this was not needed.

Funding: This research received no external funding.

Data Availability Statement: Not applicable.

Acknowledgments: The author thanks the Whitlam Government of Australia for providing support in the late 1970s through free university education and a stipend while the author was an undergraduate, which was when this work was initiated. Additionally, the author thanks Sydney University, Cornell University, Caltech, the Courant Institute, and the University of Utah, as well as the National Science Foundation, through a succession of grants, for support. C. Kern is thanked for his help with the figures, for the numerical simulations needed to produce them, and in particular, for his discovery of the example in Figure 4d. M. Milgrom is thanked for helpful comments, recently and dating back to the early 1990s, for supplying Figure 1, noticing some minor errors, and providing many useful references. The Referees, especially one referee who carefully read the paper and spotted many things in need of correction, are thanked for their comments and for providing additional references. L. Fabbri is thanked for helpful advice. G. Cope is thanked for drawing my attention to the papers [14,15].

Conflicts of Interest: The author declares no conflict of interest.

Abbreviations

The following abbreviations are used in this manuscript:

MOND	Modified Newtonian Dynamics
ΛCDM	Lambda Cold Mark Matter

References

1. Peebles, P.J.E.; Ratra, B. The cosmological constant and dark energy. *Rev. Mod. Phys.* **2003**, *75*, 559–606. [CrossRef]
2. Sahni, V. Dark Matter and Dark Energy. In *The Physics of the Early Universe*; Lecture Notes in Physics; Springer: Berlin/Heidelberg, Germany; London, UK, 2004; Volume 653, Chapter 5, pp. 141–179. [CrossRef]
3. Collin, R.E. Dynamics of Dark Energy. *Int. J. Mod. Phys. D* **2006**, *15*, 1753–1935. [CrossRef]
4. Frieman, J.A.; Turner, M.S.; Huterer, D. Dark Energy and the Accelerating Universe. *Annu. Rev. Astron. Astrophys.* **2008**, *46*, 385–432. [CrossRef]
5. Amendola, L.; Tsujikawa, S. *Dark Energy: Theory and Observations*; Cambridge University Press: Cambridge, UK, 2010; p. xiii, 491p. [CrossRef]
6. Li, M.; Li, X.D.; Wang, S.; Wang, Y. Dark Energy. *Commun. Theor. Phys.* **2016**, *56*, 525–604. [CrossRef]
7. Arun, K.; Gudennavar, S.B.; Sivaram, C. Dark matter, dark energy, and alternate models: A review. *Adv. Space Res.* **2017**, *60*, 166–186. [CrossRef]
8. Abbott, T.; Abbott, T.M.C.; Alarcon, A.; Allam, S.; Andersen, P.; Andrade-Oliveira, F.; Annis, J.; Asorey, J.; Avila, S.; Bacon, D.; et al. Cosmological Constraints from Multiple Probes in the Dark Energy Survey. *Phys. Rev. Lett.* **2019**, *122*, 171301. [CrossRef]
9. Nadler, E.O.; Drlica-Wagner, A.; Bechtol, K.; Mau, S.; Wechsler, R.H.; Gluscevic, V.; Boddy, K.; Pace, A.B.; Li, T.S.; McNanna, M. Milky Way Satellite Census. III. Constraints on Dark Matter Properties from Observations of Milky Way Satellite Galaxies. *Phys. Rev. Lett.* **2021**, *126*, 091101. [CrossRef]
10. Wong, K.C.; Suyu, S.H.; Chen, G.C.; Rusu, C.E.; Millon, M.; Sluse, D.; Bonvin, V.; Fassnacht, C.D.; Taubenberger, S.; Auger, M.W. H0LiCOW XIII. A 2.4 per cent measurement of H_0 from lensed quasars: 5.3σ tension between early and late-Universe probes. *Mon. Not. R. Astron. Soc.* **2020**, *498*, 1420–1439. [CrossRef]
11. Riess, A.G.; Casertano, S.; Yuan, W.; Macri, L.M.; Scolnic, D. Large Magellanic Cloud Cepheid Standards Provide a 1% Foundation for the Determination of the Hubble Constant and Stronger Evidence for Physics beyond ΛCDM. *Astrophys. J.* **2019**, *876*, 85. [CrossRef]
12. Chae, K.H.; Lelli, F.; Desmond, H.; McGaugh, S.S.; Li, P.; Schombert, J.M. Testing the Strong Equivalence Principle: Detection of the External Field Effect inRotationally Supported Galaxies. *Astrophys. J.* **2020**, *904*, 51. [CrossRef]
13. Pigozzo, C.; Carneiro, S.; Alcaniz, J.; Borges, H.; Fabris, J. Evidence for cosmological particle creation? *J. Cosmol. Astropart. Phys.* **2016**, *2016*, 022. [CrossRef]
14. Bizoń, P.; Rostworowski, A. Weakly Turbulent Instability of Anti-de Sitter Spacetime. *Phys. Rev. Lett.* **2011**, *107*, 031102. [CrossRef] [PubMed]
15. Krasiński, A. Accelerating expansion or inhomogeneity? A comparison of the ΛCDM and Lemaître-Tolman models. *Phys. Rev. D* **2014**, *89*, 023520; Erratum in *Phys. Rev. D* **2014**, *89*, 089901. [CrossRef]
16. Wang, B.; Abdalla, E.; Atrio-Barandela, F.; Pavón, D. Dark matter and dark energy interactions: Theoretical challenges, cosmological implications and observational signatures. *Rep. Prog. Phys.* **2016**, *79*, 096901. [CrossRef]
17. Borges, H.A.; Wands, D. Growth of structure in interacting vacuum cosmologies. *Phys. Rev. D* **2020**, *101*, 103519. [CrossRef]
18. Sabbata, V.D.; Sivaram, C. *Spin and Torsion in Gravitation*; World Scientific Publishing Co.: Singapore; Philadelphia, PA, USA; River Edge, NJ, USA, 1994; p. 328.
19. Ray, S.; Khlopov, M.; Ghosh, P.P.; Mukhopadhyay, U. Phenomenology of ΛCDM model: a possibility of accelerating Universe with positive pressure. *Int. J. Theor. Phys.* **2011**, *50*, 939–951. [CrossRef]

20. Famaey, B.; McGaugh, S.S. Modified Newtonian Dynamics (MOND): Observational Phenomenology and Relativistic Extensions. *Living Rev. Relativ.* **2012**, *15*, 10. [CrossRef]
21. Milgrom, M. A modification of the Newtonian dynamics: Implications for galaxy systems. *Astrophys. J.* **1983**, *270*, 384–389. [CrossRef]
22. Bekenstein, J.D. Relativistic gravitation theory for the modified Newtonian dynamics paradigm. *Phys. Rev. D* **2004**, *70*, 083509. [CrossRef]
23. Skordis, C. Generalizing tensor-vector-scalar cosmology. *Phys. Rev. D* **2008**, *77*, 123502. [CrossRef]
24. Merritt, D. *A Philosophical Approach to MOND: Assessing the Milgromian Research Program in Cosmology*; Cambridge University Press: Cambridge, UK, 2020; p. 285.
25. Milgrom, M. MOND vs. dark matter in light of historical parallels. *Stud. Hist. Philos. Sci. Part B Stud. Hist. Philos. Mod. Phys.* **2020**, *71*, 170–195. [CrossRef]
26. Frieman, J.A.; Turner, M.S.; Huterer, D. A unifying theory of dark energy and dark matter: Negative masses and matter creation within a modified ΛCDM framework. *Astron. Astrophys.* **2018**, *620*, A92. [CrossRef]
27. Cartan, E. *Sur les variétés à Connexion Affine et la théorie de la Relativit é Généralisée*; Gauthier Villars: Paris, France, 1955.
28. Magnon, A.; Ashtekar, A. *On Manifolds with an Affine Connection and the Theory of General Relativity*; Monographs and Textbooks in Physical Science (Book 1); Bibliopolis: Napoli, Italy, 1986; p. 199.
29. Misner, C.W.; Thorne, K.S.; Wheeler, J.A. *Gravitation*, 3rd ed.; Freeman: San Francisco, CA, USA, 1973; p. xxvi, 1279p.
30. Hehl, F.W.; von der Heyde, P.; Kerlick, G.D.; Nester, J.M. General Relativity with spin and torsion: Foundations and prospects. *Rev. Mod. Phys.* **1976**, *48*, 393–416. [CrossRef]
31. Hehl, F.W.; McCrea, J.D.; Mielke, E.W.; Ne'eman, Y. Metric-affine gauge theory of gravity: Field equations, Noether identities, world spinors, and breaking of dilation invariance. *Phys. Rep.* **1995**, *258*, 1–171. [CrossRef]
32. Shapiro, I. Physical aspects of the space–time torsion. *Phys. Rep.* **2002**, *357*, 113–214. [CrossRef]
33. Ortín, T. *Gravity and Strings*; Cambridge Monographs on Mathematical Physics: Cambridge, UK, 2004; p. xx, 684p. [CrossRef]
34. Trautman, A. *Einstein–Cartan Theory*; Encyclopedia of Mathematical Physics; Elsevier: Amsterdam, The Netherlands, 2006; Volume 2, pp. 189–195.
35. Popławski, N.J. Spacetime and fields. *arXiv* **2009**, arXiv:0911.0334.
36. Fabbri, L. Fundamental Theory of Torsion Gravity. *Universe* **2021**, *7*, 305. [CrossRef]
37. Jiméneza, J.B.; Heisenberg, L.; Koivisto, T.S. The Geometrical Trinity of Gravity. *Universe* **2019**, *5*, 173. [CrossRef]
38. Ivanov, A.N.; Wellenzohn, M. Einstein–Cartan Gravity with Torsion Field Serving as Origin for Cosmological Constant or Dark Energy Density. *Astrophys. J.* **2016**, *829*, 47. [CrossRef]
39. Sengupta, S. Gravity theory with a dark extra dimension. *Phys. Rev. D* **2020**, *101*, 104040. [CrossRef]
40. Watanabe, T.; Hayashi, M.J. General relativity with torsion. *arXiv* **2004**, arXiv:gr-qc/0409029.
41. Minkevich, A.V. Accelerating Universe with spacetime torsion but without dark matter and dark energy. *Phys. Lett. B* **2009**, *678*, 423–426. [CrossRef]
42. de Berredo-Peixoto, G.; de Freitas, E.A. On the torsion effects of a relativistic spin fluid in early cosmology. *Class. Quantum Gravity* **2009**, *26*, 175015. [CrossRef]
43. Belyaev, A.S.; Thomas, M.C.; Shapiro, I.L. Torsion as a Dark Matter Candidate from the Higgs Portal. *Phys. Rev. D* **2017**, *95*, 095033. [CrossRef]
44. Vasak, D.; Kirsch, J.; Struckmeier, J. Dark energy and inflation invoked in CCGG by locally contorted spacetime. *Eur. Phys. J. Plus* **2020**, *135*, 404. [CrossRef]
45. Nierenberg, A.M.; Gilman, D.; Treu, T.; Brammer, G.; Birrer, S.; Moustakas, L.; Agnello, A.; Anguita, T.; Fassnacht, C.D.; Motta, V.; et al. Double dark matter vision: Twice the number of compact-source lenses with narrow-line lensing and the WFC3 grism. *Mon. Not. R. Astron. Soc.* **2019**, *492*, 5314–5335. [CrossRef]
46. Milgrom, M. The a_0–cosmology connection in MOND. *arXiv* **2009**, arXiv:2001.09729.
47. Cosserat, E.M.P.; Cosserat, F. *Theory of Deformable Bodies*; NASA: Washington, DC, USA, 1968; p. vi, 220p.
48. Sciama, D.W. The analogy between charge and spin in general relativity. In *Recent Developments in General Relativity*; Festschrift for Leopold Infeld; Państwowe Wydawnictwo Naukowe: Warsaw, Poland; Pergamon Press: New York, NY, USA, 1962; pp. 415–439.
49. Kibble, T.W.B. Lorentz invariance and the gravitational field. *J. Math. Phys.* **1961**, *2*, 212–221. [CrossRef]

Article

Can a Chameleon Field Be Identified with Quintessence?

A. N. Ivanov [1,*] and M. Wellenzohn [1,2]

1 Atominstitut, Technische Universität Wien, Stadionallee 2, A-1020 Wien, Austria; max.wellenzohn@gmail.com
2 FH Campus Wien, University of Applied Sciences, Favoritenstraße 226, 1100 Wien, Austria
* Correspondence: ivanov@kph.tuwien.ac.at

Received: 18 October 2020; Accepted: 22 November 2020; Published: 26 November 2020

Abstract: In the Einstein–Cartan gravitational theory with the chameleon field, while changing its mass independently of the density of its environment, we analyze the Friedmann–Einstein equations for the Universe's evolution with the expansion parameter a being dependent on time only. We analyze the problem of an identification of the chameleon field with quintessence, i.e., a canonical scalar field responsible for dark energy dynamics, and for the acceleration of the Universe's expansion. We show that since the cosmological constant related to the relic dark energy density is induced by torsion (*Astrophys. J.* **2016**, *829*, 47), the chameleon field may, in principle, possess some properties of quintessence, such as an influence on the dark energy dynamics and the acceleration of the Universe's expansion, even in the late-time acceleration, but it cannot be identified with quintessence to the full extent in the classical Einstein–Cartan gravitational theory.

Keywords: torsion/Einstein-Cartan; gravity/chameleon

PACS: 03.50.-z; 04.25.-g; 04.25.Nx; 14.80.Va

---

1. Introduction

The chameleon field, changing its mass independently of a density of its environment [1,2], has been invented to avoid the problem of the equivalence principle violation [3]. Nowadays it is accepted that the chameleon field, identified with quintessence [4–10], i.e., a canonical scalar field, can be useful for an explanation of the late-time acceleration of the Universe's expansion [11–14] and may shed light on the origins of dark energy and dark energy dynamics [15–21]. Since the relic dark energy density is closely related to the cosmological constant [15], in contrast to such a hypothesis that the chameleon field might originate the cosmological constant proportionally to the homogeneous static dark energy density, it has been shown at the model-independent level within the Einstein–Cartan gravitational theory [22–38] that the cosmological constant or the relic dark energy density has a geometrical origin caused by torsion [39]. In this case the chameleon field is able only to evolve above the relic background of the dark energy simulating its dynamics and, of course, to make a certain influence on the acceleration of the Universe's expansion.

For the observation of torsion in the terrestrial laboratories there have been derived potentials of low-energy torsion–neutron interactions [40–42]. In terrestrial laboratories, the extreme smallness of absolute values of torsion was confirmed in different estimates of constraints on the contributions of torsion to observables of elementary particle interactions [43–48], including the qBounce experiments with ultracold neutrons (UCNs) [49–55] (see also [48]).

The chameleon–matter interactions were also intensively investigated in terrestrial laboratories [49–57] in experiments with ultracold and cold neutrons through some effective low-energy chameleon–neutron potentials [58–61] and by using cold atoms in the atom interferometry [62–66]. However, recently the importance of the chameleon field as quintessence in the late-time acceleration of the Universe has been questioned by Wang et al. [67] and Khoury [68] by pointing out that the conformal factor, relating Einstein's and Jordan's frames and defining the chameleon–matter interactions, is essentially constant over the last Hubble time. According to Wang et al. [67] and Khoury [68], this implies a negligible influence of the chameleon field on the late-time acceleration of the Universe's expansion. To some extent, this should also imply that the chameleon field cannot possess such a property of quintessence as responsibility for the late-time acceleration of the Universe's expansion [5–7].

Thus, the aim of this paper is to investigate the properties of the chameleon field in comparison to the properties of quintessence. We would like to remind readers that by definition, quintessence is a hypothetical state of dark energy described by a canonical scalar field for an explanation of the observable acceleration of the Universe's expansion. We have to also emphasize that our analysis is restricted by the classical Einstein–Cartan gravitational theory. Below we show that the chameleon field has no relation to the origin of the cosmological constant, or the relic dark energy density, which is induced by torsion [39]. However, the chameleon field can still influence on the Universe's expansion even in the late-time acceleration, caused by its evolution above the background of the relic dark energy [39]. By analyzing Einstein's equations for the flat Universe in spacetime with the Friedmann metric, dependent on the expansion parameter a [69], we show that conservation of a total energy–momentum tensor of the system, including the chameleon field, radiation and matter (dark and baryon matter), demands the conformal factor to be equal to unity if and only if the dependencies of the radiation $\rho_r(a)$ and matter $\rho_m(a)$ densities on the expansion parameter a do not deviate from their standard forms, $\rho_r(a) \sim a^{-4}$ and $\rho_m(a) \sim a^{-3}$ respectively [69]. We obtain the same result by analyzing the first order differential Friedmann–Einstein equation, relating $\dot{a}^2/a^2$ to the chameleon field, radiation and matter densities, and the second order differential Friedmann–Einstein equation, relating $\ddot{a}/a$ to the chameleon field, radiation and matter densities and their pressures, where $\dot{a}$ and $\ddot{a}$ are the first and second time derivatives of the expansion parameter. Of course, the equality of the conformal factor to unity suppresses any coupling of the chameleon field to a matter density of its environment and makes such a scalar field unhelpful for avoiding the problem of the equivalence principle violation [3]. However, it does not prevent the chameleon field, evolving above the background of the relic dark energy, from a simulation of a dark energy dynamics and having an influence on the acceleration of the Universe's expansion. Then, we show that the Friedmann–Einstein equation for $\dot{a}^2/a^2$ is the first integral of the Friedmann–Einstein equation for $\ddot{a}/a$ if and only if the total energy–momentum of the system, including the chameleon field, radiation and matter, is locally conserved. As a result we infer that (i) if the radiation and matter densities obey their standard dependence on the expansion parameter $\rho_r(a) \sim a^{-4}$ and $\rho_m(a) \sim a^{-3}$ the conformal factor is equal to unity and the chameleon field loses the possibility to couple to an environment, and (ii) if the dependencies of the radiation and matter densities deviate from their standard behavior $\rho_r(a) \sim a^{-4}$ and $\rho_m(a) \sim a^{-3}$, the conformal factor is not equal to unity and makes possible interactions of the chameleon field with its environment. In this case, usage of the chameleon field for the problem of equivalence principle violation becomes meaningful. In spite of the fact that the chameleon field does not possess the main property of quintessence in order to be a hypothetical form of dark energy [4], since the relic dark energy density or the cosmological constant has a geometrical origin related to torsion [39], the chameleon field, evolving above the relic dark energy and simulating a dark energy dynamics, might be responsible for an acceleration of the Universe's expansion.

The paper is organized as follows. In Section 2 we derive Einstein's equations in the Einstein–Cartan gravitational theory with torsion, chameleon and matter fields. Following [39] we show that the contribution of torsion to the Einstein–Hilbert action is presented in the form of the cosmological constant. Then, following Khoury and Weltman [1] we include the part of the integrand of the Einstein–Hilbert action proportional to the cosmological constant for the potential of the self-interaction of the chameleon field. This implies that the chameleon field has no relation to an origin of the cosmological constant or the relic dark energy density but can only evolve above such a relic background caused by torsion and simulate dark energy dynamics. In Section 3 in the flat Friedmann spacetime with the standard Friedmann metric $g_{\mu\nu}$, i.e., $g_{00} = 1$, $g_{0j} = 0$ and $g_{ij} = a^2(t)\,\eta_{ij}$ and $\eta_{ij} = -\delta_{ij}$, we show that the Einstein equations reduce themselves to the Friedmann–Einstein equations of the Universe's evolution with the chameleon field, radiation and matter (dark and baryon) densities. Since the Einstein tensor $G_{\mu\nu} = R_{\mu\nu} - \frac{1}{2}\,g_{\mu\nu}R$, where $R_{\mu\nu}$ and R are the Ricci tensor and scalar curvature, respectively, obey the Bianchi identity $G^{\mu\nu}{}_{;\mu} = 0$, where $G^{\mu\nu}{}_{;\mu}$ is the covariant divergence [69], the total energy–momentum tensor of the system, including the chameleon field, radiation and matter (dark and baryon), should be locally conserved. We find that local conservation of the total energy–momentum tensor imposes the evolution equations for the radiation and matter densities, where the dependence of them on the expansion parameter a is corrected by the conformal factor in comparison to the standard dependence $\rho_r(a) \sim a^{-4}$ and $\rho_a \sim a^{-3}$, respectively [69]. We show that the Friedmann–Einstein equation for $\dot{a}^2/a^2$ is the first integral of the Friedmann–Einstein equation for $\ddot{a}/a$ if and only if the total energy momentum of the system, including the chameleon field, radiation and matter, is locally conserved. In case of the standard dependence of the radiation and matter densities on the expansion parameters $\rho_r(a) \sim a^{-4}$ and $\rho_m \sim a^{-3}$ [69], local conservation of the total energy–momentum tensor of the chameleon field, radiation and matter demand the conformal factor to be equal to unity. This suppresses any interaction of the chameleon field with an ambient environment. In Section 4 we discuss experiments to probe torsion in the terrestrial laboratories through effective low-energy torsion-neutron interactions derived in [40–42]. In Section 5 we discuss the results obtained.

2. Einstein's Equations in the Einstein–Cartan Gravitational Theory with Chameleon and Matter Fields

We take the Einstein–Hilbert action of the Einstein–Cartan gravitational theory without chameleon and matter fields in the standard form [27,37,69]:

$$S_{\rm EH} = \frac{1}{2}\,M^2_{\rm Pl}\int d^4x\,\sqrt{-g}\,\mathcal{R}, \tag{1}$$

where $M_{\rm Pl} = 1/\sqrt{8\pi G_N} = 2.435 \times 10^{27}\,{\rm eV}$ is the reduced Planck mass, G_N is the Newtonian gravitational constant [70] and g is the determinant of the metric tensor $g_{\mu\nu}$. The scalar curvature $\mathcal{R}$ is defined by [27,37]

$$\mathcal{R} = g^{\mu\nu}\mathcal{R}^{\alpha}{}_{\mu\alpha\nu} = g^{\mu\nu}\Big(\frac{\partial}{\partial x^{\nu}}\Gamma^{\alpha}{}_{\alpha\mu} - \frac{\partial}{\partial x^{\alpha}}\Gamma^{\alpha}{}_{\nu\mu} + \Gamma^{\alpha}{}_{\nu\varphi}\Gamma^{\varphi}{}_{\alpha\mu} - \Gamma^{\alpha}{}_{\alpha\varphi}\Gamma^{\varphi}{}_{\nu\mu}\Big) = g^{\mu\nu}\mathcal{R}_{\mu\nu}, \tag{2}$$

where $\mathcal{R}^{\alpha}{}_{\mu\beta\nu}$ and $\mathcal{R}_{\mu\nu}$ are the Riemann and Ricci tensors in the Einstein–Cartan gravitational theory, respectively. Then, $\Gamma^{\alpha}{}_{\mu\nu}$ is the affine connection

$$\Gamma^{\alpha}{}_{\mu\nu} = \{^{\alpha}{}_{\mu\nu}\} + \mathcal{K}^{\alpha}{}_{\mu\nu} = \{^{\alpha}{}_{\mu\nu}\} + g^{\alpha\sigma}\mathcal{K}_{\sigma\mu\nu}, \tag{3}$$

where $\{^{\alpha}{}_{\mu\nu}\}$ are the Christoffel symbols [69]

$$\{^{\alpha}{}_{\mu\nu}\} = \frac{1}{2}g^{\alpha\lambda}\Big(\frac{\partial g_{\lambda\mu}}{\partial x^{\nu}} + \frac{\partial g_{\lambda\nu}}{\partial x^{\mu}} - \frac{\partial g_{\mu\nu}}{\partial x^{\lambda}}\Big) \tag{4}$$

and $\mathcal{K}_{\sigma\mu\nu}$ is the contorsion tensor, related to torsion $\mathcal{T}_{\sigma\mu\nu}$ by $\mathcal{K}_{\sigma\mu\nu} = \frac{1}{2}(\mathcal{T}_{\sigma\mu\nu} + \mathcal{T}_{\mu\sigma\nu} + \mathcal{T}_{\nu\sigma\mu})$ and $\mathcal{T}^{\alpha}{}_{\mu\nu} = g^{\alpha\sigma}\mathcal{T}_{\sigma\mu\nu} = \Gamma^{\alpha}{}_{\mu\nu} - \Gamma^{\alpha}{}_{\nu\mu}$ with the following properties: $\mathcal{K}_{\sigma\mu\nu} = -\mathcal{K}_{\nu\mu\sigma}$ and $\mathcal{T}_{\mu\nu} = -\mathcal{T}_{\nu\mu}$ [27,37]. The integrand of the Einstein–Hilbert action Equation (1) can be represented in the following form:

$$\sqrt{-g}\,\mathcal{R} = \sqrt{-g}\,R + \sqrt{-g}\,\mathcal{C} + \frac{\partial}{\partial x^{\nu}}(\sqrt{-g}\,g^{\mu\nu}\mathcal{K}^{\alpha}{}_{\alpha\mu}) - \sqrt{-g}\,g^{\mu\nu}\Big(\frac{1}{\sqrt{-g}}\frac{\partial}{\partial x^{\alpha}}(\sqrt{-g}\,\mathcal{K}^{\alpha}{}_{\nu\mu}) - \{^{\varphi}{}_{\alpha\mu}\}\,\mathcal{K}^{\alpha}{}_{\nu\varphi} - \{^{\alpha}{}_{\nu\varphi}\}\,\mathcal{K}^{\varphi}{}_{\alpha\mu}\Big), \tag{5}$$

where we have denoted

$$\mathcal{C} = g^{\mu\nu}(\mathcal{K}^{\varphi}{}_{\alpha\mu}\,\mathcal{K}^{\alpha}{}_{\nu\varphi} - \mathcal{K}^{\alpha}{}_{\alpha\varphi}\mathcal{K}^{\varphi}{}_{\nu\mu}) \tag{6}$$

and R is the Ricci scalar curvature of the Einstein gravitational theory, expressed in terms of the Christoffel symbols $\{^{\alpha}{}_{\mu\nu}\}$ [69] only. When removing in Equation (5) the total derivatives and integrating by parts, we delete the third term and transcribe the fourth term into the form $\sqrt{-g}\,g^{\mu\nu}{}_{;\alpha}\,\mathcal{K}^{\alpha}{}_{\nu\mu}$, where $g^{\mu\nu}{}_{;\alpha}$ is the covariant derivative of the metric tensor $g^{\mu\nu}$, vanishing because of the metricity condition $g^{\mu\nu}{}_{;\alpha} = 0$ [69].

Derivation of Equation (5)

Let us show that $\sqrt{-g}\,\mathcal{R}$ can be presented in the form of Equation (5). Using Equation (3) we get an obvious relation

$$\begin{aligned}\sqrt{-g}\,\mathcal{R} &= \sqrt{-g}\,R + \sqrt{-g}\,\mathcal{C} + \sqrt{-g}\,g^{\mu\nu}\Big(\frac{\partial}{\partial x^{\nu}}\mathcal{K}^{\alpha}{}_{\alpha\mu} - \frac{\partial}{\partial x^{\alpha}}\mathcal{K}^{\alpha}{}_{\nu\mu} + \{^{\alpha}{}_{\nu\varphi}\}\,\mathcal{K}^{\varphi}{}_{\alpha\mu} + \{^{\varphi}{}_{\alpha\mu}\}\,\mathcal{K}^{\alpha}{}_{\nu\varphi}\\ &- \{^{\alpha}{}_{\alpha\varphi}\}\,\mathcal{K}^{\varphi}{}_{\nu\mu} - \{^{\varphi}{}_{\nu\mu}\}\,\mathcal{K}^{\alpha}{}_{\alpha\varphi}\Big).\end{aligned} \tag{7}$$

Having replaced the first term in the brackets by the total derivative and by adding the last term in the brackets, we arrive at the expression

$$\begin{aligned}\sqrt{-g}\,\mathcal{R} &= \sqrt{-g}\,R + \sqrt{-g}\,\mathcal{C} + \tfrac{\partial}{\partial x^{\nu}}(\sqrt{-g}\,g^{\mu\nu}\,\mathcal{K}^{\alpha}{}_{\alpha\mu}) - \tfrac{\partial}{\partial x^{\nu}}(\sqrt{-g}\,g^{\mu\nu})\,\mathcal{K}^{\alpha}{}_{\alpha\mu} - \sqrt{-g}\,g^{\mu\nu}\,\{^{\varphi}{}_{\nu\mu}\}\,\mathcal{K}^{\alpha}{}_{\alpha\varphi}\\ &- \sqrt{-g}\,g^{\mu\nu}\Big(\tfrac{\partial}{\partial x^{\alpha}}\mathcal{K}^{\alpha}{}_{\nu\mu} + \{^{\alpha}{}_{\alpha\varphi}\}\,\mathcal{K}^{\varphi}{}_{\nu\mu} - \{^{\alpha}{}_{\nu\varphi}\}\,\mathcal{K}^{\varphi}{}_{\alpha\mu} - \{^{\varphi}{}_{\alpha\mu}\}\,\mathcal{K}^{\alpha}{}_{\nu\varphi}\Big).\end{aligned} \tag{8}$$

Then, the first term in the brackets we rewrite as follows:

$$\begin{aligned}&\sqrt{-g}\,\mathcal{R} = \sqrt{-g}\,R + \sqrt{-g}\,\mathcal{C} + \frac{\partial}{\partial x^{\nu}}(\sqrt{-g}\,g^{\mu\nu}\,\mathcal{K}^{\alpha}{}_{\alpha\mu}) - \frac{\partial}{\partial x^{\nu}}(\sqrt{-g}\,g^{\mu\nu})\,\mathcal{K}^{\alpha}{}_{\alpha\mu} - \sqrt{-g}\,g^{\mu\nu}\,\{^{\varphi}{}_{\nu\mu}\}\,\mathcal{K}^{\alpha}{}_{\alpha\varphi}\\ &-\sqrt{-g}\,g^{\mu\nu}\Big(\frac{1}{\sqrt{-g}}\frac{\partial}{\partial x^{\alpha}}(\sqrt{-g}\,\mathcal{K}^{\alpha}{}_{\nu\mu}) - \frac{1}{\sqrt{-g}}\frac{\partial\sqrt{-g}}{\partial x^{\alpha}}\,\mathcal{K}^{\alpha}{}_{\nu\mu} + \{^{\alpha}{}_{\alpha\varphi}\}\,\mathcal{K}^{\varphi}{}_{\nu\mu} - \{^{\alpha}{}_{\nu\varphi}\}\,\mathcal{K}^{\varphi}{}_{\alpha\mu} - \{^{\varphi}{}_{\alpha\mu}\}\,\mathcal{K}^{\alpha}{}_{\nu\varphi}\Big).\end{aligned} \tag{9}$$

Since $\sqrt{-g}\,g^{\mu\nu}\{^{\varphi}{}_{\nu\mu}\}$ and $\{^{\alpha}{}_{\alpha\varphi}\}$ are equal to [69] (see Equation (10.107) and Equation (9.56))

$$\sqrt{-g}\,g^{\mu\nu}\{^{\varphi}{}_{\nu\mu}\} = -\frac{\partial}{\partial x^{\nu}}(\sqrt{-g}\,g^{\varphi\nu}) \quad , \quad \{^{\alpha}{}_{\alpha\varphi}\} = \frac{1}{\sqrt{-g}}\frac{\partial\sqrt{-g}}{\partial x^{\varphi}}, \tag{10}$$

the fourth and fifth terms in the first line in Equation (9) and the second and third terms in the brackets are cancelled out in pairs. This reduces Equation (9) to Equation (5).

Now we may show that the contribution of the fourth term in Equation (5) to the Einstein–Hilbert action reduces to the contribution of the term $\sqrt{-g}\,g^{\mu\nu}{}_{;\alpha}\,\mathcal{K}^{\alpha}{}_{\nu\mu}$. The contribution of the fourth term in Equation (5) to the Einstein–Hilbert action is defined by the integral

$$\int d^4x\,\sqrt{-g}\,g^{\mu\nu}\Big(\frac{1}{\sqrt{-g}}\frac{\partial}{\partial x^{\alpha}}(\sqrt{-g}\,\mathcal{K}^{\alpha}{}_{\nu\mu}) - \{^{\varphi}{}_{\alpha\mu}\}\,\mathcal{K}^{\alpha}{}_{\nu\varphi} - \{^{\alpha}{}_{\nu\varphi}\}\,\mathcal{K}^{\varphi}{}_{\alpha\mu}\Big). \tag{11}$$

After the integration by parts in the first term we get

$$\oint \sqrt{-g}\,g^{\mu\nu}\mathcal{K}^{\alpha}{}_{\nu\mu}dS_{\alpha} - \int d^4x\,\sqrt{-g}\Big(\frac{\partial g^{\mu\nu}}{\partial x^{\alpha}}\,\mathcal{K}^{\alpha}{}_{\nu\mu} + g^{\mu\nu}\{^{\varphi}{}_{\alpha\mu}\}\,\mathcal{K}^{\alpha}{}_{\nu\varphi} + g^{\mu\nu}\{^{\alpha}{}_{\nu\varphi}\}\,\mathcal{K}^{\varphi}{}_{\alpha\mu}\Big). \tag{12}$$

Having omitted the surface term and having renamed some indices in the second integral in Equation (12), we arrive at the expression

$$-\int d^4x\,\sqrt{-g}\,\Big(\frac{\partial g^{\mu\nu}}{\partial x^{\alpha}} + g^{\rho\nu}\{^{\mu}{}_{\rho\alpha}\} + g^{\nu\rho}\{^{\mu}{}_{\rho\alpha}\}\Big)\,\mathcal{K}^{\alpha}{}_{\nu\mu}, \tag{13}$$

where we have used the property of the Christoffel symbols $\{^{\mu}{}_{\alpha\rho}\} = \{^{\mu}{}_{\rho\alpha}\}$. The expression in the brackets is the covariant derivative of the metric tensor $g^{\mu\nu}$ (see, for example, [27,69])

$$g^{\mu\nu}{}_{;\alpha} = \frac{\partial g^{\mu\nu}}{\partial x^{\alpha}} + g^{\rho\nu}\{^{\mu}{}_{\rho\alpha}\} + g^{\mu\rho}\{^{\nu}{}_{\rho\alpha}\}. \tag{14}$$

Thus, the contribution of the fourth term to the Einstein–Hilbert action is proportional to the integral

$$\int d^4x\,\sqrt{-g}\,g^{\mu\nu}\Big(\frac{1}{\sqrt{-g}}\frac{\partial}{\partial x^{\alpha}}(\sqrt{-g}\,\mathcal{K}^{\alpha}{}_{\nu\mu}) - \{^{\varphi}{}_{\alpha\mu}\}\,\mathcal{K}^{\alpha}{}_{\nu\varphi} - \{^{\alpha}{}_{\nu\varphi}\}\,\mathcal{K}^{\varphi}{}_{\alpha\mu}\Big) = -\int d^4x\,\sqrt{-g}\,g^{\mu\nu}{}_{;\alpha}\,\mathcal{K}^{\alpha}{}_{\nu\mu}. \tag{15}$$

This confirms our assertion concerning a vanishing contribution of the fourth term in Equation (5) to the Einstein–Hilbert action in case of the metricity condition $g^{\mu\nu}{}_{;\alpha} = 0$ [69].

Since it has been shown in [39] that $\mathcal{C} = -2\Lambda_C$, where Λ_C is the cosmological constant [69,71,72] (see also [15]) or the relic dark energy density, the Einstein–Hilbert action Equation (1) of the Einstein–Cartan gravitational theory with the scalar curvature Equation (2) can be represented in the following form [39]:

$$S_{\rm EH} = \frac{1}{2}\,M^2_{\rm Pl}\int d^4x\,\sqrt{-g}\,(R - 2\Lambda_C). \tag{16}$$

As has been shown in [39], the same result is valid for the Poincaré gauge gravitaitonal theory [73–77] (see also [31–34]). Using Equation (11) the action of the Einstein–Cartan gravitational theory with torsion, chameleon fields and matter fields we take in the form [39]

$$S_{\rm EH} = \frac{1}{2}\,M^2_{\rm Pl}\int d^4x\,\sqrt{-g}\,R + \int d^4x\,\sqrt{-g}\,\mathcal{L}[\phi] + \int d^4x\,\sqrt{-\tilde{g}}\,\mathcal{L}_m[\tilde{g}], \tag{17}$$

where $\mathcal{L}[\phi]$ is the Lagrangian of the chameleon field

$$\mathcal{L}[\phi] = \frac{1}{2}\,g^{\mu\nu}\,\partial_{\mu}\phi\partial_{\nu}\phi - V(\phi) \tag{18}$$

and $V(\phi)$ is the potential of the chameleon self-interaction. In Equation (17), following Khoury and Weltman [1], we have included additively the cosmological constant Λ_C in the form of the relic dark energy density $\rho_{\Lambda} = M^2_{\rm Pl}\Lambda_C$ into the potential $V(\phi)$ of the chameleon field self-interaction; i.e., $V(\phi) = \rho_{\Lambda} + \Phi(\phi)$. This implies that the chameleon field has no relation to the origin of the cosmological constant or the relic dark energy density. It can only evolve above the relic background of the dark energy, caused by torsion.

The matter fields and the radiation [78,79] are described by the Lagrangian $\mathcal{L}_m[\tilde{g}_{\mu\nu}]$. The interactions of the matter fields and radiation with the chameleon field are expressed in terms of the metric tensor $\tilde{g}_{\mu\nu}$ in the Jordan frame [1,2,80], which is conformally related to Einstein's frame metric tensor $g_{\mu\nu}$ by $\tilde{g}_{\mu\nu} = f^2\,g_{\mu\nu}$ (or $\tilde{g}^{\mu\nu} = f^{-2}\,g^{\mu\nu}$) and $\sqrt{-\tilde{g}} = f^4\,\sqrt{-g}$ with $f = e^{\beta\phi/M_{\rm Pl}}$, where β is the chameleon–matter coupling constant [1,2]. The factor $f = e^{\beta\phi/M_{\rm Pl}}$ can be interpreted also as a conformal coupling to matter fields and radiation [80] (see also [1,2,81]). For simplicity we have set the chameleon–photon coupling constant β_{γ} [79] to be equal to the chameleon–matter coupling constant β.

By varying the action of Equation (17) with respect to the metric tensor $\delta g^{\mu\nu}$ (see, for example, [69]), we arrive at Einstein's equations, modified by the contribution of the chameleon field. We get

$$R_{\mu\nu} - \frac{1}{2}\,g_{\mu\nu}\,R = -\frac{1}{M^2_{\rm Pl}}\Big(f^2\,\tilde{T}^{(m)}_{\mu\nu} + T^{(\phi)}_{\mu\nu}\Big), \tag{19}$$

where $R_{\mu\nu}$ is the Ricci tensor [69]; $\tilde{T}^{(m)}_{\mu\nu}$ and $T^{(\phi)}_{\mu\nu}$ are the matter (with radiation, which we treat as a radiative fluid [82–86]) and chameleon energy–momentum tensors, respectively, determined by

$$\begin{aligned} \tilde{T}^{(m)}_{\mu\nu} &= \frac{2}{\sqrt{-\tilde{g}}}\frac{\delta}{\delta\tilde{g}^{\mu\nu}}\Big(\sqrt{-\tilde{g}}\,\mathcal{L}[\tilde{g}]\Big) = (\tilde{\rho}+\tilde{p})\,\tilde{u}_\mu\tilde{u}_\nu - \tilde{p}\,\tilde{g}_{\mu\nu}, \\ T^{(\phi)}_{\mu\nu} &= \frac{2}{\sqrt{-g}}\frac{\delta}{\delta g^{\mu\nu}}\Big(\sqrt{-g}\,\mathcal{L}[\phi]\Big) = \frac{\partial\phi}{\partial x^\mu}\frac{\partial\phi}{\partial x^\nu} - g_{\mu\nu}\Big(\frac{1}{2}g^{\lambda\rho}\frac{\partial\phi}{\partial x^\lambda}\frac{\partial\phi}{\partial x^\rho} - V(\phi)\Big). \end{aligned} \tag{20}$$

The factor f^2 appears in front of $\tilde{T}^{(m)}_{\mu\nu}$ because of the relation

$$\frac{2}{\sqrt{-g}}\frac{\delta}{\delta g^{\mu\nu}}\Big(\sqrt{-\tilde{g}}\,\mathcal{L}_m[\tilde{g}]\Big) = \frac{\sqrt{-\tilde{g}}}{\sqrt{-g}}\frac{\delta\tilde{g}^{\lambda\rho}}{\delta g^{\mu\nu}}\,\tilde{T}^{(m)}_{\lambda\rho} = f^2\,\tilde{T}^{(m)}_{\mu\nu}, \tag{21}$$

where we have used that

$$\frac{\sqrt{-\tilde{g}}}{\sqrt{-g}} = f^4 \quad , \quad \frac{\delta\tilde{g}^{\lambda\rho}}{\delta g^{\mu\nu}} = f^{-2}\frac{1}{2}(g^\lambda{}_\mu g^\rho{}_\nu + g^\lambda{}_\nu g^\rho{}_\mu), \tag{22}$$

since $\tilde{g}^{\lambda\rho} = f^{-2}\,g^{\lambda\rho}$ [80] and $\tilde{T}^{(m)}_{\mu\nu} = \tilde{T}^{(m)}_{\nu\mu}$. Then, the quantities $\tilde{\rho}$, $\tilde{p}$ and $\tilde{u}_\mu$ in the Jordan frame are related to the quantities ρ, p and u_μ in Einstein's frame as [80]

$$\tilde{\rho} = f^{-3}\rho \quad , \quad \tilde{p} = f^{-3}p \quad , \quad \tilde{u}_\mu = f\,u_\mu \quad , \quad \tilde{u}^\mu = f^{-1}\,u^\mu. \tag{23}$$

This gives $\tilde{T}^{(m)}_{\mu\nu} = f^{-1}T^{(m)}_{\mu\nu}$. By plugging Equation (20) with $\tilde{T}^{(m)}_{\mu\nu} = f^{-1}T^{(m)}_{\mu\nu}$ into Equation (19), we arrive at Einstein's equations

$$R_{\mu\nu} - \frac{1}{2}g_{\mu\nu}\,R = -\frac{1}{M^2_{\rm Pl}}\,T_{\mu\nu}, \tag{24}$$

where $T_{\mu\nu}$ is the total energy–momentum tensor equal to

$$T_{\mu\nu} = \Big((\rho+p)\,u_\mu u_\nu - p\,g_{\mu\nu}\Big)\,e^{\beta\phi/M_{\rm Pl}} + \Big(\frac{\partial\phi}{\partial x^\mu}\frac{\partial\phi}{\partial x^\nu} - g_{\mu\nu}\Big(g^{\lambda\rho}\frac{1}{2}\frac{\partial\phi}{\partial x^\lambda}\frac{\partial\phi}{\partial x^\rho} - V(\phi)\Big)\Big), \tag{25}$$

where the contribution of torsion $T^{(\rm tor)}_{\mu\nu} = \rho_\Lambda g_{\mu\nu} = M^2_{\rm Pl}\Lambda_C g_{\mu\nu}$ [39] is included additively in the potential $V(\phi)$ of the self-interactions of the chameleon field. Below we analyze the Einstein equations (Equation (24)) in the cold dark matter (CDM) model [70] in the Friedmann flat spacetime with the line element [69,70]

$$ds^2 = g_{\mu\nu}(x)dx^\mu dx^\nu = dt^2 + a^2(t)\,\eta_{ij}dx^i dx^j, \tag{26}$$

where $g_{00}(x) = 1$ and $g_{ij}(x) = a^2(t)\,\eta_{ij}$ with $\eta_{ij} = -\delta_{ij}$. Then, $a(t)$ is the expansion parameter of the Universe's evolution [69]. The Christoffel symbols $\{^\alpha{}_{\mu\nu}\}$, the components of the Ricci tensor $R_{\mu\nu}$ and the scalar curvature R are equal to [69]

$$\begin{aligned} \{^0{}_{00}\} &= \{^0{}_{0j}\} = \{^j{}_{00}\} = \{^i{}_{kj}\} = 0, \,, \{^0{}_{kj}\} = -a\dot{a}\,\eta_{kj}\,, \{^i{}_{0j}\} = \frac{\dot{a}}{a}\,\delta^i{}_j, \\ R_{00} &= 3\,\frac{\ddot{a}}{a}\,,\; R_{0j} = 0\,,\; R_{ij} = \Big(\frac{\ddot{a}}{a} + 2\frac{\dot{a}^2}{a^2}\Big)\,g_{ij}\,,\; R = 6\,\Big(\frac{\ddot{a}}{a} + \frac{\dot{a}^2}{a^2}\Big), \end{aligned} \tag{27}$$

where $\eta^{i\ell}\eta_{\ell j} = \delta^i{}_j$ and $\dot{a}$ and $\ddot{a}$ are first and second derivatives with respect to time.

3. Friedmann–Einstein Equations of the Universe's Evolution

In Friedmann spacetime, Einstein's equations (Equation (24)) define the equations of the Universe's evolution, which are usually called Friedmann's equations (or the Friedmann–Einstein equations) [69]. They are given by

$$\frac{\dot{a}^2}{a^2} = \frac{1}{3M_{\rm Pl}^2}\,(\rho_\phi + (\rho_r + \rho_m)f(\phi)) \tag{28}$$

and

$$\frac{\ddot{a}}{a} = -\frac{1}{6M_{\rm Pl}^2}\,(\rho_\phi + 3p_\phi + (\rho_r + 3p_r)f(\phi) + \rho_m f(\phi)), \tag{29}$$

where ρ_r and ρ_m are the radiation and matter densities. The scalar field ϕ couples to radiation and matter densities through the conformal factor $f(\phi) = e^{\beta\phi/M_{\rm Pl}}$. Then, the radiation density ρ_r and pressure p_r are related by the equation of state $p_r = \rho_r/3$ [69]. For the description of matter we use the cold dark matter (CDM) model with the pressureless dark and baryon matter [70]. The scalar field density ρ_ϕ and pressure p_ϕ are equal to

$$\rho_\phi = \frac{1}{2}\,\dot{\phi}^2 + V(\phi) \quad , \quad p_\phi = \frac{1}{2}\,\dot{\phi}^2 - V(\phi). \tag{30}$$

Varying the action Equation (17) with respect to the scalar field ϕ and its derivative one gets the equation of motion for the scalar field [81]. In Friedmann spacetime it reads

$$\ddot{\phi} + 3\,\frac{\dot{a}}{a}\,\dot{\phi} + \frac{dV_{\rm eff}(\phi)}{d\phi} = 0, \tag{31}$$

where $V_{\rm eff}(\phi)$ is the effective potential given by

$$V_{\rm eff}(\phi) = V(\phi) + \rho_m\big(f(\phi) - 1\big). \tag{32}$$

The contribution of the radiation density comes into the effective potential in the form $(\rho_r - 3p_r)\big(f(\phi) - 1\big)$. As for the equation of state $p_r = \rho_r/3$, such a contribution vanishes. Thus, through the interaction with matter density ρ_m the scalar field can acquire a non-vanishing mass if the effective potential $V_{\rm eff}(\phi)$ obeys the constraints

$$\frac{dV_{\rm eff}(\phi)}{d\phi}\Big|_{\phi=\phi_{\rm min}} = 0 \quad , \quad \frac{d^2V_{\rm eff}(\phi)}{d\phi^2}\Big|_{\phi=\phi_{\rm min}} > 0, \tag{33}$$

i.e., the effective potential $V_{\rm eff}(\phi)$ possesses a minimum at $\phi = \phi_{\rm min}$. An important role for a dependence of a chameleon field mass on a density of an environment is the conformal factor $f(\phi)$ and its deviation from unity.

3.1. Bianchi Identity, Conservation of Total Energy–Momentum Tensor and Conformal Factor

By using Equation (27) and taking into account that in the Friedmann flat spacetime the non-vanishing components of the Einstein tensor $G^{\mu\nu} = R^{\mu\nu} - \frac{1}{2}\,g^{\mu\nu}R$ are equal to

$$G^{00} = -3\,\frac{\dot{a}^2}{a^2} \quad , \quad G^{ij} = \Big(-2\,\frac{\ddot{a}}{a} - \frac{\dot{a}^2}{a^2}\Big)\,g^{ij} \tag{34}$$

one may show that Einstein's tensor $G_{\mu\nu}$ obeys the Bianchi identity [69]

$$G^{\mu\nu}{}_{;\mu} = \frac{1}{\sqrt{-g}}\frac{\partial}{\partial x^{\rho}}\Big(\sqrt{-g}\,G^{\rho\nu}\Big) + \Gamma^{\nu}{}_{\mu\rho}G^{\mu\rho} = 0, \tag{35}$$

where $G^{\mu\nu}{}_{;\mu}$ is a covariant divergence and $\Gamma^{\nu}{}_{\mu\rho} = \{^{\nu}{}_{\mu\rho}\}$ are the Christoffel symbols [69]. As a result, the covariant divergence of the total energy–momentum tensor $T^{\mu\nu}{}_{;\mu}$ should also vanish

$$T^{\mu\nu}{}_{;\mu} = \frac{1}{\sqrt{-g}}\frac{\partial}{\partial x^{\rho}}\Big(\sqrt{-g}\,T^{\rho\nu}\Big) + \Gamma^{\nu}{}_{\mu\rho}T^{\mu\rho} = 0. \tag{36}$$

Due to time-dependence only Equation (31) takes the form

$$\frac{1}{\sqrt{-g}}\frac{\partial}{\partial t}\Big(\sqrt{-g}\,T^{00}\Big) + \Gamma^{0}{}_{ij}T^{ij} = 0, \tag{37}$$

where we have taken into account Equation (27). Using the non-vanishing components of the total energy momentum tensor

$$T^{00} = \rho_{\phi} + (\rho_r + \rho_m)f(\phi) \quad , \quad T^{ij} = -(p_{\phi} + p_r f(\phi))\,g^{ij} \tag{38}$$

we transcribe Equation (32) into the form

$$\frac{d}{dt}\Big(\rho_{\phi} + (\rho_r + \rho_m)\,f(\phi)\Big) + 3\,\frac{\dot{a}}{a}\Big(\rho_{\phi} + p_{\phi} + (\rho_r + p_r)\,f(\phi) + \rho_m\,f(\phi)\Big) = 0. \tag{39}$$

Since Equation (31) can be rewritten as follows:

$$\frac{d}{dt}\Big(\rho_{\phi} + \rho_m\,f(\phi)\Big) = \frac{d}{dt}\rho_m - 3\,\frac{\dot{a}}{a}\,\Big(\rho_{\phi} + p_{\phi}\Big), \tag{40}$$

we may remove the contribution of the chameleon field in Equation (39). As result, we get

$$\frac{d}{dt}\Big(\rho_r f(\phi) + \rho_m\Big) + \frac{\dot{a}}{a}\,\Big(4\rho_r f(\phi) + 3\rho_m f(\phi)\Big) = 0, \tag{41}$$

where we have used the equation of state $p_r = \rho_r/3$ [69]. Due to independence of radiation and matter densities, Equation (41) can be split into evolution equations of the radiation and matter densities:

$$\begin{aligned} \frac{d}{dt}(\rho_r f(\phi)) + 4\frac{\dot{a}}{a}(\rho_r f(\phi)) &= 0, \\ \frac{d}{dt}\rho_m + 3\frac{\dot{a}}{a}\rho_m f(\phi) &= 0. \end{aligned} \tag{42}$$

For the standard dependence of the radiation and matter densities on the expansion parameter $a(t)$ [69],

$$\rho_r = 3M_{\rm Pl}^2 {\rm H}_0^2 \Omega_r\,\frac{a_0^4}{a^4}, \quad \rho_m = 3M_{\rm Pl}^2 {\rm H}_0^2 \Omega_m\,\frac{a_0^3}{a^3}, \tag{43}$$

where a_0, ${\rm H}_0 = 1.438(11) \times 10^{-33}\,{\rm eV}$, Ω_r and Ω_m are the expansion parameter, the Hubble rate and the relative radiation and matter densities at our time $t_0 = 1/{\rm H}_0$ [70], the equations for the radiation and matter densities Equation (42) are satisfied identically for $f(\phi) = 1$.

Thus, if the radiation and matter densities depend on the expansion parameter a as $\rho_r(a) \sim a^{-4}$ and $\rho_m(a) \sim a^{-3}$, local conservation of the total energy–momentum in the Universe can be fulfilled if and only if the conformal factor $f(\phi)$, relating Einstein's and Jordan's frames and defining the chameleon–matter coupling, is equal to unity; i.e., $f(\phi) = 1$. However, in this case there is no influence of the chameleon field on the evolution of the radiation and matter densities and a dependence of the chameleon field mass on a density of its environment. In turn, for $f(\phi) \neq 1$ the evolution equations (Equation (37)) admit some exact solutions. It is convenient to search these solutions independently of the expansion parameter a. Treating the conformal factor $f(\phi)$ as a function of the expansion parameter a, i.e., setting $f(\phi) = f(a) \neq 1$, the solutions to Equation (37) can be given by

$$\begin{aligned} \rho_r(a) &= \rho_{r0}\,\frac{a_0^4}{a^4}\,\frac{f(a_0)}{f(a)}, \\ \rho_m(a) &= \rho_{m0}\,\frac{a_0^3}{a^3}\,\exp\Big(3\int_a^{a_0}\frac{f(a')-1}{a'}\,da'\Big), \end{aligned} \tag{44}$$

where $\rho_{r0} = 3M_{\rm Pl}^2 H_0^2 \Omega_r$ and $\rho_{m0} = 3M_{\rm Pl}^2 H_0^2 \Omega_m$ are the radiation and matter densities at out time $t_0 = 1/H_0$ and $a(t_0) = a_0$, i.e., in the era of the late-time acceleration of the Universe's expansion or the dark energy–dominated era. The integration constants of the first order differential equations (Equation (35)) are fixed by the conditions $\rho_r(a_0) = \rho_{r0}$ and $\rho_m(a_0) = \rho_{m0}$, respectively [69,70]. According to the solutions (Equation (44)), the chameleon field has an influence on the evolution of the radiation and matter densities.

As an example of the conformal factor we may use $f = e^{\beta\varphi(a)/M_{\rm Pl}}$ [1,2], where $\varphi(a)$ is the chameleon field as a function of the expansion parameter a and the solution to Equation (31), i.e., $\phi(t) = \varphi(a)$. Keeping the linear order contributions in the $\beta\varphi(a)/M_{\rm Pl}$ expansion we get

$$\begin{aligned} \rho_r(a) &= \rho_{r0}\,\frac{a_0^4}{a^4}\Big(1 + \frac{\beta}{M_{\rm Pl}}(\varphi(a_0) - \varphi(a))\Big), \\ \rho_m(a) &= \rho_{m0}\,\frac{a_0^3}{a^3}\Big(1 + 3\,\frac{\beta}{M_{\rm Pl}}\int_a^{a_0}\varphi(a')\,\frac{da'}{a'}\Big). \end{aligned} \tag{45}$$

Thus, the deviations of the radiation and matter densities from their standard behavior $\rho_r(a) \sim a^{-4}$ and $\rho_m(a) \sim a^{-3}$ are given by

$$\begin{aligned} \delta\rho_r(a) &= \frac{\beta}{M_{\rm Pl}}\,\rho_{r0}\,\frac{a_0^4}{a^4}\Big(\varphi(a_0) - \varphi(a)\Big), \\ \delta\rho_m(a) &= 3\,\frac{\beta}{M_{\rm Pl}}\,\rho_{m0}\,\frac{a_0^3}{a^3}\int_a^{a_0}\varphi(a')\,\frac{da'}{a'}. \end{aligned} \tag{46}$$

Some observations of deviations of the radiation and matter densities in the Universe from their standard form might, in principle, evidence an existence of the chameleon field. Nevertheless, we have to emphasize that the contributions of the conformal factor to the radiation and matter densities at our time are not practically observable. It is seen from the solutions (Equation (44)) that the conformal factor affects the evolution of the radiation and matter densities during the radiation and matter-dominated eras only. Of course, an influence of the chameleon field evolution on the distribution of the radiation density might seem rather questionable, since the evolution equation (Equation (37)) defines an evolution of the product $\rho_r f(\phi)$, where one may hardly separate ρ_r from $f(\phi)$. By introducing an effective radiation density $\rho_r^{(\rm eff)} = \rho_r f(\phi)$ we obtain a canonical radiation density Equation (43), where the contribution of $f(\phi)$ at $a = a_0$ is hidden very likely in Ω_r.

3.2. The Friedmann–Einstein (Equation Equation (28)) as the First Integral of the Friedmann–Einstein Equation (Equation (29))

It is well–known that without the chameleon field and for the conformal factor $f(a) = 1$ the Friedmann–Einstein differential equation for $\dot a^2/a^2$ is the first integral of the Friedmann–Einstein differential equation for $\ddot a/a$ [69]. However, such a property of Equation (28) with the chameleon field and the conformal factor $f(a) \neq 1$ to be the first integral of Equation (29) has not so far been investigated and proven in the literature. In order to prove that Equation (28) is the first integral of Equation (29) with the contributions of the chameleon field and the conformal factor $f(a) \neq 1$, we rewrite Equation (28) as follows:

$$\frac{\dot{a}^2}{a^2} = \frac{1}{3M_{\rm Pl}^2}\,(\rho_{\rm ch} + \rho_r f + \rho_m), \tag{47}$$

where $\rho_{\rm ch} = \rho_\phi + \rho_m(f-1) = \frac{1}{2}\dot{\phi}^2 + V_{\rm eff}(\phi)$ is the chameleon field density, given by Equation (30) with the replacement $V(\phi) \to V_{\rm eff}(\phi)$ (see Equation (32)). In order to find $\rho_{\rm ch}$ as a function of the expansion parameter a we use Equation (31) and transcribe it into the form

$$a\frac{d}{da}\rho_{\rm ch}(a) + 6\rho_{\rm ch}(a) = 6V_{\rm eff}(a), \tag{48}$$

where we have denoted $V_{\rm eff}(\phi) = V_{\rm eff}(a)$, assuming that ϕ is a function of a; i.e., $\phi = \phi(a)$. As a function of the expansion parameter a, the effective potential $V_{\rm eff}(a)$ is given by

$$V_{\rm eff}(a) = V(a) + \rho_m(a)(f(a) - 1), \tag{49}$$

where $V(a) = V(\phi) = V(\varphi(a))$ with the additive contribution of the relic dark energy density, induced by torsion, and $f(a) = e^{\beta\varphi(a)/M_{\rm Pl}}$. The solution to Equation (48) is equal to

$$\rho_{\rm ch}(a) = \frac{C_\phi}{a^6} + \frac{6}{a^6}\int a^5 V_{\rm eff}(a)da, \tag{50}$$

where the term C_ϕ/a^6 corresponds to the contribution of the kinetic term of a scalar field [87]. The integration constant C_ϕ, we define as follows: $C_\phi = 3M_{\rm Pl}^2 {\rm H}_0^2 \Omega_\phi a_0^6$, where Ω_ϕ is the integration constant, having the meaning of a relative density of a scalar field at time $t_0 = 1/{\rm H}_0$ [70]. As a result, Equation (47) takes the form

$$\frac{\dot{a}^2}{a^2} = \frac{1}{3M_{\rm Pl}^2}\,(\rho_{\rm ch}(a) + \rho_r(a)f(a) + \rho_m(a)), \tag{51}$$

where in the right-hand-side (r.h.s.) all densities and the conformal factor are functions of the expansion parameter a. Further, it is convenient to rewrite Equation (29) as follows:

$$\frac{\ddot{a}}{a} = -2\frac{\dot{a}^2}{a^2} + \frac{1}{3M_{\rm Pl}^2}\,\rho_r(a)f(a) + \frac{1}{2M_{\rm Pl}^2}\,\rho_m(a)f(a) + \frac{1}{M_{\rm Pl}^2}\,V(a), \tag{52}$$

where we have used Equation (51). Since the second derivative $\ddot{a}$ of the expansion parameter a with respect to time can be given by

$$\ddot{a} = \frac{1}{2}\frac{d\dot{a}^2}{da}, \tag{53}$$

one may transcribe Equation (47) into the form

$$a\frac{d}{da}\dot{a}^2 + 4\dot{a}^2 = \frac{2}{3M_{\rm Pl}^2}\,a^2\rho_r(a)f(a) + \frac{1}{M_{\rm Pl}^2}\,a^2\rho_m(a)f(a) + \frac{2}{M_{\rm Pl}^2}\,a^2 V(a). \tag{54}$$

The solution to Equation (54) amounts to

$$\dot{a}^2 = \frac{C}{a^4} + \frac{2}{3M_{\rm Pl}^2}\frac{1}{a^4}\int a^5\rho_r(a)f(a)da + \frac{1}{M_{\rm Pl}^2}\frac{1}{a^4}\int a^5\rho_m(a)f(a)da + \frac{2}{M_{\rm Pl}^2}\frac{1}{a^4}\int a^5 V(a)da, \tag{55}$$

where C is the integration constant. Dividing both sides of Equation (55) by a^2 we arrive at the equation

$$\frac{\dot{a}^2}{a^2} = \frac{1}{3M_{\rm Pl}^2}\Big(\frac{C_\phi}{a^6} + \frac{6}{a^6}\int a^5 U(a)da + \frac{2}{a^6}\int a^5\rho_r(a)f(a)da + \frac{3}{a^6}\int a^5\rho_m(a)f(a)da\Big), \quad (56)$$

where we have set $C_\phi = 3M_{\rm Pl}^2 C = 3M_{\rm Pl}^2 H_0^2\Omega_\phi$. Thus, Equation (51) is the first integral of Equation (29). Making a replacement $V(a) = V_{\rm eff}(a) - \rho_m(a)(f(a) - 1)$ we arrive at the expression

$$\frac{\dot{a}^2}{a^2} = \frac{1}{3M_{\rm Pl}^2}\Big(\rho_{\rm ch}(a) + \frac{2}{a^6}\int a^5\rho_r(a)f(a)da + \frac{6}{a^6}\int a^5\rho_m(a)da - \frac{3}{a^6}\int a^5\rho_m(a)f(a)da\Big). \quad (57)$$

Since the radiation and matter densities as functions of a obey the equations

$$\begin{aligned} a\frac{d}{da}\Big(\rho_r(a)f(a)\Big) &= -4\Big(\rho_r(a)f(a)\Big),\\ a\frac{d}{da}\rho_m(a) &= -3\rho_m(a)f(a) \end{aligned} \quad (58)$$

and that $\rho_r(a)f(a) = \rho_{r0}f(a_0)a_0^4/a^4$ (see Equation (44)), we transcribe the right–hand–side (r.h.s.) of Equation (57) into the form

$$\frac{\dot{a}^2}{a^2} = \frac{1}{3M_{\rm Pl}^2}\Big(\rho_{\rm ch}(a) + \rho_r(a)f(a) + \frac{1}{a^6}\int\frac{d}{da}\Big(a^6\rho_m(a)\Big)da\Big) = \frac{1}{3M_{\rm Pl}^2}\Big(\rho_{\rm ch}(a) + \rho_r(a)f(a) + \rho_m(a)\Big), \quad (59)$$

This proves that Equation (28) is the first integral of Equation (29) if the total energy–momentum is locally conserved. The evolution of the chameleon field density $\rho_{\rm ch}(a)$ independently of the expansion parameter a is defined by Equation (50), which we rewrite as follows:

$$\rho_{\rm ch}(a) = \rho_\Lambda + \frac{C_\phi}{a^6} + \frac{6}{a^6}\int a^5\Phi(a)da + \frac{6}{a^6}\int a^5\rho_m(a)\big(f(a) - 1\big)\,da, \quad (60)$$

where $\rho_\Lambda = M_{\rm Pl}^2\Lambda_C$. and the third term in Equation (60) is the model-dependent part of the potential of the self-interaction of the chameleon field $V(\phi) = \rho_\Lambda + \Phi(\phi)$ [4,10,16,88], taken as a function of the expansion parameter a, i.e., $\Phi(\phi) = \Phi(a)$. Such a chameleon field density may affect the acceleration of the Universe's expansion. Setting $f(a) = 1$ in Equation (60) we get

$$\rho_{\rm ch}(a) = \rho_\Lambda + \frac{C_\phi}{a^6} + \frac{6}{a^6}\int a^5\Phi(a)da, \quad (61)$$

where the second and the last terms might still provide an acceleration of the Universe's expansion additional to that caused by the first term ρ_Λ, which is induced by torsion [39].

4. Torsion–Neutron Low-Energy Interactions

Our analysis carried out above may give an impression that after the absorption of torsion by the cosmological constant, the Einstein–Cartan gravitational theory reduces to Einstein's gravitational theory [69]. Such an impression can be real only in case of the absence of fermions. As has been shown in [40–42], there is a huge variety of minimal and nonminimal low-energy torsion–neutron interactions. The torsion tensor field $\mathcal{T}_{\sigma\mu\nu}$, being a tensor of the third rank and antisymmetric with respect to indices μ and ν, i.e., $\mathcal{T}_{\sigma\mu\nu} = -\mathcal{T}_{\sigma\nu\mu}$, is defined by 24 independent components: (i) four vectors $\mathcal{E}^\mu = (\mathcal{E}^0, \vec{\mathcal{E}}\,)$, (ii) four axial vectors $\mathcal{B}^\mu = (\mathcal{B}^0, \vec{\mathcal{B}}\,)$ and (iii) 16 tensors $\mathcal{M}_{\sigma\mu\nu}$ [35,36,44] (see also [40–42]). The effective low-energy torsion–neutron potentials are presented in the form of expansion in powers of $1/m$, where m is the neutron mass, and restricted by the terms of order $O(1/m)$, by using the Foldy–Wouthuysen (FW) canonical transformations [89]. The most interesting effective low-energy torsion–neutron interactions are induced in the rotating coordinate systems, which can be used for experimental probes of torsion in terrestrial laboratories [48,60]. It is important to emphasize that a part of these effective low-energy torsion–neutron interactions provide a violation of time-reversal invariance [42], which can be probed in the terrestrial laboratories.

According to [42], in the coordinate system rotating with an angular velocity $\vec{\Omega}$ the time component $\mathcal{E}_0$ of the 4-vector $\mathcal{E}^\mu$ of the torsion field induces the T–odd, i.e., violating time reversal symmetry, optical potential

$$\Phi^{(\mathrm{T-odd})}_{\mathrm{eff}} = -i\,\frac{4}{3}\,\frac{\mathcal{E}_0}{m}\,\vec{S}\cdot\vec{\Omega}, \tag{62}$$

where $\vec{S} = \frac{1}{2}\vec{\sigma}$ is the operator of the neutron spin and $\vec{\sigma}$ are 2×2 Pauli matrices [90]. As has been shown in [47], because of the T–odd interaction (Equation (62)), the cross section for low-energy neutron–nucleus scattering, caused by the beam of polarized neutrons passing through a spinning cylinder, should acquire the correction [39]

$$\Delta\sigma_{\mathrm{TV}}(\Omega, p) = \frac{8\pi}{3\sqrt{2}}\,\mathcal{E}^0 R^2 L\,\frac{\Omega}{p}, \tag{63}$$

where R and L are the radius and length of the spinning cylinder, and p is a neutron momentum (for a detailed discussion of the p-dependence of $\Delta\sigma_{\mathrm{TV}}(\Omega, p)$ we refer to [47] below Equation (5)). The aim of the proposed experiment is a search for the Ω-dependent part of the helicity-dependent part of the difference of the cross sections for neutron–nucleus scattering, caused by neutrons polarized parallel and antiparallel to the neutron beam axis coinciding with the axis of a spinning cylinder. For contemporary experimental abilities, such a T–odd correction allows one to probe the time component of the 4-vector part of the torsion field at the level of sensitivity of about $|\mathcal{E}_0| \sim 10^{-32}\,\mathrm{GeV}$. This is a few orders of magnitude better in comparison to the estimate obtained in [44].

Another part of the effective low-energy torsion–neutron potentials, which is not proportional to $1/m$, can be used for probes of the components of the torsion field in the qBounce experiments dealing with ultracold neutrons (UCNs) bouncing in the gravitational field of the Earth [49–55] (see also [48]). As an example, we consider the effective low-energy potential of the time-component $\mathcal{B}_0$ (pseudoscalar) of the 4-axial vector $\mathcal{B}^\mu$ and the time-time-space-components $(\vec{\mathcal{M}})_k = \mathcal{M}_{00k}$ of the tensor $\mathcal{M}_{\sigma\mu\nu}$ [42]

$$\Phi_{\mathrm{eff}} = \frac{1}{3}\,\mathcal{B}_0\,\vec{S}\cdot(\vec{\Omega}_\oplus \times (\vec{R}_\oplus + \vec{r})) - \frac{1}{2}\,\vec{S}\cdot(\vec{\mathcal{M}} \times (\vec{\Omega}_\oplus \times (\vec{R}_\oplus + \vec{r}))), \tag{64}$$

where $\vec{\Omega}_\oplus$ and $\vec{R}_\oplus$ are the angular velocity and the radius vector of the Earth as they are shown in Figure 1. Then, $\vec{r}$ is the radius–vector of the UCN in the laboratory.

The experiments with UCNs, bouncing in the gravitational field of the Earth, are being performed in the laboratory at Institut Laue Langevin (ILL) in Grenoble. The ILL laboratory is fixed to the surface of the Earth in the northern hemisphere. Following [91–95] we choose the ILL laboratory or the standard laboratory frame with coordinates (t, x, y, z), where the x, y and z axes point south, east and vertically upwards, respectively, with northern and southern poles on the axis of the Earth's rotation with the Earth's sidereal frequency $\Omega_\oplus = 2\pi/(23\,\mathrm{hr}\,56\,\mathrm{min}\,4.09\,\mathrm{s} = 7.2921159\times 10^{-5}\,\mathrm{rad/s}$. The position of the ILL laboratory on the surface of the Earth is determined by the angles χ and ϕ, where $\chi = 90^0 - \theta$ is the colatitude of the laboratory, defined in terms of the latitude θ, and ϕ is the longitude of the laboratory measured east of south with the values $\theta = 45.16667^\circ\,\mathrm{N}$ and $\phi = 5.71667^\circ\,\mathrm{E}$ [96], respectively. The beam of UCNs moves from south to north antiparallel to the x–direction and with energies of UCNs quantized in the z–direction.

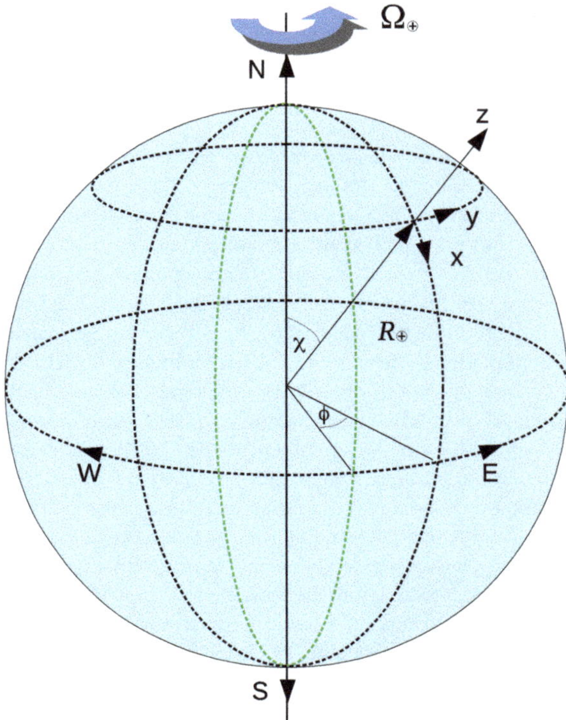

Figure 1. The position of the ILL laboratory doing the qBounce experiments on the surface of the Earth.

In the qBounce experiments the contributions of interactions beyond the gravitational interaction of the Earth are measured in terms of the transition frequencies $\omega_{pq} = E_p - E_q$ of the transitions $|q\rangle \to |p\rangle$ between two gravitational states of UCNs $|q\rangle$ and $|p\rangle$ [49–55] (see also [59,60]). As of the small values of the components of the torsion field the contribution of the $\vec{r}$-dependent part of the effective torsion–neutron potential Equation (64), where the vector $\vec{r}$ defines a location of the UCN in the coordinate system (t, x, y, z), to the transition frequencies between quantum gravitational states of UCNs can be neglected in comparison to the contributions of the terms independent of $\vec{r}$. Relative to the axes (x, y, z) the vectors $\vec{\Omega}_{\oplus}$ and $\vec{R}_{\oplus}$ are equal to $\vec{\Omega}_{\oplus} = (-\Omega_{\oplus}\sin\chi, 0, \Omega_{\oplus}\cos\chi)$ and $\vec{R}_{\oplus} = (0, 0, R_{\oplus})$, respectively. This allows one to transcribe the effective low-energy torsion–neutron potential Equation (64) into the form

$$\Phi_{\rm eff} = \Omega_{\oplus} R_{\oplus} \sin\chi \Big(\frac{1}{3}\,\mathcal{B}_0 S_y + \frac{1}{2}\,\mathcal{M}_z S_x - \frac{1}{2}\,\mathcal{M}_x S_z\Big) = 1.1 \times 10^{-6}\Big(\frac{1}{3}\,\mathcal{B}^0 S_y + \frac{1}{2}\,\mathcal{M}_z S_x - \frac{1}{2}\,\mathcal{M}_x S_z\Big), \tag{65}$$

where (S_x, S_y, S_z) are operators of the neutron spin $\vec{S}$–operator components. Thus, measuring the transition frequencies of spin-flip transitions between gravitational states $|q\downarrow\rangle \to |p\uparrow\rangle$ one may measure the contributions of the pseudoscalar $\mathcal{B}_0$ and tensor $\mathcal{M}_x = -\mathcal{M}_{00x}$ and $\mathcal{M}_z = -\mathcal{M}_{00z}$ components of the torsion field. A predictable power of the qBounce experiments we may demonstrate by example of the estimate of the contribution of the pseudoscalar component $\mathcal{B}_0$ of the torsion field coupled to UCNs. Indeed, according to Lämmerzahl [43], the value of the pseudoscalar component of the torsion field is constrained by $|\mathcal{B}_0| < 2 \times 10^{-18}\,$GeV. Its contribution to the transition frequencies between quantum gravitational states of UCNs $|q\downarrow\rangle \to |p\uparrow\rangle$ is of about $|\Delta\omega_{p\uparrow q\downarrow}| < 7 \times 10^{-16}\,$eV. This value is at the level of current experimental sensitivity $\Delta E < 10^{-15}\,$eV [55] and the sensitivity of a nearest future, which is of about $\Delta E < 10^{-17}\,$eV and even $\Delta E < 10^{-21}\,$eV [49,97]. The experiments discussed in this sections and and many others, which could be carried out by using effective low-energy potentials of torsion–neutron interactions derived in [40–42], might make reliable the geometrical origin of the cosmological constant or the relic dark energy, induced by torsion [39].

5. Discussion

We would like to emphasize that our analysis of the chameleon field as a candidate for quintessence is carried out within the classical Einstein–Cartan gravitational theory with the Einstein-Hilbert action linear in the Ricci scalar curvature. By definition [4], quintessence is a hypothetical form of dark energy described by a canonical scalar field for an explanation of the observable acceleration of the Universe's expansion. The most important that quintessence should be a hypothetical form of dark energy. In this connection in the Einstein–Cartan gravitational theory, when the cosmological constant or the relic dark energy density has the geometrical origin, caused by torsion, the chameleon field possesses no chance to be a hypothetical form of dark energy. In other words having provided a geometrical origin for the cosmological constant or the relic dark energy torsion deprives the chameleon field to have a chance to be quintessence. As a result, the chameleon field is able only to evolve above the relic background of the dark energy, caused by torsion, but not to originate it. Then, as a consequence of conservation of the total energy–momentum of the system, the chameleon field can affect the dark energy dynamics and as well as the Universe's expansion even also the late-time acceleration. We have shown that such an influence of the chameleon field on the acceleration of the Universe's expansion retains also even if the conformal factor, relating Einstein's and Jordan's frames and defining the interaction of the chameleon field with its ambient matter, is equal to unity (see Equations (60) and (61). This result is closely related to our proof that for the system, including the chameleon field, radiation and matter (dark and baryon matter), the Friedmann–Einstein equation for $\dot{a}^2/a^2$ is the first integral for the Friedmann–Einstein equation for $\ddot{a}/a$.

We have found that local conservation of the total energy–momentum of the system, including the chameleon field, radiation and matter (dark and baryon matter), leads to the equations of the evolution of the radiation and matter densities, corrected by the conformal factor. Of course, since for radiation the evolution equation defines an evolution of the product $\rho_r(a)f(\phi)$, where the conformal factor $f(\phi)$ is a function of the expansion parameter a, such a product $\rho_r(a)f(\phi)$ does not deviate from the standard behavior $\rho_r(a)f(\phi) \sim a^{-4}$. Since the radiation density $\rho_r(a)$ enters to the Friedmann–Einstein equations only in the form of the product $\rho_r(a)f(\phi)$ one may not probably separate the contribution of the conformal factor above the standard shape proportional to a^{-4}. In turn, for the matter density $\rho_m(a)$ the contribution of the conformal factor leads to a deviation from the standard behavior $\rho_m(a) \sim a^{-3}$ [69]. However, such a deviation might be, in principle, noticeable only during the matter-dominated era. In the dark energy–dominated era that is in our time of the late-time acceleration of the Universe, where the expansion parameter is equal to $a_0 = a(t_0)$ for the Hubble time $t_0 = 1/H_0$ [69], the contributions of the conformal factor to the radiation and matter densities in comparison to the standard values $\rho_r(a_0) = 3M_{\rm Pl}^2 {\rm H}_0^2 \Omega_r$ and $\rho_m(a_0) = 3M_{\rm Pl}^2 {\rm H}_0^2 \Omega_m$ are practically unobservable. This agrees well with the constraints on the deviations of the radiation and matter densities from their values at our time to a few parts per million [98], which can be obtained from the constraints on the fifth force caused by the chameleon field in the Galaxy and the Solar system.

The cosmological constant Λ_C, induced by torsion [39], we have included additively to the potential of the self-interaction of the chameleon field as a background of the relic dark energy: $V(\phi) = \rho_\Lambda + \Phi(\phi)$. In the chameleon field theory [1,2] the relic dark energy density ρ_Λ is defined as follows: $\rho_\Lambda = \Lambda^4$, where the scale $\Lambda = \sqrt[4]{3M_{\rm Pl}^2 {\rm H}_0^2 \Omega_\Lambda} = 2.24(1)\,{\rm meV}$ is calculated for the relative dark energy density $\Omega_\Lambda = 0.685(7)$ [70]. The ϕ-dependent part of the potential of the self-interaction of the chameleon field $\Phi(\phi)$ is arbitrary to some extent, i.e., model-dependent, and demands a special analysis similar to that carried out in [4,10,16,88]. However, such an analysis goes beyond the scope of our paper. We would like to emphasize that a specific analysis of a dynamics of the chameleon field such as different mechanisms of chameleon screening and a formation of a fifth force, for example, in the

Galaxy and the Solar system is related also to a special choice of the potential of the self-interaction of the chameleon field [16,98]. Such an analysis has been carried out by Brax et al. [16] and Jain et al. [98]. The repetition of such an analysis goes beyond the scope of this paper.

As regards the assertion by Wang et al. [67] and Khoury [68] that since the conformal factor is practically constant during the Hubble time, so the chameleon field is not responsible for the late-time acceleration of the Universe, one may argue that the conformal factor might be, in principle, practically constant (or better to say unity), but such a behavior of the conformal factor does not prohibit the chameleon field, evolving above the relic dark energy background induced by torsion, to take a certain part in dark energy dynamics and, correspondingly, in the acceleration of the Universe's expansion (see Equations (60) and (61)) and even so in the late-time acceleration of the Universe's expansion.

As regards another canonical scalar fields which can be introduced for an explanation of an origin of dark energy and an influence on acceleration of the Universe's expansion such as symmetron and dilaton, we may say the following. Since the dynamics of the symmetron field differs from the dynamics of the chameleon one only by the shape of the potential of self-interaction [99], our conclusion concerning an identification of the chameleon field with quintessence is fully applicable to the symmetron one. In other words the symmetron field cannot be quintessence to full extent. Moreover, we have to mention that an existence of the symmetron field and its importance for an evolution of the Universe might seem to be rather questionable after the qBounce experiments [55] on the transition frequencies between quantum gravitational states of UCNs, which have excluded the existence of the symmetron field.

Then, we have to confess that our analysis of an identification of a canonical scalar field with quintessence, carried out within the classical Einstein–Cartan gravitational theory, can say practically nothing concerning dilaton. Indeed, unlike the gravitational theories with the chameleon and symmetron fields the gravitational theories with dilaton are based on string theory in terms of the non-riemannian structure of space-time [100–107]. Of course, since an inclusion of dilaton as a canonical scalar field is closely related to a requirement of scale invariance of the action of the dynamical system under consideration, the classical Einstein–Cartan gravitational theory can be, in principle, modified by a requirement of scale invariance [108]. However, in such a modified Einstein–Cartan gravitational theory the problem of the geometrical origin of the cosmological constant or the relic dark energy caused by torsion demands a special analysis, which goes beyond the scope of this paper. Nevertheless, if in such a modified Einstein–Cartan gravitational theory torsion would be a geometrical origin of the cosmological constant or the relic dark energy density, our conclusion concerning an impossibility to identify dilation with quintessence might have been valid only within such a modified Einstein–Cartan gravitational theory.

Robust support fpr the geometrical origin of the cosmological constant or the relic dark energy could be experimental observations of torsion in terrestrial laboratories in terms of its contributions to observables of different physical processes. In Section 4 we have discussed two of these experiments, which can be carried out by using beams of polarized UCNs. We mean the contribution of the T–odd torsion–neutron low-energy interaction to the cross section for the scattering of the beam of polarized neutrons by nucleus in the end of spinning cylinder. This allows one to estimate the value of the time component $\mathcal{E}_0$o f the 4-vector part of the torsion field at the level of about $|\mathcal{E}_0| \sim 10^{-32}\,\mathrm{GeV}$. Another experiment on the probe of torsion can be carried out at ILL by the French–Austrian qBounce Collaboration by using Gravity Resonance Spectroscopy (GRS), a new measuring technique combining quantum measurements and gravity experiments [49–55]. The qBounce experiments, measuring transition frequencies between quantum gravitational states of UCNs, allow one to probe torsion with a sensitivity of about $\Delta E < 10^{-17}\,\mathrm{eV}$ and even $\Delta E \sim 10^{-21}\,\mathrm{eV}$ [49,97]. This should improve the existing upper bound on the pseudoscalar $\mathcal{B}_0$ component of the torsion field by a few orders of magnitude and give new constraints on the tensor $\mathcal{M}_{00k}$ components of the torsion field [44].

Author Contributions: A.N.I.: Conceptualization, Methodology, Data Curation, Investigation, Supervision, Writing—Original draft preparation. M.W.: Formal analysis, Investigation, Methodology, Software, Validation, Writing—Original draft preparation. All authors have read and agreed to the published version of the manuscript.

Funding: The work of A. N. Ivanov was supported by the Austrian "Fonds zur Förderung der Wissenschaftlichen Forschung" (FWF) under the contracts P31702-N27, P26636-N20 and P33279-N, and "Deutsche Förderungsgemeinschaft" (DFG) AB 128/5-2. The work of M. Wellenzohn was supported by the MA 23 (FH-Call 16) under the project "Photonik—Stiftungsprofessur für Lehre".

Acknowledgments: We are grateful to Hartmut Abele for stimulating discussions and to Philippe Brax and Alkistis Pourtsidou for fruitful discussions.

Conflicts of Interest: The authors declare no conflict of interest.

References

1. Khoury, J.; Weltman, A. Chameleon cosmology. *Phys. Rev. D* **2004**, *69*, 044026. [CrossRef]
2. Mota, D.F.; Shaw, D.J. Evading equivalence principle violations, cosmological, and other experimental constraints in scalar field theories with a strong coupling to matter. *Phys. Rev. D* **2007**, *75*, 063501. [CrossRef]
3. Will, C.M. *Theory and Experiment in Gravitational Physics*; Cambridge University Press: Cambridge, UK, 1993.
4. Ratra, B.; Peebles, P.J.E. Cosmological consequences of a rolling homogeneous scalar field. *Phys. Rev. D* **1988**, *37*, 3406. [CrossRef] [PubMed]
5. Peebles, P.J.E.; Vilenkin, A. Quintessential inflation. *Phys. Rev. D* **1999**, *59*, 063505. [CrossRef]
6. Zlatev, I.; Wang, L.; Steinhardt, P.J. Quintessence, cosmic coincidence, and the cosmological constant. *Phys. Rev. Lett.* **1999**, *82*, 896. [CrossRef]
7. Steinhardt, P.J. Quintessence and cosmic acceleration. *NATO Sci. Ser. C* **2001**, *565*, 143.
8. Steinhardt, P.J. The quintessential Universe. *AIP Conf. Proc.* **2001**, *586*, 279.
9. Steinhardt, P.J. A quintessential introduction to dark energy. *Philos. Trans. R. Soc. Lond. A* **2003**, *361*, 2497. [CrossRef]
10. Tsujikawa, S. Quintessence: A review *Class. Quantum Gravity* **2013**, *30*, 21400. [CrossRef]
11. Perlmutter, S.; Aldering, G.; Deustua, S.; Fabbro, S.; Goldhaber, G.; Groom, D.E.; Kim, A.G.; Kim, M.Y.; Knop, R.A.; Nugent, P.; et al. Cosmology from type Ia supernovae. *Bull. Am. Astron. Soc.* **1997**, *29*, 1351.
12. Riess, A.G.; Filippenko, A.V.; Challis, P.; Clocchiatti, A.; Diercks, A.; Garnavich, P.M.; Gilliland, R.L.; Hogan, C.J.; Jha, S.; Kirshner, R.P.; et al. Observational evidence from supernovae for an accelerating universe and a cosmological constant. *Astron. J.* **1998**, *116*, 1009–1038. [CrossRef]
13. Perlmutter, S.; Aldering, G.; Goldhaber, G.; Knop, R.A.; Nugent, P.; Castro, P.G.; Deustua, S.; Fabbro, S.; Goobar, A.; Groom, D.E.; et al. Measurements of Ω and Λ from 42 high redshift supernovae. *Astron. J.* **1999**, *517*, 565. [CrossRef]
14. Goobar, A.; Perlmutter, S.; Goldhaber, G.; Knop, R.A.; Nugent, P.; Castro, P.G.; Deustua, S.; Fabbro, S.; Groom, D.E.; Hook, I.M.; et al. The acceleration of the universe: Measurements of cosmological parameters from type Ia supernovae. *Phys. Scr. T* **2000**, *85*, 47. [CrossRef]
15. Peebles, P.J.E.; Ratra, B. The cosmological constant and dark energy. *Rev. Mod. Phys.* **2003** *75*, 559–606. [CrossRef]
16. Brax, P.; van de Bruck, C.; Davis, A.C.; Khoury, J.; Weltman, A. Detecting dark energy in orbit: The cosmological chameleon. *Phys. Rev. D* **2004**, *70*, 123518. [CrossRef]
17. Copeland, E.J.; Sami, M.; Tsujikawa, S. Dynamics of dark energy. *Int. J. Mod. Phys. D* **2006**, *15*, 1753–1936. [CrossRef]
18. Frieman, J.A.; Turner, M.S.; Huterer, D. Dark energy and the accelerating Universe. *Annu. Rev. Astron. Astrophys.* **2008**, *46*, 385. [CrossRef]
19. Jain, B.; Joyce, A.; Thompson, R.; Upadhye, A.; Battat, J.; Brax, P.; Davis, A.; de Rham, C.; Dodelson, S.; Erickcek, A.; et al. Novel probes of gravity and dark energy. *arXiv* **2013**, arXiv:1309.5389.
20. Brax, P.; Davis, A.-C. Casimir, gravitational, and neutron tests of dark energy. *Phys. Rev. D* **2015**, *91*, 063503. [CrossRef]
21. Pignol, G. Probing dark energy models with neutrons. *Int. J. Mod. Phys. A* **2015**, *30*, 1530048. [CrossRef]
22. Cartan, É. Sur une généralisation de la notion de courbure de Riemann et les espaces à torsion. *C. R. Acad. Sci.* **1922**, *174*, 593.

23. Cartan, É. Sur les variétés à connexion affine et la relativité généralisée (prémière partie). *Ann. Ec. Norm.* **1923**, *40*, 325. [CrossRef]
24. Cartan, É. Sur les variétés à connexion affine et la relativité généralisée (prémière partie). *Ann. Ec. Norm.* **1924**, *41*, 1. [CrossRef]
25. Cartan, É. Sur les variétés à connexion affine et la relativité généralisée (deuxième partie). *Ann. Ec. Norm.* **1925**, *42*, 17. [CrossRef]
26. Cartan, É.; Einstein, A. *Letters of Absolute Parallelism*; Princeton University Press: Princeton, NJ, USA, 1975.
27. Schrödinger, E. *Space-Time Structure*; Cambridge at the University Press: Cambridge, UK, 1950.
28. Hehl, F.W.; Kerlick, G.D.; van der Heyde, P. General relativity with spin and torsion and its deviations from Einstein's theory. *Phys. Rev. D* **1974**, *10*, 1066–1069. [CrossRef]
29. Hehl, F.W.; van der Heyde, P.; Kerlick, G.D.; Nester, J.M. General relativity with spin and torsion: Foundations and prospects. *Rev. Mod. Phys.* **1976**, *48*, 393–416. [CrossRef]
30. Hehl, F.W.; Obukhov, Y.N. Elie Cartan's torsion in geometry and in field theory, an essay. *Ann. Fond. Louis Broglie* **2007**, *32*, 157–194.
31. Hehl, F.W. Gauge Theories of Gravity and Spacetime. *arXiv* **2012**, arXiv:1204.3672v2.
32. Blagojević, M.; Hehl, F.W. (Eds.) *Gauge Theories of Gravitation: A Reader with Commentaries*; Imperial College Press: London, UK, 2013.
33. Hehl, F.W.; Obukhov, Y.N. Conservation of energy-momentum of matter as the basis for the gauge theory of gravitation. *arXiv* **2019** arXiv:1909.01791.
34. Obukhov, Y.N.; Hehl, F.W. General relativity as a special case of Poincaré gauge gravity. *Phys. Rev. D* **2020**, *102*, 044058. [CrossRef]
35. Shapiro, I.L. Physical aspects of the space-time torsion. *Phys. Rep.* **2002**, *357*, 113–213. [CrossRef]
36. Hammond, R.T. Torsion gravity. *Rep. Prog. Phys.* **2002**, *65*, 599–649. [CrossRef]
37. Kostelecký, V.A. Gravity, Lorentz violation, and the standard model. *Phys. Rev. D* **2004**, *69*, 105009. [CrossRef]
38. Ni, W.-T. Reports on Progress in Physics Searches for the role of spin and polarization in gravity. *Rep. Prog. Phys.* **2010**, *73*, 056901. [CrossRef]
39. Ivanov, A.N.; Wellenzohn, M. Einstein–Cartan gravity with torsion field serving as an origin for the cosmological constant or dark energy density. *Astrophys. J.* **2016**, *829*, 47. [CrossRef]
40. Ivanov, A.N.; Wellenzohn, M. Nonrelativistic approximation of the Dirac equation for slow fermions coupled to the chameleon and torsion fields in the gravitational field of the Earth. *Phys. Rev. D* **2015**, *92*, 065006. [CrossRef]
41. Ivanov, A.N.; Wellenzohn, M. Effective low-energy potential for slow Dirac fermions in Einstein-Cartan Gravity with torsion and chameleon field. *Phys. Rev. D* **2015**, *92*, 125004. [CrossRef]
42. Ivanov, A.N.; Wellenzohn, M. Spin precession of slow neutrons in Einstein-Cartan gravity with torsion, chameleon, and magnetic field. *Phys. Rev. D* **2016**, *93*, 045031. [CrossRef]
43. Lämmerzahl, C. Constraints on space-time torsion from Hughes-Drever experiments. *Phys. Lett. A* **1997**, *228*, 223–231. [CrossRef]
44. Kostelecký, V.A.; Russell, N.; Tasson, J.D. Constraints on torsion from bounds on Lorentz violation. *Phys. Rev. Lett.* **2008**, *100*, 111102. [CrossRef]
45. Obukhov, Y.N.; Silenko, A.J.; Teryaev, O.V. Spin-torsion coupling and gravitational moments of Dirac fermions: Theory and experimental bounds. *Phys. Rev. D* **2014**, *90*, 124068. [CrossRef]
46. Lehnert, R.; Snow, W.M.; Yan, H. A first experimental limit on in-matter torsion from neutron spin rotation in liquid $^4He^4$. *Phys. Lett. B* **2014**, *730*, 353–356; Erratum in **2015**, *744*, 415. [CrossRef]
47. Ivanov, A.N.; Snow, W.M. Parity-even and time-reversal-odd neutron optical potential in spinning matter induced by gravitational torsion. *Phys. Lett. B* **2017**, *764*, 186–189. [CrossRef]
48. Ivanov, A.N. Presented at the Workshop Dark Energy in the Laboratory, Chicheley Hall, UK, 20–22 April 2016.
49. Abele, H.; Jenke, T.; Leeb, H.; Schmiedmayer, J. Ramsey's method of separated oscillating fields and its application to gravitationally induced quantum phase shifts. *Phys. Rev. D* **2010**, *81*, 065019. [CrossRef]
50. Jenke, T.; Geltenbort, P.; Lemmel, H.; Abele, H. Realization of a gravity-resonance-spectroscopy technique. *Nat. Phys.* **2011**, *7*, 468–472. [CrossRef]
51. Abele, H.; Jenke, T.; Stadler, D.; Geltenbort, P. QuBounce: The dynamics of ultra-cold neutrons falling in the gravity potential of the Earth. *Nucl. Phys. A* **2009**, *827*, 593c–595c. [CrossRef]

52. Jenke, T.; Stadler, D.; Abele, H.; Geltenbort, P. Q-BOUNCE—Experiments with quantum bouncing ultracold neutrons. In *Nuclear Instruments and Methods in Physics Research Section A: Accelerators, Spectrometers, Detectors and Associated Equipment*; Particle Physics with Slow Neutrons; Elsevier: Amsterdam, The Netherlands, 2009; Volume 611, pp. 318–321.
53. Abele, H.; Leeb, H. Gravitation and quantum interference experiments with neutrons. *New J. Phys.* **2012**, *14*, 055010. [CrossRef]
54. Jenke, T.; Cronenberg, G.; Bürgdorfer, J.; Chizhova, L.A.; Geltenbort, P.; Ivanov, A.N.; Lauer, T.; Lins, T.; Rotter, S.; Saul, H.; et al. Gravity resonance spectroscopy constrains dark energy and dark matter scenarios. *Phys. Rev. Lett.* **2014**, *112*, 151105. [CrossRef]
55. Cronenberg, G.; Brax, P.; Filter, H.; Geltenbort, P.; Jenke, T.; Pignol, G.; Pitschmann, M.; Thalhammer, M.; Abele, H. Acoustic Rabi oscillations between gravitational quantum states and impact on symmetron dark energy. *Nat. Phys.* **2018**, *14*, 1022–1026. [CrossRef]
56. Lemmel, H.; Brax, P.; Ivanov, A.N.; Jenke, T.; Pignol, G.; Pitschmann, M.; Potocar, T.; Wellenzohn, M.; Zawisky, M.; Abele, H. Neutron interferometry constrains dark energy chameleon fields. *Phys. Lett. B* **2015**, *743*, 310–314. [CrossRef]
57. Li, K.; Arif, M.; Cory, D.G.; Haun, R.; Heacock, B.; Huber, M.G.; Nsofini, J.; Pushin, D.A.; Saggu, P.; Sarenac, D.; et al. Neutron limit on the strongly-coupled chameleon field. *Phys. Rev. D* **2016**, *93*, 062001. [CrossRef]
58. Brax, P.; Pignol, G. Strongly coupled chameleons and the neutronic quantum bouncer. *Phys. Rev. Lett.* **2011**, *107*, 111301. [CrossRef] [PubMed]
59. Ivanov, A.N.; Höllwieser, R.; Jenke, T.; Wellenzohn, M.; Abele, H. Influence of the chameleon field potential on transition frequencies of gravitationally bound quantum states of ultracold neutrons. *Phys. Rev. D* **2013**, *87*, 105013. [CrossRef]
60. Ivanov, A.N.; Cronenberg, G.; Höllwieser, R.; Jenke, T.; Pitschmann, M.; Wellenzohn, M.; Abele, H. Exact solution for chameleon field, self-coupled through the Ratra-Peebles potential with $n = 1$ and confined between two parallel plates. *Phys. Rev. D* **2016**, *94*, 085005. [CrossRef]
61. Ivanov, A.N.; Pitschmann, M. Nonrelativistic approximation of the Dirac equation for slow fermions in static metric spacetimes. *Phys. Rev. D* **2014**, *90*, 045040. [CrossRef]
62. Burrage, C.; Copeland, E.J.; Hinds, E.A. Probing dark energy with atom interferometry. *JCAP* **2015**, *2015*, 42. [CrossRef]
63. Burrage, C.; Copeland, E.J. Using atom interferometry to detect dark energy. *Contemp. Phys.* **2016**, *57*, 164–176. [CrossRef]
64. Hamilton, P.; Jaffe, M.; Haslinger, P.; Simmons, Q.; Müller, H.; Khoury, J. Atom-interferometry constraints on dark energy. *Science* **2015**, *349*, 849–851. [CrossRef]
65. Elder, B.; Khoury, J.; Haslinger, P.; Jaffe, M.; Müller, H.; Hamilton, P. Chameleon dark energy and atom interferometry. *Phys. Rev. D* **2016**, *94*, 044051. [CrossRef]
66. Burrage, C.; Sakstein, J. Tests of chameleon gravity. *Living Rev. Relat.* **2018**, *21*, 1. [CrossRef]
67. Wang, J.; Hui, L.; Khoury, J. No-go theorems for generalized chameleon field theories. *Phys. Rev. Lett.* **2012**, *109*, 241301. [CrossRef] [PubMed]
68. Khoury, J. Chameleon field theories. *Class. Quantum Gravity* **2013**, *30*, 214004. [CrossRef]
69. Rebhan, E. *Theoretische Physik: Relativitätstheorie und Kosmologie*; Springer: Berlin/Heidelberg, Germany, 2012.
70. Particle Data Group. Review of Particle Physics. *Prog. Theor. Exp. Phys.* **2020**, *2020*, 083C01. [CrossRef]
71. Weinberg, S. The cosmological constant problem. *Rev. Mod. Phys.* **1989**, *61*, 1. [CrossRef]
72. Peebles, P.J.E. Evolution of the cosmological constant. *Nature* **1999**, *398*, 25. [CrossRef]
73. Kibble, T.W.B. Lorentz invariance and the gravitational field. *J. Math. Phys.* **1961**, *2*, 212. [CrossRef]
74. Utiyama, R. Invariant theoretical interpretation of interaction. *Phys. Rev.* **1956**, *101*, 1597. [CrossRef]
75. Sciama, D.W. On the interpretation of the Einstein-Schrodinger unified field theory. *J. Math. Phys.* **1961**, *2*, 472. [CrossRef]
76. Sciama, D.W. The physical structure of general relativity. *Rev. Mod. Phys.* **1964**, *36*, 463–469. [CrossRef]
77. Blagojević, M. *Gravitation and Gauge Symmetries*; Series in High–Energy Physics, Cosmology and Gravitation; Foster, B., Grishchuk, L., Kolb, E.W., MacCallum, M.A.H., Perkins, D.H., Schutz, B.F., Eds.; Institute of Physics Publishing: Bristol, UK; Philadelphia, PA, USA, 2001.

78. Davis, A.-C.; Schelpe, C.A.O.; Shaw, D.J. Effect of a chameleon scalar field on the cosmic microwave background. *Phys. Rev. D* **2009**, *80*, 064016. [CrossRef]
79. Baum, S.; Cantatore, G.; Hoffmann, D.H.H.; Karuza, M.; Semertzidis, Y.K.; Upadhye, A.; Zioutas, K. Detecting solar chameleons through radiation pressure. *Phys. Lett. B* **2014**, *739*, 167–173. [CrossRef]
80. Dicke, R.H. Mach's principle and invariance under transformation of units. *Phys. Rev.* **1962**, *125*, 2163. [CrossRef]
81. Ivanov, A.N.; Wellenzohn, M. Standard electroweak interactions in gravitational theory with chameleon field and torsion. *Phys. Rev. D* **2015**, *91*, 085025. [CrossRef]
82. Tomas, L.H. The radiation field in a fluid in motion. *Q. J. Math.* **1930**, *1*, 239. [CrossRef]
83. Weinberg, S. Entropy generation and the survival of photo-galaxies in an expanding Universe. *Astrophys. J.* **1971**, *168*, 175. [CrossRef]
84. Straumann, N. On radiative fluids. *Helv. Phys. Acta* **1976**, *49*, 269.
85. Schweizer, M.A. Transient and transport coefficients for radiative fluids. *Astrophys. J.* **1982**, *258*, 798–811. [CrossRef]
86. Schweizer, M.A. Relativistic radiative hydrodynamics. *Ann. Phys.* **1988**, *183*, 80. [CrossRef]
87. Steinhardt, P.J.; Turok, N. Cosmic evolution in a cyclic universe. *Phys. Rev. D* **2002**, *65*, 126003. [CrossRef]
88. Copeland, E.J.; Mizuno, S.; Shaeri, M. Dynamics of a scalar field in Robertson-Walker spacetimes. *Phys. Rev. D* **2009**, *79*, 103515. [CrossRef]
89. Foldy, L.L.; Wouthuysen, S.A. On the Dirac theory of spin 1/2 particle and its nonrelativistic limit. *Phys. Rev.* **1950**, *78*, 29. [CrossRef]
90. Itzykson, C.; Zuber, J.-B. *Quantum Field Theory*; McGraw-Hill Inc.: New York, NY, USA, 1980.
91. Bluhm, R.; Kostelecký, V.A.; Lane, C.D.; Russell, N. Clock-comparison tests of Lorentz and CPT symmetry in space. *Phys. Rev. Lett.* **2002**, *88*, 090801. [CrossRef] [PubMed]
92. Kostelecký, V.A.; Mewes, M. Signals for Lorentz violation in electrodynamics. *Phys. Rev. D* **2002**, *66*, 056005. [CrossRef]
93. Bluhm, R.; Kostelecký, V.A.; Lane, C.D.; Russell, N. Probing Lorentz and CPT violation with space-based experiments. *Phys. Rev. D* **2003**, *68*, 125008. [CrossRef]
94. Kostelecký, V.A.; Mewes, M. Electrodynamics with Lorentz-violating operators of arbitrary dimension. *Phys. Rev. D* **2009**, *80*, 015020. [CrossRef]
95. Ding, Y.; Kostelecký, V.A. Lorentz-violating spinor electrodynamics and Penning traps. *Phys. Rev. D* **2016**, *94*, 056008. [CrossRef]
96. GPS Coordinates of Grenoble. Available online: https://latitude.to/map/fr/france/cities/grenoble (accessed on 26 November 2020).
97. Ivanov, A.N.; Wellenzohn, M.; Abele, H. Probing of violation of Lorentz invariance by ultracold neutrons in the Standard Model Extension. *Phys. Lett. B* **2019**, *797*, 134819. [CrossRef]
98. Jain, B.; Vikram, V.; Sakstein, J. Astrophysical tests of modified gravity: Constraints from distance indicators in the nearby Universe. *Astrophys. J.* **2013**, *779*, 39. [CrossRef]
99. Hinterbichler, K.; Khoury, J. Symmetron fields: Screening Long-range forces through local symmetry restoration. *Phys. Rev. Lett.* **2010**, *104*, 231301. [CrossRef]
100. Damour, T.; Polyakov, A.M. The string dilation and a least coupling principle. *Nucl. Phys. B* **1994**, *423*, 532–558.
101. Gasperini, M.; Piazza, F.; Veneziano, G. Quintessence as a runaway dilaton. *Phys. Rev. D* **2001**, *65*, 023508. [CrossRef]
102. Damour, T.; Piazza, F.; Veneziano, G. Violations of the equivalence principle in a dilaton-runaway scenario. *Phys. Rev. D* **2002**, *66*, 046007. [CrossRef]
103. Damour, T.; Piazza, F.; Veneziano, G. Runaway dilaton and equivalence principle violations. *Phys. Rev. Lett.* **2002**, *89*, 081601. [CrossRef] [PubMed]
104. Fujii, Y. Mass of the dilaton and the cosmological constant. *Prog. Theor. Phys.* **2003**, *110*, 433–439. [CrossRef]
105. Gasperini, M. Dilaton cosmology and phenomenology. *Lect. Notes Phys.* **2008**, *737*, 787–844.
106. Guendelman, E.I.; Kaganovich, A.B. Absence of the fifth force problem in a model with spontaneously broken dilatation symmetry. *Ann. Phys.* **2008**, *323*, 866–882. [CrossRef]

107. Brax, P.; van de Bruck, C.; Davis, A.-C. The dilaton and modified gravity. *Phys. Rev. D* **2010**, *82*, 063519. [CrossRef]
108. Saa, A. Einstein-Cartan theory of gravity revisited. *arXiv* **1993**, arXiv:9309027.

Article

Big Bang Nucleosynthesis Constraints on $f(T, T_G)$ Gravity

Petros Asimakis [1], **Emmanuel N. Saridakis** [2,3,4,*], **Spyros Basilakos** [2,5,6] **and Kuralay Yesmakhanova** [7,8]

1 Department of Physics, School of Applied Mathematical and Physical Sciences, National Technical University of Athens, 9 Iroon Polytechniou Street, Zografou Campus, 15780 Athens, Greece
2 National Observatory of Athens, Lofos Nymfon, 11852 Athens, Greece
3 CAS Key Laboratory for Researches in Galaxies and Cosmology, Department of Astronomy, University of Science and Technology of China, Hefei 230026, China
4 School of Astronomy, School of Physical Sciences, University of Science and Technology of China, Hefei 230026, China
5 Academy of Athens, Research Center for Astronomy and Applied Mathematics, Soranou Efesiou 4, 11527 Athens, Greece
6 School of Sciences, European University Cyprus, Diogenes Street, Engomi, Nicosia 1516, Cyprus
7 Ratbay Myrzakulov Eurasian International Centre for Theoretical Physics, Nur-Sultan 010009, Kazakhstan
8 Physics Department, Eurasian National University, Nur-Sultan 010008, Kazakhstan
* Correspondence: msaridak@phys.uoa.gr

Abstract: We confront $f(T, T_G)$ gravity, with big bang nucleosynthesis (BBN) requirements. The former is obtained using both the torsion scalar, as well as the teleparallel equivalent of the Gauss–Bonnet term, in the Lagrangian, resulting to modified Friedmann equations in which the extra torsional terms constitute an effective dark energy sector. We calculate the deviations of the freeze-out temperature T_f, caused by the extra torsion terms in comparison to ΛCDM paradigm. Then, we impose five specific $f(T, T_G)$ models and extract the constraints on the model parameters in order for the ratio $|\Delta T_f/T_f|$ to satisfy the observational BBN bound. As we find, in most of the models the involved parameters are bounded in a narrow window around their general relativity values as expected, asin the power-law model, where the exponent n needs to be $n \lesssim 0.5$. Nevertheless, the logarithmic model can easily satisfy the BBN constraints for large regions of the model parameters. This feature should be taken into account in future model building.

Keywords: modified gravity; nucleosynthesis; torsional gravity

PACS: 98.80.−k; 04.50.Kd; 26.35.+c; 98.80.Es

Citation: Asimakis, P.; Saridakis, E.N.; Basilakos, S.; Yesmakhanova, K. Big Bang Nucleosynthesis Constraints on $f(T, T_G)$ Gravity. *Universe* **2022**, *8*, 486. https://doi.org/10.3390/universe8090486

Academic Editor: Luca Fabbri

Received: 25 June 2022
Accepted: 2 September 2022
Published: 14 September 2022

1. Introduction

There are two motivations that lead to the construction of modifications of gravity. The first is purely theoretical, namely, to construct gravitational theories that do not suffer from the renormalizability problems of general relativity and thus are closer to a quantum description [1,2]. The second is cosmological, namely, to construct gravitational theories that at a cosmological framework can describe the early and late accelarating eras [3–7], as well as to alleviate various observational tensions [8].

There is a rich literature on modified and extended theories of gravity. One may start from the Einstein–Hilbert Lagrangian and add extra terms, resulting in $f(R)$ gravity [9–11], in $f(G)$ gravity [12–14], in $f(G, \mathcal{T})$ theories [15], in $f(P)$ gravity [16–18] in Lovelock gravity [19,20], in Weyl gravity [21], in Horndeski/Galileon scalar-tensor theories [22,23], etc. Nevertheless, one can follow a different approach and add new terms to the equivalent torsional formulation of gravity, resulting in $f(T)$ gravity [24,25], in $f(T, T_G)$ gravity [26–28], in $f(T, B)$ gravity [29,30], in scalar-torsion theories [31], etc. Torsional gravity has been proven to exhibit interesting phenomenology, both at the cosmological framework [32–57] and at the level of local, spherically symmetric solutions [58–75].

One crucial test that every modification of gravity should pass that is usually underestimated in the literature is the confrontation with Big Bang Nucleosynthesis (BBN) data [76–80]. Specifically, the amount of modification needed in order to fulfill the late-time cosmological requirements must not at the same time spoil the successes of early-time cosmology, and among them the BBN phase. Hence, whatever are the advantages of a specific modified theory of gravity, if it cannot satisfy the BBN constraints it must be excluded [81–84].

In the present manuscript, we are interested in investigating the BBN epoch in a universe governed by $f(T, T_G)$ gravity. In particular, we desire to study various specific models that are known to lead to viable phenomenology and extract constraints on the involved model parameters. The plan of the article is as follows: In Section 2, we briefly present $f(T, T_G)$ gravity, extracting the field equations and applying them to a cosmological framework. In Section 3, we summarize the BBN formalism and provide the difference in the freeze-out temperature caused by the extra torsion terms. Then, in Section 4, we investigate five specific $f(T, T_G)$ models, confronting them with the observational BBN bounds. Finally, Section 5 is devoted to the Conclusions.

2. $f(T, T_G)$ Gravity

In this section, we briefly review $f(T, T_G)$ gravity [26–28]. As usual in torsional formulation of gravity, we use the tetrad field as the dynamical variable, which forms an orthonormal basis at the tangent space. In a coordinate basis, one can relate it with the metric through $g_{\mu\nu}(x) = \eta_{AB} e^A_\mu(x) e^B_\nu(x)$, where $\eta_{AB} = \text{diag}(-1, 1, 1, 1)$, and with Greek and Latin letters, denoting coordinate and tangent indices, respectively. Applying the Weitzenböck connection $W^\lambda_{\nu\mu} \equiv e^\lambda_A \, \partial_\mu e^A_\nu$ [25], the corresponding torsion tensor is

$$T^\lambda_{\mu\nu} \equiv W^\lambda_{\nu\mu} - W^\lambda_{\mu\nu} = e^\lambda_A \left(\partial_\mu e^A_\nu - \partial_\nu e^A_\mu\right), \tag{1}$$

and then the torsion scalar is obtained through the contractions

$$T \equiv \frac{1}{4} T^{\rho\mu\nu} T_{\rho\mu\nu} + \frac{1}{2} T^{\rho\mu\nu} T_{\nu\mu\rho} - T_{\rho\mu}^{\ \ \rho} T^{\nu\mu}_{\ \ \ \nu}, \tag{2}$$

and incorporates all information of the gravitational field. Used as a Lagrangian, the torsion scalar gives rise to exactly the same equations with general relativity, which is why the theory was named the teleparallel equivalent of general relativity (TEGR).

Similarly to curvature gravity, where one can construct higher-order invariants such as the Gauss–Bonnet one, in torsional gravity one may construct higher-order torsional invariants, too. In particular, since the curvature (Ricci) scalar and the torsion scalar differ by a total derivative, in [26] the authors followed the same recipe and extracted a higher-order torsional invariant that differs from the Gauss–Bonnet one by a boundary term, namely

$$\begin{aligned} T_G = \Big(& K^{\kappa}{}_{\varphi\pi} K^{\varphi\lambda}{}_{\rho} K^{\mu}{}_{\chi\sigma} K^{\chi\nu}{}_{\tau} - 2K^{\kappa\lambda}{}_{\pi} K^{\mu}{}_{\varphi\rho} K^{\varphi}{}_{\chi\sigma} K^{\chi\nu}{}_{\tau} \\ & + 2K^{\kappa\lambda}{}_{\pi} K^{\mu}{}_{\varphi\rho} K^{\varphi\nu}{}_{\chi} K^{\chi}{}_{\sigma\tau} + 2K^{\kappa\lambda}{}_{\pi} K^{\mu}{}_{\varphi\rho} K^{\varphi\nu}{}_{\sigma,\tau} \Big) \delta^{\pi\rho\sigma\tau}_{\kappa\lambda\mu\nu}, \end{aligned} \tag{3}$$

where $K^{\mu\nu}{}_{\rho} \equiv -\frac{1}{2}\left(T^{\mu\nu}{}_{\rho} - T^{\nu\mu}{}_{\rho} - T_{\rho}{}^{\mu\nu}\right)$ is the contortion tensor, and the generalized $\delta^{\pi\rho\sigma\tau}_{\kappa\lambda\mu\nu}$ denotes the determinant of the Kronecker deltas. Note that similarly to the Gauss–Bonnet term, the teleparallel equivalent of the Gauss–Bonnet term T_G is also a topological invariant in four dimensions.

Using the above torsional invariants, one can construct the new class of $f(T, T_G)$ gravitational modifications, characterized by the action [26]

$$S = \frac{M_P^2}{2} \int d^4x \, e \, f(T, T_G), \tag{4}$$

with M_P^2 the reduced Planck mass. The general field equations of the above action can be found in [26], where one can clearly see that the theory is different from $f(R)$, $f(R,G)$, and $f(T)$ gravitational modifications, and thus it corresponds to a novel class of modified gravity.

In this work, we are interested in the cosmological applications of $f(T,T_G)$ gravity. Hence, we consider a spatially flat Friedmann–Robertson–Walker (FRW) metric of the form

$$ds^2 = -dt^2 + a^2(t)\delta_{ij}dx^i dx^j\,, \tag{5}$$

with $a(t)$ the scale factor, which corresponds to the diagonal tetrad

$$e^A_{\ \mu} = \mathrm{diag}(1, a(t), a(t), a(t)). \tag{6}$$

In this case, the torsion scalar (2) and the teleparallel equivalent of the Gauss–Bonnet term (3) become

$$T = 6H^2 \tag{7}$$
$$T_G = 24H^2(\dot{H} + H^2), \tag{8}$$

with $H = \frac{\dot{a}}{a}$ the Hubble parameter and where dots denote derivatives with respect to t.

The general field equations for the FRW geometry are [27]

$$f - 12H^2 f_T - T_G f_{T_G} + 24H^3 \dot{f_{T_G}} = 2M_P^{-2}(\rho_r + \rho_m) \tag{9}$$
$$f - 4(3H^2 + \dot{H})f_T - 4H\dot{f_T} - T_G f_{T_G} + \frac{2}{3H}T_G \dot{f_{T_G}} + 8H^2 \ddot{f_{T_G}} = -2M_P^{-2}(p_r + p_m)\,, \tag{10}$$

with $\dot{f_T} = f_{TT}\dot{T} + f_{TT_G}\dot{T}_G$, $\dot{f_{T_G}} = f_{TT_G}\dot{T} + f_{T_GT_G}\dot{T}_G$, and $\ddot{f_{T_G}} = f_{TTT_G}\dot{T}^2 + 2f_{TT_GT_G}\dot{T}\dot{T}_G + f_{T_GT_GT_G}\dot{T}_G^2 + f_{TT_G}\ddot{T} + f_{T_GT_G}\ddot{T}_G$, and where f_{TT}, f_{TT_G},... denote multiple partial differentiations with respect to T and T_G. Note that in the above equations, we have also introduced the radiation and matter sectors, corresponding to perfect fluids with energy densities ρ_r, ρ_m and pressures p_r, p_m, respectively. Lastly, we mention that the above equations for $f(T,T_G) = -T + \Lambda$ recover the TEGR and general relativity equations, where Λ is the cosmological constant.

As we can see, we can re-write the Friedmann Equations (9) and (10) in the usual form

$$3M_P^2 H^2 = (\rho_r + \rho_m + \rho_{DE}) \tag{11}$$
$$-2M_P^2\dot{H} = (\rho_r + p_r + \rho_m + p_m + \rho_{DE} + p_{DE}), \tag{12}$$

where we have defined the effective dark energy density and pressure as

$$\rho_{DE} \equiv \frac{M_P^2}{2}\left(6H^2 - f + 12H^2 f_T + T_G f_{T_G} - 24H^3 \dot{f_{T_G}}\right), \tag{13}$$

$$p_{DE} \equiv \frac{M_P^2}{2}\Big[-2(2\dot{H} + 3H^2) + f - 4(\dot{H} + 3H^2)f_T - 4H\dot{f_T} - T_G f_{T_G} + \frac{2}{3H}T_G \dot{f_{T_G}} + 8H^2 \ddot{f_{T_G}}\Big], \tag{14}$$

of gravitational origin.

3. Big Bang Nucleosynthesis Constraints

Big bang nucleosynthesis (BBN) was a process that took place during radiation era. Let us first present the framework, which provides the BBN constraints through standard cosmology [76–80]. The first Friedmann equation from Einstein–Hilbert action can be written as

$$3H^2 = M_P^{-2}\rho, \tag{15}$$

where $\rho = \rho_r + \rho_m$. In the radiation era, the radiation sector dominates; hence, we can write

$$H^2 \approx \frac{M_P^{-2}}{3}\rho_r \equiv H_{GR}^2. \tag{16}$$

In addition, it is known that the energy density of relativistic particles is

$$\rho_r = \frac{\pi^2}{30} g_* T^4, \tag{17}$$

where $g_* \sim 10$ is the effective number of degrees of freedom and T is the temperature. Thus, if we combine (16) with (17) we obtain

$$H(T) \approx \left(\frac{4\pi^3 g_*}{45}\right)^{1/2} \frac{T^2}{M_{Pl}}, \tag{18}$$

where $M_{Pl} = (8\pi)^{\frac{1}{2}} M_P = 1.22 \times 10^{19}$ GeV is the Planck mass.

During the radiation era, the scale factor evolves as $a(t) \sim t^{1/2}$. Therefore, using the relation of the Hubble parameter with the scale factor, we find that in the radiation era the Hubble parameter evolves as $H(t) = \frac{1}{2t}$. Combining the last one with (18), we find the relation between temperature and time. Thus, we have $\frac{1}{t} \simeq \left(\frac{32\pi^3 g_*}{90}\right)^{1/2} \frac{T^2}{M_{Pl}}$ (or $T(t) \simeq (t/\text{sec})^{-1/2}$ MeV).

During the BBN, we have interactions between particles. For example, we have interactions between neutrons, protons, electrons, and neutrinos, namely, $n + \nu_e \to p + e^-$, $n + e^+ \to p + \bar{\nu}_e$, and $n \to p + e^- + \bar{\nu}_e$. We name the conversion rate from a particle A to particle B as λ_{BA}. Hence, the conversion rate from neutrons to protons is λ_{pn}, and it is equal to the sum of the three interaction conversion rates written above. Therefore, the calculation of the neutron abundance arises from the protons-neutron conversion rate [78,79]

$$\lambda_{pn}(T) = \lambda_{(n+\nu_e \to p+e^-)} + \lambda_{(n+e^+ \to p+\bar{\nu}_e)} + \lambda_{(n \to p+e^- +\bar{\nu}_e)} \tag{19}$$

and its inverse $\lambda_{np}(T)$, and therefore for the total rate we have $\lambda_{tot}(T) = \lambda_{np}(T) + \lambda_{pn}(T)$. Now, we assume that the various particle (neutrino, electron, and photon) temperatures are the same and low enough in order to use the Boltzmann distribution instead of the Fermi-Dirac one, and we neglect the electron mass compared to the electron and neutrino energies. The final expression for the conversion rate is [81–84]

$$\lambda_{tot}(T) = 4A\, T^3(4!T^2 + 2 \times 3!QT + 2!Q^2), \tag{20}$$

where $Q = m_n - m_p = 1.29 \times 10^{-3}$ GeV is the mass difference between neutron and proton and $A = 1.02 \times 10^{-11}$ GeV^{-4}.

We proceed in calculating the corresponding freeze-out temperature. This will arise comparing the universe expansion rate $\frac{1}{H}$ with $\lambda_{tot}(T)$. In particular, if $\frac{1}{H} \ll \lambda_{tot}(T)$, namely, if the expansion time is much smaller than the interaction time, we can consider thermal equilibrium [76,77]. On the contrary, if $\frac{1}{H} \gg \lambda_{tot}(T)$ then particles do not have enough time to interact so they decouple. The freeze-out temperature T_f, in which the decoupling takes place, corresponds to $H(T_f) = \lambda_{tot}\left(T_f\right) \simeq c_q\, T_f^5$, with $c_q \equiv 4A\,4! \simeq 9.8 \times 10^{-10}$ GeV^{-4} [81–84]. Now, if we use (18) and $H(T_f) = \lambda_{tot}\left(T_f\right) \simeq c_q\, T_f^5$, we acquire

$$T_f = \left(\frac{4\pi^3 g_*}{45 M_{Pl}^2 c_q^2} \right)^{1/6} \sim 0.0006 \text{ GeV}. \tag{21}$$

Using modified theories, we obtain extra terms in energy density due to the modification of gravity. The first Friedmann Equation (11) during radiation era becomes

$$3M_P^2 H^2 = \rho_r + \rho_{DE}, \tag{22}$$

where ρ_{DE} must be very small compared to ρ_r in order to be in accordance with observations. Hence, we can write (22) using (16) as

$$H = H_{GR}\sqrt{1 + \frac{\rho_{DE}}{\rho_r}} = H_{GR} + \delta H, \tag{23}$$

where H_{GR} is the Hubble parameter of standard cosmology. Thus, we have $\Delta H = \left(\sqrt{1 + \frac{\rho_{DE}}{\rho_r}} - 1\right) H_{GR}$, which quantifies the deviation from standard cosmology, i.e., form H_{GR}. This will lead to a deviation in the freeze-out temperature ΔT_f. Since $H_{GR} = \lambda_{tot} \approx c_q T_f^5$ and $\sqrt{1 + \frac{\rho_{DE}}{\rho_r}} \approx 1 + \frac{1}{2}\frac{\rho_{DE}}{\rho_r}$, we easily find

$$\left(\sqrt{1 + \frac{\rho_{DE}}{\rho_r}} - 1\right) H_{GR} = 5c_q T_f^4 \Delta T_f, \tag{24}$$

and finally

$$\frac{\Delta T_f}{T_f} \simeq \frac{\rho_{DE}}{\rho_r} \frac{H_{GR}}{10 c_q T_f^5}, \tag{25}$$

where we used that $\rho_{DE} << \rho_r$ during BBN era. This theoretically calculated $\frac{\Delta T_f}{T_f}$ should be compared with the observational bound

$$\left| \frac{\Delta T_f}{T_f} \right| < 4.7 \times 10^{-4}, \tag{26}$$

which is obtained from the observational estimations of the baryon mass fraction converted to ^4He [85–91].

4. BBN Constraints on $f(T, T_G)$ Gravity

In this section, we will apply the BBN analysis in the case of $f(T, T_G)$ gravity. Let us mention here that in general, in modified gravity, inflation is not straightaway driven by an inflaton field, but the inflaton is hidden inside the gravitational modification, i.e., it is one of the extra scalar degrees of freedom of the modified graviton. Hence, in such frameworks reheating is usually performed gravitationally, and the reheating and BBN temperatures may differ from standard ones. Nevertheless, in the present work we make the assumption that we do not deviate significantly from the successful concordance scenario, in order to examine whether $f(T, T_G)$ gravity can at first pass BBN constraints or not. Clearly a more general analysis should be performed in a separate project, to cover more radical cases too. In the following, we will examine five specific models that are considered to be viable in the literature.

4.1. *Model I:* $f = -T + \beta_1\sqrt{T^2 + \beta_2 T_G}$

Firstly, we investigate the model $f = -T + \beta_1\sqrt{T^2 + \beta_2 T_G}$ [28]. Since in our analysis we focus on the radiation era where the Hubble parameter $H(t) = \frac{1}{2t}$, we can express the derivatives of the Hubble parameter as powers of the Hubble parameter itself, e.g.,

$\dot{H} = -2H^2$ and $\ddot{H} = 8H^3$. Additionally, in order to eliminate one model parameter we will apply the Friedmann equation at present time, requiring

$$\Omega_{DE0} \equiv \rho_{DE0}/(3M_P^2 H_0^2), \tag{27}$$

where Ω_{DE} is the dark energy density parameter and with the subscript "0" denoting the value of a quantity at present time. Doing so, and inserting $f = -T + \beta_1\sqrt{T^2 + \beta_2 T_G}$ into (13) and then into (25), we finally find

$$\frac{\Delta T_f}{T_f} = (10c_q T_f^3)^{-1}\zeta H_0 \Omega_{DE0}(3 - 2\beta_2)^{-3/2} \cdot \left(9 - 15\beta_2 + 6\beta_2^2\right)\left[(3 + 2\beta_2)H_0^2 + 2\beta_2 \dot{H}_0\right]^{3/2} \cdot \left[\left(9 + 3\beta_2 - 2\beta_2^2\right)H_0^4 + 9\beta_2 H_0^2 \dot{H}_0 + \beta_2^2 H_0 \ddot{H}_0\right]^{-1}, \tag{28}$$

where

$$\zeta \equiv \left(\frac{4\pi^3 g_*}{45}\right)^{\frac{1}{2}} M_{Pl.}^{-1}. \tag{29}$$

In this expression, we insert [92]

$$\Omega_{DE0} \approx 0.7, \quad H_0 = 1.4 \times 10^{-42}\ \text{GeV}, \tag{30}$$

and the derivatives of the Hubble function at present are calculated through $\dot{H}_0 = -H_0^2(1 + q_0)$ and $\ddot{H}_0 = H_0^3(j_0 + 3q_0 + 2)$ with $q_0 = -0.503$ the current deceleration parameter of the Universe [92], and $j_0 = 1.011$ the current jerk parameter [93,94]. Hence, $\dot{H}_0 \approx -9.7 \times 10^{-85}\ \text{GeV}^2$ and $\ddot{H}_0 \approx 4.1 \times 10^{-126}\ \text{GeV}^3$.

Using the BBN constraint (26), we conclude that $\beta_2 \in (-2.98, -2.93) \cup (0.99, 1.01)$, where we have used (27) to find

$$\beta_1 = \sqrt{3} H_0 \Omega_{DE0}\left[(3 + 2\beta_2)H_0^2 + 2\beta_2 \dot{H}_0\right]^{3/2} \cdot \left[\left(9 + 3\beta_2 - 2\beta_2^2\right)H_0^4 + 9\beta_2 H_0^2 \dot{H}_0 + \beta_2^2 H_0 \ddot{H}_0\right]^{-1}. \tag{31}$$

Using the above range of β_2, we find that $\beta_1 \in (2.09 \times 10^{-26}, 0.001) \cup (1.380, 1.384)$.

In Figure 1, we depict $|\Delta T_f / T_f|$ appearing in (28) versus the model parameter β_2. As we can see, the allowed range is within the vertical dashed lines.

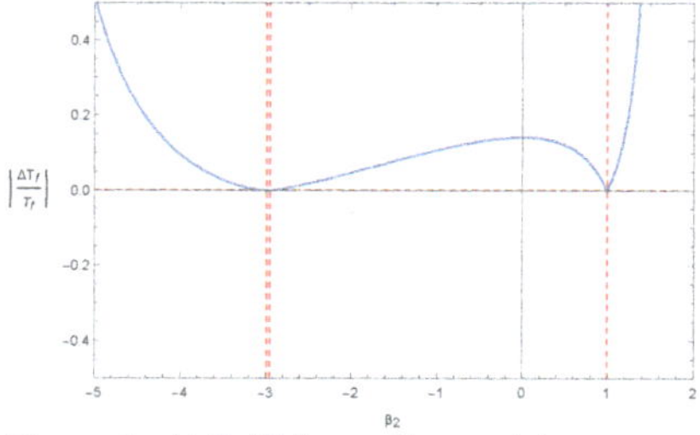

Figure 1. $|\Delta T_f / T_f|$ vs. the model parameter β_2 (blue solid curve), for Model I: $f = -T + \beta_1\sqrt{T^2 + \beta_2 T_G}$. The allowed range of β_2, where (26) is satisfied (horizontal red dashed line), is within the vertical dashed lines.

4.2. Model II: $f = -T + a_1 T^2 + a_2 T\sqrt{|T_G|}$

Let us now study the case $f = -T + a_1 T^2 + a_2 T\sqrt{|T_G|}$, where a_1, a_2 are the free parameters of the theory [28]. In this case, we find

$$\frac{\Delta T_f}{T_f} = \frac{3}{10}c_q^{-1}\zeta^3 T_f\left\{\frac{\Omega_{DE0}}{3H_0^2} - \sqrt{6}a_2\left[\frac{\sqrt{H_0^2+\dot{H}_0}}{6H_0}\left(6 - \frac{2\dot{H}_0^2 - H_0\ddot{H}_0}{H_0^2+\dot{H}_0}\right) - 1\right]\right\}. \tag{32}$$

Using the constraint (26), and according to (32), $\Delta T_f/T_f$ is linear in a_2; we deduce that (32) is valid for a small region around 2.7×10^{83} GeV^{-2}, where we have used the constraint from current cosmological era (27)

$$a_1 = \frac{\Omega_{DE0}}{18H_0^2} - \sqrt{6}a_2\frac{\sqrt{H_0^2+\dot{H}_0}}{36H_0}\left[6 - \frac{2\dot{H}_0^2 - H_0\ddot{H}_0}{(H_0^2+\dot{H}_0)^2}\right]. \tag{33}$$

Using the above value of a_2, we find that $a_1 = -1.1 \times 10^{83}$ GeV^{-2}.

4.3. Model III: $f = -T + \beta_1\sqrt{T^2+\beta_2 T_G} + a_1 T^2 + a_2 T\sqrt{|T_G|}$

Now, we analyze the model $f = -T + \beta_1\sqrt{T^2+\beta_2 T_G} + a_1 T^2 + a_2 T\sqrt{|T_G|}$, where we have four free parameters, namely, $\beta_1, \beta_2, a_1, a_2$ [28]. In order to simplify the analysis, we will impose the constraint $-2.99 < \beta_2 < \frac{3}{2}$, obtained above.

In this case, we find

$$\begin{aligned}\frac{\Delta T_f}{T_f} = -\left(60c_q T_f^3\right)^{-1}\Big\{&3\sqrt{12}\beta_1(3-2\beta_2)^{-1/2}(1+\beta_2-2\beta_1)\\ &-18\Big\{\frac{\Omega_{DE0}}{3H_0^2} + \frac{\sqrt{12}\beta_1}{18H_0^3}\left[(3+2\beta_2)H_0^2 + 2\beta_2\dot{H}_0\right]^{-1/2}\\ &\cdot\left[(3-6\beta_1+2\beta_2)H_0^2 + 2\beta_2\dot{H}_0\right]\\ &-\frac{a_2}{\sqrt{6}H_0}\sqrt{H_0^2+\dot{H}_0}\left[6 - \frac{2\dot{H}_0^2 - H_0\ddot{H}_0}{(H_0^2+\dot{H}_0)^2}\right] + \sqrt{6}a_2\\ &-\frac{\sqrt{12}\beta_1\beta_2}{18H_0^3}\left[(3+2\beta_2)H_0^2 + 2\beta_2\dot{H}_0\right]^{-3/2}\\ &\cdot\Big[(3+2\beta_2)H_0^4 + (9+8\beta_2)H_0^2\dot{H}_0\\ &+\beta_2\left(4\dot{H}_0^2 + H_0\ddot{H}_0\right)\Big]\Big\}\zeta^2 T_f^4\Big\}\zeta.\end{aligned}$$

Observing that expression (34) is linear in a_2, and using the constraint (26) and two values for β_1 from the aforementioned range we extracted in model I, i.e., $\beta_1 = 1.4$ and $\beta_2 = 1$, we find that (32) is valid for a small region around the point -3.5×10^{83} GeV^{-2}. Using another set of values ($\beta_1 = 0.001$, $\beta_2 \approx -2.96$), we find that (32) is valid for a small region around the point -5.3×10^{83} GeV^{-2}, where we have used

$$
\begin{aligned}
a_1 = \frac{\Omega_{DE0}}{18H_0^2} + \frac{\sqrt{12}\beta_1}{108H_0^3}\Big[(3+2\beta_2)H_0^2 + 2\beta_2\dot{H}_0\Big]^{-1/2} \\
\Big[(3-6\beta_1+2\beta_2)H_0^2 + 2\beta_2\dot{H}_0\Big] \\
- \frac{\sqrt{6}}{36}\frac{a_2}{H_0}\sqrt{H_0^2+\dot{H}_0}\left(6 - \frac{2\dot{H}_0^2 - H_0\ddot{H}_0}{(H_0^2+\dot{H}_0)^2}\right) \\
- \frac{\sqrt{12}\beta_1\beta_2}{108H_0^3}\Big[(3+2\beta_2)H_0^2 + 2\beta_2\dot{H}_0\Big]^{-3/2} \\
\times\Big[(3+2\beta_2)H_0^4 + (9+8\beta_2)H_0^2\dot{H}_0 + \beta_2\Big(4\dot{H}_0^2 + H_0\ddot{H}_0\Big)\Big], \quad (34)
\end{aligned}
$$

from (27). Imposing the above range of a_2, we find that $a_1 = 1.4 \times 10^{83}$ GeV^{-2} for the first case and $a_1 = 2.2 \times 10^{83}$ GeV^{-2} for the second.

4.4. Model IV: $f = -T + \beta_1(T^2 + \beta_2 T_G)^n$

As a next model, we consider the power-law model $f = -T + \beta_1(T^2 + \beta_2 T_G)^n$, where the free parameters are β_1, β_2, n. In this model, we use values of β_1, β_2 in order to constrain the power n. In this case, repeating the above steps, we find

$$
\begin{aligned}
\frac{\Delta T_{\dot{f}}}{T_f} = (10c_q)^{-1}\Omega_{DE0}H_0^{2(1-n)}\zeta^{4n-1}T_f^{8n-7}(3-2\beta_2)^{n-2} \\
\cdot\Big[(3+2\beta_2)H_0^2 + 2\beta_2\dot{H}_0\Big]^{2-n}\Big[\Big(9-12\beta_2+4\beta_2^2\Big) \\
-2n\Big(18-39\beta_2+18\beta_2^2\Big) + 16n^2\beta_2(2\beta_2-3)\Big] \\
\cdot\Big\{\Big(9+12\beta_2+4\beta_2^2\Big)H_0^4 + 4\beta_2(3+2\beta_2)H_0^2\dot{H}_0 + 4\beta_2^2\dot{H}_0^2 \\
-2n\Big[\Big(18+15\beta_2+2\beta_2^2\Big)H_0^4 + \beta_2(27+12\beta_2)H_0^2\dot{H}_0 \\
+6\beta_2^2\dot{H}_0^2 + 2\beta_2^2H_0\ddot{H}_0\Big] + 2n^2\beta_2\Big[4(3+2\beta_2)H_0^2\dot{H}_0 \\
+4\beta_2\dot{H}_0^2 + 2\beta_2H_0\ddot{H}_0\Big]\Big\}^{-1}. \quad (35)
\end{aligned}
$$

We use the constraint (26) and four values for β_2 from the range we extracted in model I above. For $\beta_2 \approx -2.9$, we find that the constraint (26) is valid for $n \lesssim 0.5$. Similarly, using the value $\beta_2 = -2$, we find $n \lesssim 0.47$, while for $\beta_2 = -1$ we find $n \lesssim 0.46$. Finally, for $\beta_2 = 1$, we find $n \lesssim 0.47$. We mention that we have used the relation

$$
\begin{aligned}
\beta_1 = -6(12)^{-n}H_0^{2(1-n)}\Omega_{DE0}\Big[(3+2\beta_2)H_0^2+2\beta_2\dot{H}_0\Big]^{2-n} \\
\cdot\Big\{\Big(9+12\beta_2+4\beta_2^2\Big)H_0^4 + 4\beta_2(3+2\beta_2)H_0^2tH_0 + 4\beta_2^2\dot{H}_0^2 \\
-2n\Big[\Big(18+15\beta_2+2\beta_2^2\Big)H_0^4 + \beta_2(27+12\beta_2)H_0^2\dot{H}_0 \\
+6\beta_2^2\dot{H}_0^2 + 2\beta_2^2H_0\ddot{H}_0\Big] + 2n^2\beta_2\Big[4(3+2\beta_2)H_0^2\dot{H}_0 \\
+4\beta_2\dot{H}_0^2 + 2\beta_2H_0\ddot{H}_0\Big]\Big\}^{-1}, \quad (36)
\end{aligned}
$$

which arises from (27).

Now, taking $\beta_2 \approx -2.9$, $n \lesssim 0.5$ we find $\beta_1 \in \left[-6.1 \times 10^{-82}, 0.0007\right]$ GeV$^{2(1-2n)}$. Similarly, for $\beta_2 = -2$, $n \lesssim 0.47$ we find $\beta_1 \in \left[-3.5 \times 10^{-74}, 5.9 \times 10^{-6}\right]$ GeV$^{2(1-2n)}$, while using $\beta_2 = -1$, $n \lesssim 0.46$ we find $\beta_1 \in \left[-4.4 \times 10^{-58}, 1.2 \times 10^{-6}\right]$ GeV$^{2(1-2n)}$. Finally, for $\beta_2 = 1$, $n \lesssim 0.47$ we find $\beta_1 \in \left[-6.4 \times 10^{-8}, 9.0 \times 10^{-6}\right]$ GeV$^{2(1-2n)}$.

In order to provide the above results in a more transparent way, in Figure 2, we present $|\Delta T_f/T_f|$ from (35) in terms of the model parameter n. As we observe, n needs to be $n \lesssim 0.5$ to pass the BBN constraint (26).

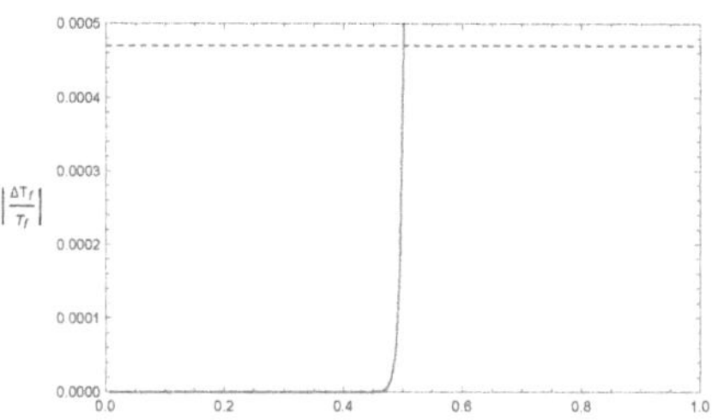

Figure 2. $|\Delta T_f/T_f|$ vs. the model parameter n (blue solid curve), for Model IV: $f = -T + \beta_1(T^2 + \beta_2 T_G)^n$ with $\beta_2 \approx -2.90$, and the upper bound for $|\Delta T_f/T_f|$ from (26) (red dashed line). As we observe, constraints from BBN require $n \lesssim 0.5$.

4.5. Model V: $f = -T + \alpha \ln \beta_1 (T^2 + \beta_2 T_G)^n$

In the last model we examine is the logarithmic one, characterized by $f = -T + \alpha \ln \beta_1 (T^2 + \beta_2 T_G)^n$, where β_1, β_2, n are the free parameters. Repeating the above analysis, we find

$$\begin{aligned}\frac{\Delta T_f}{T_f} = \left(10 c_q \zeta T_f^7\right)^{-1} H_0^2 \Omega_{DE0} \Big\{ & \ln \beta_1 + n[\ln 12 \\ & +4\ln\left(\zeta T_f^2\right) + \ln(3-2\beta_2) \\ & -2(3-2\beta_2)^{-2}\left(18-39\beta_2+18\beta_2^2\right)\Big]\Big\} \\ \cdot \Big\{ & \ln \beta_1 + n\Big\{\ln 12 + 2\ln(H_0) + \ln[(3+2\beta_2)H_0^2 + 2\beta_2 \dot{H}_0] \\ & -2\Big[(3+2\beta_2)H_0^2 + 2\beta_2\dot{H}_0\Big]^{-2}\Big[\left(18+15\beta_2+2\beta_2^2\right)H_0^4 \\ & +\beta_2(27+12\beta_2)H_0^2\dot{H}_0 + 6\beta_2^2\dot{H}_0^2 + 2\beta_2^2 H_0 \ddot{H}_0\Big]\Big\}\Big\}^{-1},\end{aligned} \quad (37)$$

where using relation (27) we find

$$\begin{aligned}\alpha = -6H_0^2\Omega_{DE0}\Big\{ & \ln\beta_1 + n\Big\{\ln 12 + 2\ln(H_0) \\ & + \ln\Big[(3+2\beta_2)H_0^2 + 2\beta_2\dot{H}_0\Big] \\ & -2\Big[(3+2\beta_2)H_0^2 + 2\beta_2\dot{H}_0\Big]^{-2}\Big[(18+15\beta_2+2\beta_2^2)H_0^4 \\ & + \beta_2(27+12\beta_2)H_0^2\dot{H}_0 + 6\beta_2^2\dot{H}_0^2 + 2\beta_2^2 H_0\ddot{H}_0\Big]\Big\}\Big\}^{-1}.\end{aligned} \quad (38)$$

We consider the values $\beta_1 = 0.001$ GeV^{-4n}, $\beta_2 \approx -2.9$, and we find that n is allowed to take every value apart from -0.0003 and a very small region around it since (37) diverges. Moreover, α is allowed to take every value apart from 0, which is the value it obtains using the above narrow window for n. Using the same considerations as the above models, we find that for $\beta_1 = 0.001$ GeV^{-4n}, $\beta_2 = -2$ the value of n is allowed to take every value apart from -0.012 and a every value but 0. Similarly, for $\beta_1 = 0.001$ GeV^{-4n}, $\beta_2 = -1$ we find that $n \neq -0.018$ and $a \neq 0$, while for $\beta_1 = 0.001$ GeV^{-4n}, $\beta_2 = 1$ we find $n \neq -0.018$ and $a \neq 0$.

As an example, in Figure 3 we present $|\Delta T_f/T_f|$ from (37) as a function of the model parameter n. The model parameter n is allowed to take all possible values except those values around a very small region centered at -0.0003, in which (37) diverges. Hence, we conclude that the logarithmic $f(T, T_G)$ model can easily satisfy the BBN bounds.

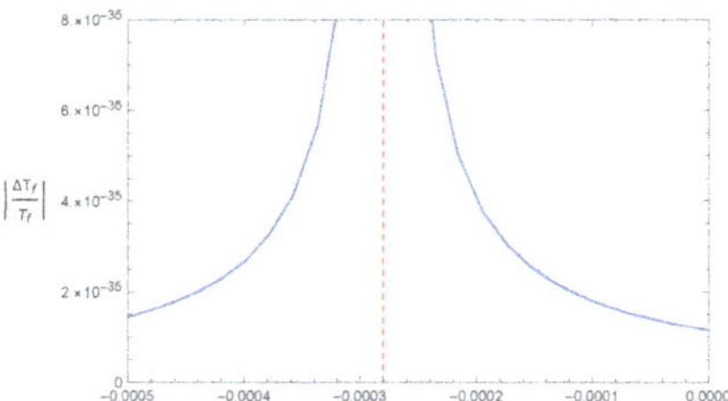

Figure 3. $|\Delta T_f/T_f|$ vs. the model parameter n (blue solid curve), for Model V: $f = -T + \alpha \ln \beta_1 (T^2 + \beta_2 T_G)^n$, choosing $\beta_1 = 0.001\ \text{GeV}^{-4n}$, $\beta_2 \approx -2.7$. The vertical dashed line at $n - 0.0003$ denotes the point where (37) diverges.

5. Conclusions

Modified gravity aims to provide explanations for various epochs of the universe evolution, and at the same time to improve the renormalizability issues of general relativity. Nevertheless, despite the specific advantages at a given era of cosmological evolution, one should be very careful not to spoil the other, well understood and significantly constrained, phases, such as the big bang nucleosynthesis (BBN) one.

In particular, there are many modified gravity models, which are constructed phenomenologically in order to be able to describe the late-time universe evolution at both the background and perturbation level. Typically, these models are confronted with observational data such as Supernovae Type Ia (SNIa), Baryonic Acoustic Oscillations (BAO), cosmic microwave background (CMB), cosmic chronometers (CC), gamma-ray bursts (GRB), growth data, etc. The problem is that although modified gravity scenarios, through the extra terms they induce, are very efficient in describing the late-time universe, quite often they induce significant terms at early times too, thus spoiling the early-time evolution, such as the BBN phase, in which the concordance cosmological paradigm is very successful. Hence, independently of the late-universe successes that a modified gravity model may have, one should always examine whether the model can pass the BBN constraints too.

In the present work, we confronted one interesting class of gravitational modification, namely, $f(T, T_G)$ gravity, with BBN requirements. The former is obtained using both the torsion scalar, as well as the teleparallel equivalent of the Gauss–Bonnet term, in the Lagrangian. Hence, one obtains modified Friedmann equations in which the extra torsional terms constitute an effective dark energy sector.

We started by calculating the deviations of the freeze-out temperature T_f, caused by the extra torsion terms, in comparison to ΛCDM paradigm. We imposed five specific $f(T, T_G)$ models that have been proposed in the literature in phenomenological grounds, i.e., in order to be able to describe the late-time evolution and lead to acceleration without an explicit cosmological constant. Hence, we extracted the constraints on the model parameters in order for the ratio $|\Delta T_f/T_f|$ to satisfy the BBN bound $\left|\frac{\Delta T_f}{T_f}\right| < 4.7 \times 10^{-4}$. As we found, in most of the models the involved parameters are bounded in a narrow window around their general relativity values, as expected. However, the logarithmic model can easily satisfy the BBN constraints for large regions of the model parameters, which acts as an advantage for this scenario.

We stress here that we did not fix the cosmological parameters to their general relativity values; on the contrary, we left them completely free and we examined which parameter regions are allowed if we want the models to pass the BBN constraints. The fact that in most models the parameter regions are constrained to a narrow window around their general relativity values was in some sense expected, but in general is not guaranteed or known a priori since many modified gravity models are completely excluded under the BBN analysis since for all parameter regions their early-universe effect is huge.

In conclusion, $f(T, T_G)$ gravity, apart from having interesting cosmological implications both in the inflationary and late-time phase, possesses particular sub-classes that can safely pass BBN bounds; nevertheless, the torsional modification is constrained in narrow windows around the general relativity values. This feature should be taken into account in future model building.

Author Contributions: Conceptualization, E.N.S.; methodology, E.N.S., P.A.; software, P.A.; validation, E.N.S., S.B.; formal analysis, P.A.; investigation, P.A.; data curation, P.A.; writing—original draft preparation, E.N.S.; writing—review and editing, P.A., S.B., K.Y.; visualization, P.A.; supervision, E.N.S., S.B.; project administration, P.A. All authors have read and agreed to the published version of the manuscript.

Funding: This research received no external funding.

Institutional Review Board Statement: Not applicable.

Informed Consent Statement: Not applicable.

Data Availability Statement: Not applicable.

Acknowledgments: This research is co-financed by Greece and the European Union (European Social Fund-ESF) through the Operational Programme "Human Resources Development, Education and Lifelong Learning" in the context of the project "Strengthening Human Resources Research Potential via Doctorate Research" (MIS-5000432), implemented by the State Scholarships Foundation (IKY). The work of N.E.M is supported in part by the UK Science and Technology Facilities research Council (STFC) under the research grant ST/T000759/1. S.B., N.E.M., and E.N.S. also acknowledge participation in the COST Association Action CA18108 "Quantum Gravity Phenomenology in the Multimessenger Approach (QG-MM)".

Conflicts of Interest: The authors declare no conflict of interest.

References

1. Stelle, K.S. Renormalization of Higher Derivative Quantum Gravity. *Phys. Rev. D* **1977**, *16*, 953. [CrossRef]
2. Addazi, A.; Alvarez-Muniz, J.; Batista, R.A.; Amelino-Camelia, G.; Antonelli, V.; Arzano, M.; Asorey, M.; Atteia, J.L.; Bahamonde, S.; Bajardi, F.; et al. Quantum gravity phenomenology at the dawn of the multi-messenger era—A review. *arXiv* **2021**. [CrossRef]
3. Nojiri, S.; Odintsov, S.D. Unified cosmic history in modified gravity: From F(R) theory to Lorentz non-invariant models. *Phys. Rep.* **2011**, *505*, 59–144. [CrossRef]
4. Clifton, T.; Ferreira, P.G.; Padilla, A.; Skordis, C. Modified Gravity and Cosmology. *Phys. Rep.* **2012**, *513*, 1–189. [CrossRef]
5. Nojiri, S.; Odintsov, S.D.; Oikonomou, V.K. Modified Gravity Theories on a Nutshell: Inflation, Bounce and Late-time Evolution. *Phys. Rep.* **2017**, *692*, 1–104. [CrossRef]
6. Saridakis, E.N.; Lazkoz, R.; Salzano, V.; Moniz, P.V.; Capozziello, S.; Jiménez, J.B.; Laurentis, M.D.; Olmo, G.J.; Akrami, Y.; Bahamonde, S.; et al. Modified Gravity and Cosmology: An Update by the CANTATA Network. *arXiv* **2021**. [CrossRef]
7. Ishak, M. Testing General Relativity in Cosmology. *Living Rev. Relativ.* **2019**, *22*, 1–204. [CrossRef]
8. Abdalla, E.; Abellán, G.F.; Aboubrahim, A.; Agnello, A.; Akarsu, Ö.; Akrami, Y.; Alestas, G.; Aloni, D.; Amendola, L.; Anchordoqui, L.A.; et al. Cosmology intertwined: A review of the particle physics, astrophysics, and cosmology associated with the cosmological tensions and anomalies. *J. High Energy Astrophys.* **2022**, *34*, 49–211. [CrossRef]
9. Starobinsky, A.A. A New Type of Isotropic Cosmological Models Without Singularity. *Phys. Lett.* **1980**, *91*, 99–102. [CrossRef]
10. Capozziello, S. Curvature Quintessence. *Int. J. Mod. Phys.* **2002**, *11*, 483–491 [CrossRef]
11. Felice, A.D.; Tsujikawa, S. $f(R)$ theories. *Living Rev. Relativ.* **2010**, *13*, 3. [CrossRef] [PubMed]
12. Antoniadis, I.; Rizos, J.; Tamvakis, K. Singularity—Free cosmological solutions of the superstring effective action. *Nucl. Phys. B* **1994**, *415*, 497–514. [CrossRef]
13. Nojiri, S.; Odintsov, S.D. Modified Gauss-Bonnet theory as gravitational alternative for dark energy. *Phys. Lett. B* **2005**, *631*, 1–6. [CrossRef]
14. Felice, A.D.; Tsujikawa, S. Construction of cosmologically viable f(G) dark energy models. *Phys. Lett. B* **2009**, *675*, 1–8. [CrossRef]
15. Yousaf, Z.; Bhatti, M.Z.; Khan, S.; Sahoo, P.K. f(G,T$\alpha\beta$T$\alpha\beta$) theory and complex cosmological structures. *Phys. Dark Univ.* **2022**, *36*, 101015. [CrossRef]
16. Erices, C.; Papantonopoulos, E.; Saridakis, E.N. Cosmology in cubic and $f(P)$ gravity. *Phys. Rev. D* **2019**, *99*, 1235277. [CrossRef]
17. Marciu, M. Note on the dynamical features for the extended $f(P)$ cubic gravity. *Phys. Rev. D* **2020**, *101*, 10. [CrossRef]

18. Jiménez, J.B.; Jiménez-Cano, A. On the strong coupling of Einsteinian Cubic Gravity and its generalisations. *J. Cosmol. Astropart. Phys.* **2021**, *1*, 69. [CrossRef]
19. Lovelock, D. The Einstein tensor and its generalizations. *J. Math. Phys.* **1971**, *12*, 498–501. [CrossRef]
20. Deruelle, N.; Farina-Busto, L. The Lovelock Gravitational Field Equations in Cosmology. *Phys. Rev. D* **1990**, *41*, 3696. [CrossRef]
21. Mannheim, P.D.; Kazanas, D. Exact Vacuum Solution to Conformal Weyl Gravity and Galactic Rotation Curves. *Astrophys. J.* **1989**, *342*, 635–638. [CrossRef]
22. Horndeski, G.W. Second-order scalar-tensor field equations in a four-dimensional space. *Int. J. Theor. Phys.* **1974**, *10*, 363–384. [CrossRef]
23. Deffayet, C.; Esposito-Farese, G.; Vikman, A. Covariant Galileon. *Phys. Rev. D* **2009**, *79*, 84003. [CrossRef]
24. Bengochea, G.R.; Ferraro, R. Dark torsion as the cosmic speed-up. *Phys. Rev. D* **2009**, *79*, 124019. [CrossRef]
25. Cai, Y.F.; Capozziello, S.; Laurentis, M.D.; Saridakis, E.N. $f(T)$ teleparallel gravity and cosmology. *Rep. Prog. Phys.* **2016**, *79*, 106901. [CrossRef]
26. Kofinas, G.; Saridakis, E.N. Teleparallel equivalent of Gauss-Bonnet gravity and its modifications. *Phys. Rev. D* **2014**, *90*, 84044. [CrossRef]
27. Kofinas, G.; Leon, G.; Saridakis, E.N. Dynamical behavior in $f(T, T_G)$ cosmology. *Class. Quantum Gravity* **2014**, *31*, 175011. [CrossRef]
28. Kofinas, G.; Saridakis, E.N. Cosmological applications of $F(T, T_G)$ gravity. *Phys. Rev. D* **2014**, *90*, 84045. [CrossRef]
29. Bahamonde, D.; Böhmer, C.G.; Wright, M. Modified teleparallel theories of gravity. *Phys. Rev. D* **2015**, *92*, 104042. [CrossRef]
30. Bahamonde, S.; Capozziello, S. Noether Symmetry Approach in $f(T, B)$ teleparallel cosmology. *Eur. Phys. J. C* **2017**, *77*, 107. [CrossRef]
31. Geng, C.Q.; Lee, C.C.; Saridakis, E.N.; Wu, Y.P. "Teleparallel" dark energy. *Phys. Lett. B* **2011**, *704*, 384–387. [CrossRef]
32. Wu, P.; Yu, H.W. Observational constraints on $f(T)$ theory. *Phys. Lett. B* **2010**, *693*, 415–420. [CrossRef]
33. Myrzakulov, R. Accelerating universe from $F(T)$ gravity. *Eur. Phys. J. C* **2011**, *71*, 1752. [CrossRef]
34. Zheng, R.; Huang, Q.G. Growth factor in $f(T)$ gravity. *J. Cosmol. Astropart. Phys.* **2011**, *3*, 2. [CrossRef]
35. Tamanini, N.; Boehmer, C.G. Good and bad tetrads in $f(T)$ gravity. *Phys. Rev. D* **2012**, *86*, 44009. [CrossRef]
36. Bamba, K.; Myrzakulov, R.; Nojiri, S.; Odintsov, S.D. Reconstruction of $f(T)$ gravity: Rip cosmology, finite-time future singularities and thermodynamics. *Phys. Rev. D* **2012**, *85*, 104036. [CrossRef]
37. Dong, H.; Wang, Y.B.; Meng, X.H. Extended Birkhoff's Theorem in the $f(T)$ Gravity. *Eur. Phys. J. C* **2012**, *72*, 2002. [CrossRef]
38. Karami, K.; Abdolmaleki, A. Generalized second law of thermodynamics in $f(T)$-gravity. *J. Cosmol. Astropart. Phys.* **2012**, *4*, 7. [CrossRef]
39. Liu, D.; Reboucas, M.J. Energy conditions bounds on $f(T)$ gravity. *Phys. Rev. D* **2012**, *86*, 83515. [CrossRef]
40. Otalora, G. Scaling attractors in interacting teleparallel dark energy. *J. Cosmol. Astropart. Phys.* **2013**, *7*, 44. [CrossRef]
41. Ong, Y.C.; Izumi, K.; Nester, J.M.; Chen, P. Problems with Propagation and Time Evolution in $f(T)$ Gravity. *Phys. Rev. D* **2013**, *88*, 024019. [CrossRef]
42. Chen, P.; Izumi, K.; Nester, J.M.; Ong, Y.C. Remnant Symmetry, Propagation and Evolution in f(T) Gravity. *Phys. Rev. D* **2015**, *91*, 064003. [CrossRef]
43. Farrugia, G.; Said, J.L. Stability of the flat FLRW metric in $f(T)$ gravity. *Phys. Rev. D* **2016**, *94*, 124054. [CrossRef]
44. Bejarano, C.; Ferraro, R.; Guzmán, M.J. McVittie solution in $f(T)$ gravity. *Eur. Phys. J. C* **2017**, *77*, 825. [CrossRef]
45. Hohmann, M.; Jarv, L.; Ualikhanova, U. Dynamical systems approach and generic properties of $f(T)$ cosmology. *Phys. Rev. D* **2017**, *96*, 43508. [CrossRef]
46. Bahamonde, S.; Böhmer, C.G.; Krššák, M. New classes of modified teleparallel gravity models. *Phys. Lett. B* **2017**, *775*, 37–43. [CrossRef]
47. Abedi, H.; Capozziello, S.; D'Agostino, R.; Luongo, O. Effective gravitational coupling in modified teleparallel theories. *Phys. Rev. D* **2018**, *97*, 84008. [CrossRef]
48. Golovnev, A.; Koivisto, T. Cosmological perturbations in modified teleparallel gravity models. *J. Cosmol. Astropart. Phys.* **2018**, *11*, 12. [CrossRef]
49. Krssak, M.; van den Hoogen, R.J.; Pereira, J.G.; Böhmer, C.J.; Coley, A.A. Teleparallel theories of gravity: Illuminating a fully invariant approach. *Class. Quantum Gravity* **2019**, *36*, 183001. [CrossRef]
50. Deng, X.M. Probing $f(T)$ gravity with gravitational time advancement. *Class. Quantum Gravity* **2018**, *35*, 175013. [CrossRef]
51. Caruana, M.; Farrugia, G.; Said, J.L. Cosmological bouncing solutions in $f(T, B)$ gravity. *Eur. Phys. J. C* **2020**, *80*, 640. [CrossRef]
52. Ren, X.; Wong, T.H.T.; Cai, Y.F.; Saridakis, E.N. Data-driven Reconstruction of the Late-time Cosmic Acceleration with $f(T)$ Gravity. *Phys. Dark Univ.* **2021**, *32*, 100812. [CrossRef]
53. Briffa, R.; Escamilla-Rivera, C.; Levi, J.S.; Mifsud, J.; Pullicino, N.L. Impact of H_0 priors on $f(T)$ late time cosmology. *Eur. Phys. J. Plus* **2022**, *137*, 532. [CrossRef]
54. Benisty, D.; Guendelman, E.I.; van de Venn, A.; Vasak, D.; Struckmeier, J.; Stoecker, H. The dark side of the torsion: Dark energy from propagating torsion. *Eur. Phys. J. C* **2022**, *82*, 264. [CrossRef]
55. Dialektopoulos, K.F.; Said, J.L.; Oikonomopoulou, Z. Classification of teleparallel Horndeski cosmology via Noether symmetries. *Eur. Phys. J. C* **2022**, *82*, 259. [CrossRef]
56. Papagiannopoulos, G.; Basilakos, S.; Saridakis, E.N. Dynamical system analysis of Myrzakulov gravity. *arXiv* **2022**. [CrossRef]

57. Papanikolaou, T.; Tzerefos, C.; Basilakos, S.; Saridakis, E.N. No constraints for $f(T)$ gravity from gravitational waves induced from primordial black hole fluctuations. *arXiv* **2022**. [CrossRef]
58. Wang, T. Static Solutions with Spherical Symmetry in $f(T)$ Theories. *Phys. Rev. D* **2011**, *84*, 24042. [CrossRef]
59. Boehmer, C.G.; Mussa, A.; Tamanini, N. Existence of relativistic stars in $f(T)$ gravity. *Class. Quantum Gravity* **2011**, *28*, 245020. [CrossRef]
60. Ferraro, R.; Fiorini, F. Spherically symmetric static spacetimes in vacuum $f(T)$ gravity. *Phys. Rev. D* **2011**, *84*, 83518. [CrossRef]
61. Meng, X.H.; Wang, Y.B. Birkhoff's theorem in the $f(T)$ gravity. *Eur. Phys. J. C* **2011**, *71*, 1755. [CrossRef]
62. Rodrigues, M.E.; Houndjo, M.J.S.; Saez-Gomez, D.; Rahaman, F. Anisotropic Universe Models in $f(T)$ Gravity. *Phys. Rev. D* **2012**, *86*, 104059. [CrossRef]
63. Rodrigues, M.E.; Houndjo, M.J.S.; Tossa, J.; Momeni, D.; Myrzakulov, R. Charged Black Holes in Generalized Teleparallel Gravity. *J. Cosmol. Astropart. Phys.* **2013**, *11*, 24. [CrossRef]
64. Nashed, G.G.L. Spherically symmetric charged-dS solution in $f(T)$ gravity theories. *Phys. Rev. D* **2013**, *88*, 104034. [CrossRef]
65. Bejarano, C.; Ferraro, R.; Guzmán, M.J. Kerr geometry in $f(T)$ gravity. *Eur. Phys. J. C* **2015**, *75*, 77. [CrossRef]
66. Das, A.; Rahaman, F.; Guha, B.K.; Ray, S. Relativistic compact stars in $f(T)$ gravity admitting conformal motion. *Astrophys. Space Sci.* **2015**, *358*, 36. [CrossRef]
67. Mai, Z.F.; Lu, H. Black Holes, Dark Wormholes and Solitons in $f(T)$ Gravities. *Phys. Rev. D* **2017**, *95*, 124024. [CrossRef]
68. Mustafa, G.; Abbas, G.; Xia, T. Wormhole solutions in $F(T, T_G)$ gravity under Gaussian and Lorentzian non-commutative distributions with conformal motions. *Chin. J. Phys.* **2019**, *60*, 362–378. [CrossRef]
69. Nashed, G.G.L.; Capozziello, S. Stable and self-consistent compact star models in teleparallel gravity. *Eur. Phys. J. C* **2020**, 80, 969. [CrossRef]
70. Pfeifer, C.; Schuster, S. Static spherically symmetric black holes in weak f(T)-gravity. *Universe* **2021**, *7*, 153. [CrossRef]
71. Ren, X.; Zhao, Y.; Saridakis, E.N.; Cai, Y.F. Deflection angle and lensing signature of covariant $f(T)$ gravity. *J. Cosmol. Astropart. Phys.* **2021**, *10*, 62. [CrossRef]
72. Bahamonde, S.; Golovnev, A.; Guzmán, M.J.; Said, J.L.; Pfeifer, C. Black holes in f(T,B) gravity: Exact and perturbed solutions. *J. Cosmol. Astropart. Phys.* **2022**, *1*, 37. [CrossRef]
73. Bahamonde, S.; Ducobu, L.; Pfeifer, C. Scalarized black holes in teleparallel gravity. *J. Cosmol. Astropart. Phys.* **2022**, *4*, 18. [CrossRef]
74. Huang, Y.; Zhang, J.; Ren, X.; Saridakis, E.N.; Cai, Y.F. N-body simulations, halo mass functions, and halo density profile in $f(T)$ gravity. *arXiv* **2022**. [CrossRef]
75. Zhao, Y.; Ren, X.; Ilyas, A.; Saridakis, E.N.; Cai, Y.F. Quasinormal modes of black holes in $f(T)$ gravity. *arXiv* **2022**. [CrossRef]
76. Bernstein, J.; Brown, L.S.; Feinberg, G. Cosmological helium production simplified. *Rev. Mod. Phys.* **1989**, *61*, 25. [CrossRef]
77. Kolb, E.W.; Turner, M.S. The Early Universe. *Front. Phys.* **1990**, *69*, 1–547.
78. Olive, K.A.; Steigman, G.; Walker, T.P. Primordial nucleosynthesis: Theory and observations. *Phys. Rep.* **2000**, *333*, 389-407. [CrossRef]
79. Cyburt, R.H.; Fields, B.D.; Olive, K.A.; Yeh, T.H. Big Bang Nucleosynthesis: 2015. *Rev. Mod. Phys.* **2016**, *88*, 15004. [CrossRef]
80. Asimakis, P.; Basilakos, S.; Mavromatos, N.E.; Saridakis, E.N. Big bang nucleosynthesis constraints on higher-order modified gravities. *Phys. Rev. D* **2022**, *105*, 8. [CrossRef]
81. Torres, D.F.; Vucetich, H.; Plastino, A. Early universe test of nonextensive statistics. *Phys. Rev. Lett.* **1997**, *79*, 1588–1590. [CrossRef]
82. Lambiase, G. Lorentz invariance breakdown and constraints from big-bang nucleosynthesis. *Phys. Rev. D* **2005**, *72*, 87702. [CrossRef]
83. Lambiase, G. Dark matter relic abundance and big bang nucleosynthesis in Horava's gravity. *Phys. Rev. D* **2011**, *83*, 107501. [CrossRef]
84. Anagnostopoulos, F.K.; Gakis, V.; Saridakis, E.N.; Basilakos, S. New models and Big Bang Nucleosynthesis constraints in $f(Q)$ gravity. *arXiv* **2022**. [CrossRef]
85. Coc, A.; Vangioni-Flam, E.; Descouvemont, P.; Adahchour, A.; Angulo, C. Updated Big Bang nucleosynthesis confronted to WMAP observations and to the abundance of light elements. *Astrophys. J.* **2004**, 600, 544-552. [CrossRef]
86. Olive, K.A.; Skillman, E.; Steigman, G. The Primordial abundance of He-4: An Update. *Astrophys. J.* **1997**, *483*, 788. [CrossRef]
87. Izotov, Y.I.; Thuan, T.X. The Primordial abundance of 4-He revisited. *Astrophys. J.* **1998**, *500*, 188. [CrossRef]
88. Fields, B.D.; Olive, K.A. On the evolution of helium in blue compact galaxies. *Astrophys. J.* **1998**, *506*, 177. [CrossRef]
89. Izotov, Y.I.; Chaffee, F.H.; Foltz, C.B.; Green, R.F.; Guseva, N.G.; Thuan, T.X. Helium abundance in the most metal-deficient blue compact galaxies: I zw 18 and sbs 0335-052. *Astrophys. J.* **1999**, *527*, 757–777. [CrossRef]
90. Kirkman, D.; Tytler, D.; Suzuki, N.; O'Meara, J.M.; Lubin, D. The Cosmological baryon density from the deuterium to hydrogen ratio towards QSO absorption systems: D/H towards Q1243+3047. *Astrophys. J. Suppl.* **2003**, *149*, 1. [CrossRef]
91. Izotov, Y.I.; Thuan, T.X. Systematic effects and a new determination of the primordial abundance of He-4 and dY/dZ from observations of blue compact galaxies. *Astrophys. J.* **2004**, *602*, 200–230. [CrossRef]
92. Aghanim, N. et al. [Planck Collaboration]. Planck 2018 results. VI. Cosmological parameters. *Astron. Astrophys.* **2020**, *641*, A6.
93. Visser, M. Jerk and the cosmological equation of state. *Class. Quantum Gravity* **2004**, *21*, 2603–2616. [CrossRef]
94. Al Mamon, A.; Bamba, K. Observational constraints on the jerk parameter with the data of the Hubble parameter. *Eur. Phys. J. C* **2018**, *78*, 862. [CrossRef]

MDPI
St. Alban-Anlage 66
4052 Basel
Switzerland
Tel. +41 61 683 77 34
Fax +41 61 302 89 18
www.mdpi.com

Universe Editorial Office
E-mail: universe@mdpi.com
www.mdpi.com/journal/universe

www.ingramcontent.com/pod-product-compliance
Lightning Source LLC
LaVergne TN
LVHW070101170726
843515LV00007B/1589